IONS IN SOLUTION AND THEIR SOLVATION

IONS IN SOLUTION AND THEIR SOLVATION

Yizhak Marcus

Hebrew University of Jerusalem

WILEY

Published by John Wiley & Sons, Inc., Hoboken, New Jersey
Published simultaneously in Canada

For general information on our other products and services or for technical support, please contact our Customer Care Department within the United States at (800) 762-2974, outside the United States at (317) 572-3993 or fax (317) 572-4002.

Wiley also publishes its books in a variety of electronic formats. Some content that appears in print may not be available in electronic formats. For more information about Wiley products, visit our web site at www.wiley.com.

Library of Congress Cataloging-in-Publication Data:

Marcus, Y.
   Ions in solution and their solvation / Yizhak Marcus.
      pages   cm
   Includes indexes.
   ISBN 978-1-118-88914-5 (cloth)
1. Ionic solutions.   2. Ions.   3. Solution (Chemistry)   4. Solvation.   I. Title.
   QD561 .M3365
   541′.372–dc23

                                                             2015020563

Cover credits: Art courtesy of the author & Background Image: Roman Okopny/iStockphoto

10 9 8 7 6 5 4 3 2 1

1   2015

CONTENTS

PREFACE

Thirty years ago, my book *Ion Solvation* was published, which became very popular and has been since then extensively cited, in particular for the numerical data concerning the properties of ions. However, much has happened during these years regarding experimental and computer simulation results pertinent to this subject as well as regarding physicochemical insights of the interactions that occur between ions and their surrounding solvents and other solutes too. It is therefore high time for an updated and revised edition of the early book to be available to scientists and engineers who deal with ions in solution. It is expected that the present book fulfills its mission to provide the needed up-to-date information.

It is appropriate to quote here some passages from the preface of my earlier book. "My main purpose ... on the one hand is to look back on what has been accomplished in a given field, organize it in an orderly fashion, and present it in a comprehensible and unified manner. On the other hand, I use this opportunity to locate gaps in our knowledge, and either fill these with new research while the book is being written, or to do this to the best of my ability in the course of time." The latter purposes I still pursue in spite of a rather advanced age.

A further quotation from the preface of my earlier book is as true at the present time as, if not even more so than, at the time my earlier book was written. "University research nowadays ... is not conducive to the carrying out of extensive and systematic sets of measurements of high accuracy on the properties of systems. Both from the stand point of student interest and from that of the necessary financial support, work on systems that are not of immediate practical importance ... is not encouraged." Still, impressive amounts of work of high quality are produced in universities too, and where relevant have found their way into the present book.

Numerical tables were an essential part of my earlier book on *ion solvation* and are also necessarily so in the present book on ions in solution and their solvation. These are interspersed in the places in the text where the quantities are being discussed, rather than being relegated to appendices, although the tables may disrupt the smooth flow of the text. Their provision, it is hoped, should make the present book as useful as a source book of data as was its predecessor. However, it must be stressed that there are some newer and corrected numerical data, for example, the CODATA thermochemical compilations or data examined by the European Commission on Atomic Energy that could not find their way into the preset book. Readers are also encouraged to look up recent volumes of the *Journal of Physical and Chemical Reference Data* for the latest critically examined values.

The significance of ions in solution and of their solvation is dealt with in the introductory chapter, which also describes the organization of the material in the chapters, as is also reflected in the list of contents.

It remains for me to express my appreciation of my university, of which I am now an emeritus professor for 15 years, which has provided me with the facilities required for doing nonlaboratory research and for writing reviews and books. The academic environment and discussions with colleagues here as well as at conferences that I attend, being supported by my university, are invaluable for doing what I have undertaken in this book.

Jerusalem, August 2014

1

INTRODUCTION

1.1 THE SIGNIFICANCE AND PHENOMENOLOGY OF IONS IN SOLUTION

Chemistry is for a large part conducted in solutions involving ions and such solutions are ubiquitous in nature. Oceans are vast aqueous solutions of salts, consisting mainly of sodium chloride, but other salts and minor components are also present in ocean water. Lakes, rivers, and brackish water are dilute solutions of ions and are essential to survival, since they provide drinking water and water for irrigation. Rain and other precipitates may remove ionic species from the atmosphere that arrived there as spray from oceans and seas or from human activities, for example, acid rain. Physiological fluids consist mostly of water in which colloidal substances, but also ions essential to their function, are dissolved.

It appears from the above that water is the only medium in which ions play a role, but this picture is too narrow because human endeavors utilize many other liquid media in which ions are present and have an active role. The manufacture of organic substances, as raw materials or intermediates in many industries, such as textiles, drugs, and food additives, generally involves reactions carried out in mixed aqueous-organic or completely nonaqueous liquid media in which ions participate. In chemical analysis, such media have long been of invaluable use, for instance in electroanalytical measurements or chromatographic separations. Industrial uses of nonaqueous media involving ions include solvent extraction in hydrometallurgy or in nuclear fuel reprocessing and nuclear waste disposal (Chapter 8).

Ions in Solution and Their Solvation, First Edition. Yizhak Marcus.
© 2015 John Wiley & Sons, Inc. Published 2015 by John Wiley & Sons, Inc.

The ions involved in these as well as other systems and applications interact with each other and with nonionized solutes that may be present. These interactions are of prime interest to the chemist, but the extent, intensity, and rate of proceeding of these interactions depend heavily on the solvent or solvent mixture present, a fact that is not always clearly recognized by the operator. The ion–solvent interactions should be understood in order to make the best use of the solutions of the ions, since it is the solvated ions that take part in the interactions of interest. If a free choice of the solvent or the solvent mixture to be used is possible, the most suitable one for the purpose should be selected on the basis of the knowledge available on the interactions that take place, bearing in mind also costs, ecology, and hazards. If the solvent is prescribed, this knowledge is still needed in order to select the proper reaction conditions or the additives that could be useful. So-called "bare" or nonsolvated ions occur in gas-phase reactions (Chapter 2) but not in condensed media, that is, in solutions. A seeming exception to this generalization is the use of room temperature ionic liquids (RTILs) as reaction media, where all the ions are surrounded by ions of the opposite charge sign rather than by a nonionic liquid medium. Whether the ions of RTILs are called "bare" or "solvated" is a semantic question. In common situations, which are the subject of this book, there is always an excess of a nonionic liquid medium in which the ions find themselves, the molecules of which surround the ions more or less completely, unless some other species, be it another ion (of opposite sign) or a solute molecule (a ligand) replaces some of the solvent molecules in the ionic solvation shell.

Ions cannot be added individually to any major extent to a solvent or a solution; it is always an electrolyte consisting of ions of both signs in a combination that makes the electrolyte electrically neutral, which is added to form a solution involving ions. Many commonly used and studied electrolytes are crystalline solids, such as NaCl or $(C_4H_9)_4NClO_4$. The electrostatic energy that holds the ions constituting such crystals together, the lattice energy that must be invested in order to separate the ions in the solution, is compensated by the solvation energy that is gained in the process of dissolution, with some effect also of the entropic changes encountered in the process. Some potential electrolytes are gaseous, for example, HCl, but they produce ions only on reaction with the solvent in which the covalent H–Cl bonds are broken and replaced with others to compensate for the energy involved. On the other hand, ions may leave the solution, if not individually then as a small combination of ions, in electrospray experiments, in which they are then monitored in the gas phase by mass spectrometry. The results of such experiments have some bearing on the state of the liquid ionic solutions, but this subject is outside the scope of this book.

It should be kept in mind that, connected with such ion solvation reactions with crystalline or gaseous electrolytes, a further reaction takes place, which is not always recognized, namely the breaking of some solvent–solvent molecular interactions, required to produce the space to accommodate the ions in the solution. It is the balance of all the (Gibbs) energies that have to be invested and those that are gained that determine the extent to which an electrolyte will dissolve in a given solvent (Chapter 4).

Some electrolytes are completely dissociated into "free," that is, solvated, positively charged cations, and negatively charged anions. Other electrolytes are only partly so dissociated, depending on the concentration and on the nature of the solvent. Some substances are ionogenic, in the sense that some dissociation into ions occurs only under specific conditions, and these include also so-called "weak electrolytes" such as many acids and essentially basic substances in aqueous solutions.

A solution of a single, individual ion in a very large amount of solvent could presumably be the basis of a study of ion solvation which is not encumbered by other interactions. This situation cannot be achieved in the laboratory but can be dealt with as a thought process and now for many years also in computer simulations. The results of the latter have by now consolidated into a large body of knowledge that is constantly not only extended but also improved by the level of sophistication that can nowadays be achieved in such simulations. Such results are incorporated into the discussions in the present book, where they are compared with laboratory experimental data obtained on electrolytes, extrapolated to infinite dilution (Chapters 4 and 5). At such high dilutions, each ion is surrounded by solvent molecules only and does not interact with the very remote ions of the opposite charge sign that must be present somewhere in the solution. Still, the allocation of the extrapolated values of the electrolyte properties to its constituent individual ions is a problem that must be solved.

For the purpose of only reducing the number of items in the properties list from the many electrolytes (combination of cations and anions), that have been derived at infinite dilution to the much smaller number of individual ions that constitute them, it is sufficient to employ the so-called "conventional" values. These are based on assigning to one ion, say the solvated hydrogen ion, an arbitrary value (generally zero) and rely on the additivity of individual infinite dilution ionic values to derive values of all other ions. The sum of the conventional ionic values, weighted according to the stoichiometric coefficients (the numbers of ions of each kind constituting the electrolyte), expresses correctly the infinite dilution property of the electrolyte. Within a given charge sign series of ions, say cations only, comparisons between conventional values of diverse ions can throw some light on the effects of the individual ionic properties, such as size and valency, but the cation and the anion series cannot be compared with each other.

The problem of assignment of the so-called "absolute" individual ionic values to these infinite dilution electrolyte data is solved mainly on the basis of chemical intuition (Chapter 4) that can be assisted by the results from computer simulations. Once individual ionic values of their properties in a given solvent (or solvent mixture) at a given thermodynamic state [temperature and pressure, usually specified as 298.15 K (25°C) and 0.1 MPa (less commonly now 1 atm = 0.101325 MPa)] have been established, they may be compared with other properties of the ions (e.g., their sizes) or with theoretical expectations (models). The latter are the main incentives to obtaining the absolute values. Such comparisons and correlations provide insights into the ion–solvent interactions that take place and form the basis for understanding interactions of ions with other solutes, be they ionic themselves or nonionic.

There are some experimental measurements that can be made on solutions of ions that pertain directly to individual ions. These include transport properties, such as the ionic conductivities that are obtained from specific conductivities of electrolytes in conjunction with transport number measurements. Diffusivities of individual ionic species can also be measured by the use of isotopically labeled ions and should be compatible with the mobilities deduced from the ionic conductivities. Spectroscopic data can also in certain cases be due to individual ionic species, such as NMR chemical shifts and relaxation rates of the signals from appropriate nuclei (e.g., ^{7}Li or ^{27}Al). Such information may be used as a guide for the "chemical intuition" mentioned earlier needed for obtaining absolute individual ionic values from measurements on electrolytes extrapolated to infinite dilution.

As already mentioned at the beginning, nonaqueous and mixed aqueous-organic solvents play important roles in chemistry, but water is still the most studied solvent for ions, not only at ambient conditions but also under diverse conditions of temperature, pressure, and the existence of external fields, such as electrical ones. Their solvation by water and their properties in aqueous solutions are therefore useful as a reference basis for evaluation of the effects occurring on the exchange of the water, in part or completely, by another solvent. The situation is complicated by the fact that water is a unique solvent in many respects, and effects encountered in aqueous solutions of ions may be absent in other media. For instance, the effects that ions have on the hydrogen-bonded structure of water, breaking it or enhancing it, have no counterpart in most other solvents, but there are exceptions. The so called "hydrophobic interactions" applicable to ionic species with extended alkyl chains or aromatic rings are practically unique for aqueous solutions. These interactions do not allude only to "model" ions such as $(C_4H_9)_4N^+$ but more importantly to biomolecules in general and specifically also to side chains of proteins that are ionized due to the $-CO_2^-$ or $-NH_3^+$ groups that they carry.

The ion solvation efficacy of nonaqueous solvents, alone or mixed with water, has beneficial results for some uses, such as enhancement of the solubility of ionic substances, but may be detrimental in other aspects, for example, in the availability of ions as reaction partners. Poorly solvated ions are more reactive than strongly solvated ones and this is manifested in the rates of organic reactions. Aprotic solvents (such that do not provide hydrogen bonding) have been found as optimal media for reactions involving anions that are only poorly solvated by them. Nonpolar solvents, on the other hand, have a low solvating power for the ions and therefore play a very minor role in solutions of ions, because of the low solubility of electrolytes in such solvents. Polar solvents, whether protic or aprotic, interact with ions of both charge signs by means of their dipole moments. Protic solvents carry a hydrogen atom capable of hydrogen-bonding to anions whereas both protic and aprotic ones provide a pair of nonbonded electrons to form coordinate bonds with mainly the cations (Chapter 3). The electron pair donor–acceptor properties of the solvents are generally of more importance than the dipole moments of their molecules. The permittivity of the solvent, important for the ionic dissociation of the electrolytes, plays a minor role in the solvation of the ions or properties of the solvated ions, once formed in the solution.

Mixture of solvents poses additional problems to the study of ion solvation and the properties of the solvated ions: the interactions between the molecules of the different solvents with each other besides those of molecules of each solvent among themselves are significant. Also, the selective solvation of the ions by the molecules of the component solvents may affect the solvated ions profoundly: the ions may "see" around them nearly only molecules of the favored component (selective solvation). The preferential solvation of ions in solvent mixture needs to be described quantitatively, and once this is done should be explained in terms of both the solvating properties of the solvents and the properties of the ions (Chapter 6).

Much emphasize has been provided in this introduction to solution of electrolytes at infinite dilution, where individual ionic properties are manifested and may be obtained and discussed. However, "real life" encounters solutions of ions that involve finite, sometimes quite large, concentrations of ions, where interactions of the ions among themselves are important. The number of solvent molecules per formula unit of the electrolyte diminishes with increasing concentrations, and hence the integrity of the ionic solvation shells is eventually broken. The ion–ion interactions then compete with the ion–solvent interactions. Ion pairs, consisting of couples of ions of opposite charge sign, may have transient existence, but an equilibrium concentration of them may result and needs to be taken into account. Similarly, the presence of non-ionic solutes (already referred to earlier text, where solvent mixtures are considered), which may be solid, liquid, or gaseous solutes, poses new forms of interactions that have to be dealt with (Chapter 7).

Several aspects of ion solvation and the properties of solvated ions are dealt with in this introduction. On the whole, one goal of the discussion of ion solvation is the provision of small sets of properties of the ions and of the solvents, from which the solvation can be predicted for any ion/solvent combination. This has been attempted in this book, with a view to be useful for the many applications of ions in solution, examples of which are shown in Chapter 8, with no attempt to exhaust this subject.

1.2 LIST OF SYMBOLS AND ABBREVIATIONS

Chemical species and units of physical quantities are denoted by Roman type characters, whereas physical quantities that can be expressed by numerical values are denoted by Greek or *italic* characters. Mathematical symbols have their usual meaning and are not listed here. The same symbol is used for an extensive property of a system and for the molar quantity of a constituent of the system. The SI system of physical units is used throughout, but some extra SI units commonly used in the physicochemical literature are also included where they simplify the notation. These include the symbols °C for centigrade temperatures ($T/K - 273.15$), M for mol·dm^{-3}, and m for mol (kg·solvent)$^{-1}$.

PRINCIPAL LATIN CHARACTERS

A^{z-} a generalized anion
A coefficient in the Debye–Hückel expression for activity coefficients
AN acceptor number of a solvent
a distance of closest approach of ions in solution (nm)

a_X thermodynamic activity of species X

(aq)   an ion in aqueous solution, generally at infinite dilution

B coefficient in the Debye–Hückel expression for activity coefficients

B_η B-coefficient of the Jones–Dole viscosity expression (M^{-1})

b' coefficient of the expression of the electric field dependence of the permittivity

b parameter in the Bjerrum expression for ion pairing

C^{z+} a generalized cation

C_P molar heat capacity at constant pressure ($J·K^{-1}·mol^{-1}$)

c_X molar concentration of species X (M)

(cr)   crystalline phase

D   Debye unit of dipole moments (3.33564×10^{-30} C·m)

D diffusion coefficient ($m^2·s^{-1}·mol^{-1}$)

DN donor number of a solvent

d interatomic distance (nm)

E   generalized electrolyte

E energy, molar energy ($J·mol^{-1}$)

E electric field strength ($V·m^{-1}$)

E electromotive force of an electrochemical cell (V)

E_j liquid junction potential

E_T polarity index of a solvent (kca·mol^{-1}, 1 cal = 4.184 J)

e elementary charge (1.6022×10^{-19} C)

F Faraday constant (9.6485×10^4 C mol^{-1})

f_X fraction of species X

G Gibbs energy, molar Gibbs energy ($J·mol^{-1}$)

g Kirkwood dipole orientation parameter

(g)   gas phase

$g(r)$ pair correlation function

H enthalpy, molar enthalpy ($J·mol^{-1}$)

h solvation (hydration) number

$I^{z\pm}$ generalized ion

I ionic strength (M or m)

K equilibrium constant

K_a, K_b acid, base dissociation constant in aqueous solutions

K_{ass} ion pair association constant (M)

K_W ion product of water (M^2)

k_B Boltzmann constant (1.3807×10^{-23} J·K^{-1})

k rate constant of specified reaction (s^{-1} for unimolecular reactions, $M^{-1}·s^{-1}$ for bimolecular reactions)

(l)   liquid phase

M^{z+} metal ion of charge $z+$

M_X molar mass of species X (in kg·mol^{-1})

m_X molal concentration of species X (m)

N   generalized nonelectrolyte solute

N_A Avogadro's number ($6.0221 \times 10^{23} mol^{-1}$)

N_X number of particles of species X in the system
n_C number of carbon atoms in an alkyl chain
n_D refractive index at the sodium D line
n_X amount of substance of species X (in mol)
P pressure (Pa)
p vapor pressure (Pa)
$Q(b)$ integral in the Bjerrum theory of ion pairing
R gas constant $(8.3145\,\text{J·K}^{-1}\text{·mol}^{-1})$
R_X molar refractivity of species X $(\text{m}^3\text{·mol}^{-1})$
r_X radius of a particle of species X
S generalized solvent
S entropy, molar entropy $(\text{J·K}^{-1}\text{·mol}^{-1})$
$S(k)$ structure factor in k space
s_X molar solubility of species X (M)
(s) solid phase
T temperature (in K)
t centigrade temperature (°C)
t_b normal boiling point of liquid (°C at 0.101325 MPa)
t_m melting (freezing) temperature (°C)
U Potential interaction energy in the system (J)
u speed of sound in a liquid (m·s^{-1})
V volume, molar volume $(\text{m}^3\text{·mol}^{-1})$
W water, a generalized reference solvent
w_X mass fraction of species X
X generalized solute
x_X mole fraction of species X
Y generalized molar thermodynamic quantity $(G, H, S, V, …)$
y packing fraction of a solvent
y_X molar activity coefficient of species X
Z lattice parameter
z_X charge number of ionic species X (taken algebraically)

PRINCIPAL GREEK CHARACTERS

α fraction of electrolyte dissociated into ions
α Kamlet–Taft hydrogen bond donation ability of solvent
α_P isobaric thermal expansibility (K^{-1})
α_X polarizability of species X (m^{-3})
β Kamlet–Taft electron pair donation ability of solvent
$\gamma_\pm$ mean ionic molal activity coefficient of electrolyte
δ chemical shift of NMR signal (ppm)
δ_H Hildebrand Solubility Parameter (in $\text{Pa}^{1/2}$)
ε_0 permittivity of empty space $(8.8542 \times 10^{-12}\,\text{C}^2\text{·J}^{-1}\text{·m}^{-1})$
ε relative permittivity

η	dynamic viscosity (Pa·s)
κ	specific conductance (S·m^{-1})
κ_S, κ_T	adiabatic (isentropic), isothermal compressibility (Pa^{-1})
Λ_E	molar conductivity of an electrolyte E (S·m^{-1}·dm^3·mol^{-1})
λ_I	molar conductivity of ion I (S·m^{-1}·dm^3·mol^{-1})
μ	dipole moment (D)
μ_X	chemical potential of species X (J·mol^{-1})
ν	wave number (cm^{-1})
ν	stoichiometric coefficient (number of ions per formula)
π^*	Kamlet–Taft polarity/polarizability of solvent
ρ	density (kg·m^{-3})
σ	surface tension (N·m^{-1})
σ	molecular collision diameter (nm)
τ	relaxation time, mean residence time (s)
φ_X	volume fraction of species X
χ	molar (diamagnetic) susceptibility (m^3·mol^{-1})
χ	surface potential of a liquid against another phase (V)
ω	frequency of an electromagnetic wave (s^{-1})

PRINCIPAL SUBSCRIPTS

ad	pertaining to the process of adsorption
cav	pertaining to cavity formation
dip	contribution from dipole interactions
disp	contribution from dispersion interactions
E	pertaining to an electrolyte
el	contribution from electrostatic interactions
els	contribution from electrostriction
f	pertaining to the process of formation
hyd	pertaining to hydration
I	pertaining to the ion I
intr	intrinsic value of solute
neut	pertaining to a neutral species
S	pertaining to the solvent S
soln	for the process of dissolution
solv	for the process of solvation
str	structural contribution
tr	of transfer
vdW	van der Waals radius or volume

PRINCIPAL SUPERSCRIPTS

conv	conventional
E	excess extensive property
F	of fusion

L   local
N   normalized
V   of vaporization
*   standard state of a pure substance
o   standard thermodynamic function
∞   standard state of infinite dilution
≠   of activation
φ   (presuperscript) apparent molar

A chemical substance or ion is generally referred to in the text by its name or formula, but in tables and as subscripts, abbreviations are generally employed. The common abbreviations of alkyl chains employed are as follows: Me, methyl; Et, ethyl; Pr, 1-propyl; Bu, 1-butyl; Pe, 1-pentyl; Hx, 1-hexyl; Oc, 1-octyl; and Ph, phenyl. Common solvents have the following abbreviations: EG, 1,2-ethanediol; THF, tetrahydrofuran; Diox, 1,4-dioxane; PC, propylene carbonate; FA, formamide; DMF, N,N-dimethylformamide; NMPy, N-methyl-2-pyrrolidinone; Py, pyridine; DMSO, dimethylsulfoxide; TMS, tetramethylenesulfone (sulfolane), and HMPT, hexamethyl phosphoric triamide. The names of other solvents are occasionally abbreviated as noted in the footnotes of tables.

2

IONS AND THEIR PROPERTIES

2.1   IONS AS ISOLATED PARTICLES

Ions are defined as particles that carry electrical charges. They may exist in gaseous phases as individual ions, but in condensed phases (solids and liquids), they exist as electrically neutral combinations of cations and anions, and in solutions they exist as electrolytes, the ions of which may be bound or relatively free to migrate.

Many chemical species are ionic, and the ions may be monatomic, such as Ca^{2+} or F^-; they may consist of a few atoms, such as uranyl(VI), UO_2^{2+}, or phosphate, PO_4^{2-}; or even considerably more than a few, such as trifluoromethylsulfonate, $CF_3SO_3^-$, or tetraphenylarsonium, $(C_6H_5)_4As^+$. They may be the constituents of so-called room temperature ionic liquids, such as 1-ethyl-3-methylimidazolium, $1\text{-}C_2H_5\text{-}3\text{-}CH_3\text{-}c\text{-}1,3N_2(CH)_3^+$ or *bis*(trifluoromethylsulfonyl)amide, $(CF_3SO_2)_2N^-$. Polyions, constituting the dissociated part of polyelectrolytes, consist of many atoms carrying many charges dispersed along polymeric chains. Polypeptides, proteins, nucleic acids, and polyphosphate detergents are examples of polyelectrolytes.

Ions in an ideal gaseous state, termed isolated or bare ions, are devoid of interactions with other particles or their surroundings and are commonly monatomic or consist of relatively few atoms. They may also be the centers of clusters consisting of the ion proper surrounded by a small number of solvent molecules.

Ions in Solution and Their Solvation, First Edition. Yizhak Marcus.
© 2015 John Wiley & Sons, Inc. Published 2015 by John Wiley & Sons, Inc.

2.1.1 Bare Ions

The primary characteristics of isolated ions are the amount of electrical charge they carry and their mass. The amount of charge is given in terms of $z_I e$, where z_I is an integral positive or negative number and $e = 1.60218 \times 10^{-19}$ C is the elementary unit of the charge.

The masses of ions are generally specified as their molar mass, M_I, that is, the mass of Avogadro's number, $N_A = 6.02214 \times 10^{23}$ mol^{-1}, of ions. The units of the molar mass are therefore kg·mol^{-1} (see Table 2.1 for a large number of representative ions).

An isolated ion is formed from an atom, a radical, or a molecule by either losing one or more electrons in an ionization process to form a cation or gaining an electron in an electron capture process to form an anion. The ionization process may proceed in several successive stages and requires the investment of energy. This energy is expressed by the ionization potential, $\sum I_p$, the sum being over the successive ionization stages. The electron capture by a neutral species releases energy, the amount of which is expressed as its electron affinity, EA. The values of the EA are generally based on the appearance potentials in mass spectroscopy. Electrostatic repulsion between an anion that already carries a negative charge and an incoming electron makes electron capture by an anion an unlikely event. Multivalent anions, such as SO_4^{2-} and PO_4^{3-}, are unstable in the isolated state and are treated as virtual ions. Therefore, the EA to form a multicharged anion is based on thermochemical cycles rather than on mass spectrometry and are negative (energy has to be invested to form them).

The energies involved in the ionization $\left(\sum I_p \right)$ or electron capture (EA) process are generally reported in electronvolt units (1 eV/particle = 96.4853 kJ·mol^{-1}). However, for the purpose of this book, where thermochemical cycles involving the solvation of the ions are of consequence, it is better to convert the energies to enthalpies by adding $z_I RT$ to the $\sum I_p$ values and $-|z_I| RT$ to the $-EA$ values for the cations and anions respectively and report the values in kJ·mol^{-1}. In fact, for metallic elements M forming cations M^{z+} the ionization potentials are the differences in the standard enthalpies of formation of the cation (see below) and the metal atoms in the ideal gas phase:

$$\sum I_p = \Delta_f H°(M^{z+}, ig) - \Delta_f H°(M, ig) \tag{2.1}$$

The $\sum I_p$ and EA values pertaining to the standard conditions, $T° = 298.15$ K and $P° = 0.1$ MPa, are recorded in Table 2.2 for many ions. Another quantity that characterizes isolated anions is their proton affinity, PA. It describes the enthalpy released when the anion binds a hydrogen ion (a proton) in the ideal gas phase. These values are also shown in Table 2.2.

Thermodynamic quantities that pertain to the formation of isolated ions from the elements in their standard states are well defined. The standard molar Gibbs energy and the enthalpy of formation, $\Delta_f G°(I^{z\pm}, ig)$ and $\Delta_f H°(I^{z\pm}, ig)$, in kJ·mol^{-1} of many ions are recorded in Table 2.3 for the standard temperature $T° = 298.15$ K and pressure $P° = 0.1$ MPa. The values of the Gibbs energies were obtained from the enthalpies and entropies: $\Delta_f G°(I^{z\pm}, ig) = \Delta_f H°(I^{z\pm}, ig) - T°[S°(I^{z\pm}, ig) - \sum S°(\text{element}, ig)]$ in a

TABLE 2.1 The Names, Formulas, Charge Numbers, z_i, and Molar Masses, M_i, of Common Ions

Cation	z_i	$M/\text{kg·mol}^{-1}$	Anion	z_i	$M/\text{kg·mol}^{-1}$
Hydrogen, H^+	+1	0.001008	Hydride, H^-	−1	0.001008
Deuterium, D^+	+1	0.002016	Fluoride, F^-	−1	0.01899
Lithium, Li^+	+1	0.006941	Chloride, Cl^-	−1	0.03545
Sodium, Na^+	+1	0.02294	Bromide, Br^-	−1	0.07991
Potassium, K^+	+1	0.03910	Iodide, I^-	−1	0.12691
Rubidium, Rb^+	+1	0.08547	Hydroxide, OH^-	−1	0.01701
Cesium, Cs^+	+1	0.13291	Hydrosulfide, SH^-	−1	0.03307
Copper(I), Cu^+	+1	0.06355	Hypochlorite, ClO^-	−1	0.05145
Silver, Ag^+	+1	0.10787	Hypobromide, BrO^-	−1	0.09591
Gold(I), Au^+	+1	0.19697	Hypoiodide, IO^-	−1	0.14291
Thallium(I), Tl^+	+1	0.20438	Cyanide, CN^-	−1	0.02602
Hydronium, H_3O^+	+1	0.01902	Cyanate, NCO^-	−1	0.04203
Ammonium, NH_4^+	+1	0.01804	Thiocyanate, SCN^-	−1	0.05808
Hydroxylaminium, $HONH_3^+$	+1	0.03404	Azide, N_3^-	−1	0.04202
Hydrazinium, $H_2NNH_3^+$	+1	0.03305	Hydrogenfluoride, HF_2^-	−1	0.03901
Guanidinium, $C(NH_2)_3^+$	+1	0.06008	Hydroperoxide, HO_2^-	−1	0.03301
Tetramethylammonium, Me_4N^+	+1	0.07415	Triiodide, I_3^-	−1	0.38071
Tetraethylammonium, Et_4N^+	+1	0.13025	Metaborate, BO_2^-	−1	0.04281
Tetra-n-propylammonium, Pr_4N^+	+1	0.18636	Chlorite, ClO_2^-	−1	0.06745
Tetra-n-butylammonium, Bu_4N^+	+1	0.24247	Nitrite, NO_2^-	−1	0.04601
Tetraphenylphosphonium, Ph_4P^+	+1	0.33939	Nitrate, NO_3^-	−1	0.06201
Tetraphenylarsonium, Ph_4As^+	+1	0.38334	Chlorate, ClO_3^-	−1	0.08345
Nitrosyl, NO^+	+1	0.03001	Bromate, BrO_3^-	−1	0.12761
Nitroxyl, NO_2^+	+1	0.04601	Iodate, IO_3^-	−1	0.17491
Beryllium, Be^{2+}	+2	0.009012	Perchlorate, ClO_4^-	−1	0.09945
Magnesium, Mg^{2+}	+2	0.02431	Permanganate, MnO_4^-	−1	0.11894
Calcium, Ca^{2+}	+2	0.04008	Pertechnetate, TcO_4^-	−1	0.16301
Strontium, Sr^{2+}	+2	0.08762	Perrhenate, ReO_4^-	−1	0.25002
Barium, Ba^{2+}	+2	0.13733	Tetrafluoroborate, BF_4^-	−1	0.08681
Radium, Ra^{2+}	+2	0.226	Formate, HCO_2^-	−1	0.04502
Vanadium(II), V^{2+}	+2	0.05094	Acetate, $CH_3CO_2^-$	−1	0.05904
Chromium(II), Cr^{2+}	+2	0.05201	Benzoate, $PhCO_2^-$	−1	0.12112
Manganese(II), Mn^{2+}	+2	0.05494	Trifluoroacetate, $CF_3CO_2^-$	−1	0.11302
Iron(II), Fe^{2+}	+2	0.05585	Trifluoromethylsulfonate, $CF_3SO_3^-$	−1	0.14906
Cobalt, Co^{2+}	+2	0.05893	Tetraphenylborate, BPh_4^-	−1	0.31923
Nickel, Ni^{2+}	+2	0.05869	Bicarbonate, HCO_3^-	−1	0.06102
Copper(II), Cu^{2+}	+2	0.06355	Bisulfate, HSO_4^-	−1	0.09707
Zinc, Zn^{2+}	+2	0.06539	Dihydrogenphosphate, $H_2PO_4^-$	−1	0.09699
Palladium(II), Pd^{2+}	+2	0.10642	Hexafluorophosphate, PF_6^-	−1	0.14496
Cadmiun, Cd^{2+}	+2	0.11241	Hexafluoroantimonate, SbF_6^-	−1	0.23574
Tin(II), Sn^{2+}	+2	0.11871	Oxide, O^{2-}	−2	0.01600

TABLE 2.1 (Continued)

Cation	z_i	M/kg·mol^{-1}	Anion	z_i	M/kg·mol^{-1}
Samarium(II), Sm^{2+}	+2	0.15036	Sulfide, S^{2-}	−2	0.03207
Europium(II) Eu^{2+}	+2	0.15197	Carbonate, CO_3^{2-}	−2	0.06001
Ytterbium(II), Yb^{2+}	+2	0.17304	Oxalate, $C_2O_4^{2-}$	−2	0.08802
Platinum(II), Pt^{2+}	+2	0.19308	Sulfite, SO_3^{2-}	−2	0.08007
Mercury, Hg^{2+}	+2	0.20059	Sulfate, SO_4^{2-}	−2	0.09607
Dimercury(I), Hg_2^{2+}	+2	0.40118	Selenate, SeO_4^{2-}	−2	0.14297
Lead, Pb^{2+}	+2	0.2072	Chromate, CrO_4^{2-}	−2	0.11599
Uranyl(VI), UO_2^{2+}	+2	0.27003	Molybdate, MoO_4^{2-}	−2	0.15994
Aluminium, Al^{3+}	+3	0.02698	Tungstate, WO_4^{2-}	−2	0.24785
Scandium, Sc^{3+}	+3	0.04496	Thiosulfate, $S_2O_3^{2-}$	−2	0.11212
Vanadium(III), V^{3+}	+3	0.05094	Hexafluorosilicate, SiF_6^{2-}	−2	0.14208
Chromium(III), Cr^{3+}	+3	0.05201	Dichromate, $Cr_2O_7^{2-}$	−2	0.21599
Iron(III), Fe^{3+}	+3	0.05585	Hydrogenphosphate, HPO_4^{2-}	−2	0.09598
Cobalt(III), Co^{3+}	+3	0.05893	Phosphate, PO_4^{3-}	−3	0.09497
Gallium, Ga^{3+}	+3	0.06971	Hexacyanoferrate(III), $Fe(CN)_6^{3-}$	−3	0.21195
Yttrium, Y^{3+}	+3	0.08891	Hexacyanocobaltate(III), $Co(CN)_6^{3-}$	−3	0.21611
Indium, In^{3+}	+3	0.11482	Hexacyanoferrate(II), $Fe(CN)_6^{4-}$	−4	0.21195
Antimony(III), Sb^{3+}	+3	0.12176			
Lanthanum, La^{3+}	+3	0.13891			
Cerium(III), Ce^{3+}	+3	0.14012			
Praseodymium, Pr^{3+}	+3	0.14091			
Neodymium, Nd^{3+}	+3	0.14424			
Promethium, Pm^{3+}	+3	0.147			
Samarium(III), Sm^{3+}	+3	0.15036			
Europium(III), Eu^{3+}	+3	0.15197			
Gadolinium, Gd^{3+}	+3	0.15725			
Terbium, Tb^{3+}	+3	0.15893			
Dysprosium, Dy^{3+}	+3	0.16251			
Holmium, Ho^{3+}	+3	0.164936			
Erbium, Er^{3+}	+3	0.16726			
Thulium, Tm^{3+}	+3	0.16893			
Ytterbium(III), Yb^{3+}	+3	0.17304			
Lutetium, Lu^{3+}	+3	0.17497			
Thallium(III), Tl^{3+}	+3	0.20438			
Bismuth, Bi^{3+}	+3	0.20898			
Actinium, Ac^{3+}	+3	0.227			
Uranium(III), U^{3+}	+3	0.23803			
Americium, Am^{3+}	+3	0.241			
Curium, Cm^{3+}	+3	0.244			
Zirconium, Zr^{4+}	+4	0.09122			
Tin(IV), Sn^{4+}	+4	0.11871			
Cerium(IV), Ce^{4+}	+4	0.14012			
Hafnium, Hf^{4+}	+4	0.17849			
Thorium, Th^{4+}	+4	0.23204			
Uranium(IV), U^{4+}	+4	0.23803			
Neptunium(IV), Np^{4+}	+4	0.237			
Plutonium(IV), Pu^{4+}	+4	0.239			

TABLE 2.2 Ionization Potentials Leading to Cations, Electron Affinities Leading to Anions, and Proton Affinities of Anions, all Converted to Enthalpies

Cation	$\sum I_p + z_i RT$/kJ·mol^{-1}	Reference	Anion	$-EA - |z_i| RT$/kJ·mol^{-1}	Reference	PA/kJ·mol^{-1}	Reference
H^+	1318	1	H^-	73	2	1649	2
D^+	1331	1	F^-	328	2	1530	3
Li^+	526	1	Cl^-	349	2	1372	3
Na^+	502	1	Br^-	324	2	1326	3
K^+	425	1	I^-	295	2	1294	4
Rb^+	409	1	OH^-	176	2	1607	2
Cs^+	382	1	SH^-	223	2	1443	2
Cu^+	752	1	ClO^-	209	2	1474	2
Ag^+	737	1	BrO^-	227	5		
Au^+	896	1	IO^-	229	5		
Tl^+	596	1	CN^-	369	2	1447	2
H_3O^+	456	2	NCO^-	346	2	1415	2
NH_4^+			SCN^-	207	2	1343	2
NO^+	900	6	N_3^-	266	2	1414	2
NO_2^+	944	6	HF_2^-	>460	6		
Be^{2+}	2669	1	HO_2^-	103	6	1542	2
Mg^{2+}	2201	1	I_3^-	350	7		
Ca^{2+}	1748	1	BO_2^-	393	6		
Sr^{2+}	1626	1	ClO_2^-	207	5		
Ba^{2+}	1480	1	NO_2^-	222	2	1389	2
Ra^{2+}	1501	1	NO_3^-	378	2	1330	2
V^{2+}	2077	1	ClO_3^-	410	8		
Cr^{2+}	2259	1	BrO_3^-	444	9		
Mn^{2+}	2239	1	IO_3^-	453	9		
Fe^{2+}	2334	1	ClO_4^-	506	8	1192	4
Co^{2+}	2420	1	MnO_4^-	480	7		

Ni²⁺	2502	1					
Cu²⁺	2716	1					
Zn²⁺	2652	1					
Pd²⁺	2691	1					
Cd²⁺	2512	1					
Sn²⁺	2133	1					
Sm²⁺	1625	1					
Eu²⁺	1644	1					
Yb²⁺	1791	1					
Pt²⁺	2634	1	TcO_4^-	470	7		
Hg²⁺	2829	1	ReO_4^-	430	6		
Pb²⁺	2178	1	BF_4^-	540	7		
UO_2^{2+}	1642	11	HCO_2^-	343	10	1416	2
Al³⁺	5157	1	$CH_3CO_2^-$	326	10	1430	2
Sc³⁺	4275	1	$PhCO_2^-$		10	1393	2
V³⁺	4910	1	$CF_3CO_2^-$	433	10	1323	2
Cr³⁺	5252	1	$CF_3SO_3^-$	>470	6	1318	4
Fe³⁺	5296	1	BPh_4^-				
Co³⁺	5658	1	HSO_4^-	430	2	1264	4
Ga³⁺	5540	1	O^{2-}	−640	6		
Y³⁺	3795	1	S^{2-}	−390	6		
In³⁺	5079	1	CO_3^{2-}	−226	6		
Sb³⁺	4889	1	SO_3^{2-}	−270	6		
La³⁺	3437	1	SO_4^{2-}	−1660	6		
Ce³⁺	3549	1	SiF_6^{2-}	740	7		
Pr³⁺	3650	1	PO_4^{3-}	−657	6		
Nd³⁺	3719	1	$Fe(CN)_6^{3-}$	520	7		
Pm³⁺	3777	1	$Fe(CN)_6^{4-}$	−40	7		
Sm³⁺	3904	1					
Eu³⁺	4055	1					

(continued)

TABLE 2.2 (Continued)

Cation	$\Sigma I_p + z_r RT/\text{kJ·mol}^{-1}$	Reference	Anion	$-EA - \lvert z_1 \rvert RT/\text{kJ·mol}^{-1}$	Reference	$PA/\text{kJ·mol}^{-1}$	Reference
Gd^{3+}	3769	1					
Tb^{3+}	3810	1					
Dy^{3+}	3927	1					
Ho^{3+}	3949	1					
Er^{3+}	3953	1					
Tm^{3+}	4046	1					
Yb^{3+}	4215	1					
Lu^{3+}	3924	1					
Tl^{3+}	5457	1					
Bi^{3+}	4797	1					
Ac^{3+}	3522	1					
U^{3+}	3626	12					
Am^{3+}	3863	12					
Cm^{3+}	3793	12					
Zr^{4+}	7483	1					
Sn^{4+}	9004	6					
Ce^{4+}	7101	1					
Hf^{4+}	7491	1					
Th^{4+}	6425	1					
U^{4+}	6771	12					
Np^{4+}	6935	12					
Pu^{4+}	7130	12					

TABLE 2.3 The Standard Molar Gibbs Energy and Enthalpy of formation [1] and the Standard Molar Entropy [13] and Constant-pressure Molar Heat Capacity [14] of Isolated Ions at the Standard Temperature $T° = 298.15\,K$ and Pressure $P° = 0.1\,MPa$

Ion	$\Delta_f G°(I^{z\pm}, ig)/$ kJ·mol^{-1}	$\Delta_f H°(I^{z\pm}, ig)/$ kJ·mol^{-1}	$S°(I^{z\pm}, ig)/$ J·K^{-1}·mol^{-1}	$C_p°(I^{z\pm}, ig)/$ J·K^{-1}·mol^{-1}
H$^+$	1523.2	1536.2	108.9	20.8
D$^+$	1520.7	1540.32 [15]	117.8 [15]	20.8
Li$^+$	654.8	685.78	133.0	20.8
Na$^+$	580.5	609.36	148.0	20.8
K$^+$	487.3	514.26	154.6	20.8
Rb$^+$	464	490.1	164.4	20.8
Cs$^+$	432.7	457.96	169.9	20.8
Cu$^+$	1051.9	1089.99	161.1	20.8
Ag$^+$	984.5	1021.73	167.4	20.8
Au$^+$	1224.4	1262.44	174.6 [12]	20.8
Tl$^+$	739	772.2	175.3	20.8
H$_3$O$^+$	602.2	570.7 [16]	192.8	34.9
NH$_4^+$	681	630 [6]	186.3	34.9
HONH$_3^+$		160.7 [17]	235.4	43.5
H$_2$NNH$_3^+$	793	707 [6]	230.5	43.5
C(NH$_2$)$_3^+$		462 [18]	264.5 [19]	77.9 [18]
Me$_4$N$^+$	73.4	537 [20]	331.9 [14]	109.6
Et$_4$N$^+$	712	411 [20]	483	201 [21]
Pr$_4$N$^+$		307 [20]	641	294 [21]
Bu$_4$N$^+$		221 [20]		386 [21]
Ph$_4$P$^+$			651.0 [14]	366.3
Ph$_4$As$^+$			650.0 [14]	368.6
NO$^+$	959.6	989.9 [6]	198.4	29.1
NO$_2^+$	993.7	967.8	214.1	38.2
Be^{2+}	2955.4	2993.23	136.3	20.8
Mg^{2+}	2300.3	2348.5	148.7	20.8
Ca^{2+}	1892.1	1925.9	154.9	20.8
Sr^{2+}	1757	1790.54	164.7	20.8
Ba^{2+}	1628.3	1660.38	170.4	20.8
Ra^{2+}		1659.79	176.6	20.8
V^{2+}	2545.2	2590.86	182.1	29.6
Cr^{2+}	2592	2655.71	181.7	31.2
Mn^{2+}	2477.4	2519.69	173.8	20.8
Fe^{2+}	2689.6	2749.93	180.3	25.9
Co^{2+}	2785.7	2844.2	179.5	22.9
Ni^{2+}	2873.4	2931.39	178.1	21.7
Cu^{2+}	3011.5	3054.07	176.0	20.8
Zn^{2+}	2747.2	2782.78	161.1	20.8
Pd^{2+}	3024.9	3069.4	185.4	20.8
Cd^{2+}	2588.9	2623.54	167.8	20.8
Sn^{2+}	2399.9	2434.8	168.5	20.8
Sm^{2+}	1799.2	1833	183.1	20.8

(*continued*)

TABLE 2.3   (Continued)

Ion	$\Delta_f G°(I^{z\pm}, ig)/$ kJ·mol^{-1}	$\Delta_f H°(I^{z\pm}, ig)/$ kJ·mol^{-1}	$S°(I^{z\pm}, ig)/$ J·K^{-1}·mol^{-1}	$C_p°(I^{z\pm}, ig)/$ J·K^{-1}·mol^{-1}
Eu^{2+}	1786.9	1820	188.9	20.8
Yb^{2+}	1909.8	1943.64	172.2	20.8
Pt^{2+}	3153.5	3199.1	194.6	20.8
Hg^{2+}	2860.9	2890.47	273.0	20.8
Hg$_2^{2+}$			175.1	36.9
Pb^{2+}	2328.6	2373.33	175.5	20.8
UO$_2^{2+}$		1210 [11]	257.8 [14]	48.4
Al^{3+}	5446.9	5483.17	150.2	20.8
Sc^{3+}	4616	4652.31	156.4	20.8
V^{3+}	5380.2	5424.6	177.8	20.8
Cr^{3+}	5602.1	5648.4	179.0	20.8
Fe^{3+}	5669.1	5712.8	174.0	20.8
Co^{3+}	6038.2	6082.7	179.3	20.8
Ga^{3+}	5780.5	5816.6	161.9	20.8
Y^{3+}	4163.9	4199.86	164.9	20.8
In^{3+}	5289.1	5322	168.1	20.8
Sb^{3+}	5114	5151	169.0 [1]	20.8
La^{3+}	3871	3904.9	170.5	20.8
Ce^{3+}	3936.8	3970.6	185.5	20.8
Pr^{3+}	3971.3	4005.8	188.9	20.8
Nd^{3+}	4014.6	4050	191.6	20.8
Pm^{3+}	4044	4079 [6]	191.0	20.8
Sm^{3+}	4065.1	4100	189.0	20.8
Eu^{3+}	4199.4	4230	181.0	20.8
Gd^{3+}	4126.9	4163	189.3	20.8
Tb^{3+}	41614.1	41967	193.5	20.8
Dy^{3+}	4169	4205	195.5	20.8
Ho^{3+}	4207	4243	196.2	20.8
Er^{3+}	4231.4	4268	195.9	20.8
Tm^{3+}	4261.2	4297	194.3	20.8
Yb^{3+}	4328.3	4367.3	190.5	20.8
Lu^{3+}	4293.5	4350	173.4	20.8
Tl^{3+}	5606.1	5639.2	175.3	20.8
Bi^{3+}	4968.1	5004	175.9	20.8
Ac^{3+}	4000.8	3885 [12]	176.6	20.8
U^{3+}	4132.4	4176 [12]	195.5	20.8
Am^{3+}	4128.4	4165 [12]	177.4	20.8
Cm^{3+}	4162.4	4199 [12]	194.5	20.8
Zr^{4+}	8223.4	8261 [6]	165.2	20.8
Sn^{4+}	9285.8	9320.7	168.5	20.8
Ce^{4+}	7493.6	7523	170.6	20.8
Hf^{4+}	8153.2	8192	173.6	20.8
Th^{4+}	6984.2	7021	176.9	20.8
U^{4+}	7286.4	7327 [12]	186.4	20.8

TABLE 2.3 (Continued)

Ion	$\Delta_f G°(I^{z\pm}, ig)/$ kJ·mol^{-1}	$\Delta_f H°(I^{z\pm}, ig)/$ kJ·mol^{-1}	$S°(I^{z\pm}, ig)/$ J·K^{-1}·mol^{-1}	$C_P°(I^{z\pm}, ig)/$ J·K^{-1}·mol^{-1}
Np^{4+}	7383.8	7425 [12]	188.7	20.8
Pu^{4+}	7457.9	7498 [12]	190.7	20.8
H$^-$	132.2	139.03 [15]	108.8 [15]	20.8
F$^-$	−268.6	−255.39	145.6	20.8
Cl$^-$	−241.4	−233.13	154.4	20.8
Br$^-$	−245.1	−219.07	163.6	20.8
I$^-$	−230.2	−197	169.4	20.8
OH$^-$	−144.8	−143.5	172.3	29.1
SH$^-$	−146.6	−120 [22]	186.2	29.1
ClO$^-$	−108.5	−108 [6]	215.7	32.5
BrO$^-$			227.2	33.2
IO$^-$		−48 [6]		
CN$^-$	−59.2	36 [22]	196.7	29.1
NCO$^-$	−196.4	−192 [6]	218.9	38.0
SCN$^-$		−49	232.5	43.2
N$_3^-$	203.1	180.7	212.25	37.9
HF$_2^-$	−666	−683 [6]	211.3	34.0
HO$_2^-$	−81.5	−94 [6]	228.6	36.2
I$_3^-$	−529.9	−482 [6]	334.7	61.7
BO$_2^-$	−668	−667 [6]	215.8	38.4
ClO$_2^-$	−11.2	−29 [6]	257.0	44.7
NO$_2^-$	−182.7	−202 [22]	236.2	37.1
NO$_3^-$	−272.8	−320 [22]	245.2	44.7
ClO$_3^-$	−153.8	−200 [22]	264.3	57.6
BrO$_3^-$	−113.7	−145 [22]	278.7	60.4
IO$_3^-$	−184.9	−208 [22]	288.2	62.0
ClO$_4^-$	−266.8	−344 [22]	263.0	62.0
MnO$_4^-$	−674.8	−723.8	277.8	72.4
TcO$_4^-$			288.5	75.0
ReO$_4^-$	−930.4	−976 [2]	284.1	74.7
BF$_4^-$	−1644.2	−1687 [6]	267.9	67.8
HCO$_2^-$	−452.7	−460 [23]	238.2	38.8
CH$_3$CO$_2^-$	−464.1	−504.2 [23]	278.2	61.4
PhCO$_2^-$	−330.6	−400.4 [23]	338.0 [19]	117.9
CF$_3$CO$_2^-$	−1138	−1194 [6]	331.0 [14]	89.3
CF$_3$SO$_3^-$			346.0 [14]	107.6
BPh$_4^-$			656.0 [14]	363.7 [14]
HCO$_3^-$	−702	−738 [22]	257.9	50.6
HSO$_4^-$	−886.1	−953 [6]	283.0	70.9
H$_2$PO$_4^-$	−1190	−1280 [6]	280.7	62.5
PF$_6^-$	−2005.6	−2109.9 [6]	299.6 [19]	104.7
SbF$_6^-$	−1901	−1993 [6]	345.5 [19]	124.0

(continued)

TABLE 2.3 (Continued)

Ion	$\Delta_f G°(I^{z\pm}, ig)/$ kJ·mol⁻¹	$\Delta_f H°(I^{z\pm}, ig)/$ kJ·mol⁻¹	$S°(I^{z\pm}, ig)/$ J·K⁻¹·mol⁻¹	$C_p°(I^{z\pm}, ig)/$ J·K⁻¹·mol⁻¹
O^{2-}	939.7	950 [6]	143.3	20.8
S^{2-}			152.1	20.8
CO_3^{2-}	−300.9	−321 [22]	246.1	44.4
$C_2O_4^{2-}$			295.1	76.0
SO_3^{2-}	−1035.5		264.3	52.6
SO_4^{2-}	−704.8	−758 [22]	263.6	62.4
SeO_4^{2-}			281.2	73.5
CrO_4^{2-}	−659.5	−705 [22]	281.4	74.8
MoO_4^{2-}			291.1	77.0
WO_4^{2-}			296.6	76.6
$S_2O_3^{2-}$			291.1	71.0
SiF_6^{2-}	−2183.4	−2161 [6]	309.9	113.1
$Cr_2O_7^{2-}$			379.7	140.8
HPO_4^{2-}			283.0	67.8
PO_4^{3-}			266.4	65.4
$Fe(CN)_6^{3-}$			491.6 [14]	217.1
$Co(CN)_6^{3-}$			464.8	210.2
$Fe(CN)_6^{4-}$			469.8	210.6

References are provided for data not found in those given in this caption.

thermodynamically consistent manner. The standard molar entropy and constant-pressure heat capacity, $S°(I^{z\pm}, ig)$ and $C_p°(I^{z\pm}, ig)$, in J·K⁻¹·mol⁻¹ of isolated ions are also well-defined quantities and are recorded in Table 2.3. For monatomic ions with no unpaired electrons, the standard molar entropy reflects the translational entropy alone and depends only on the mass of the ion. At $T° = 298.15$ K and $P° = 0.1$ MPa $S°(I^{z\pm}, ig) = 108.85 + (3/2)\ln(M/M°)$J·K⁻¹·mol⁻¹, where $M/M°$ is the relative molar mass $(M° = 1$·kg·mol⁻¹$)$. For monatomic ions with no unpaired electrons, the standard molar heat capacity depends on the translational degrees of freedom alone, and hence is common to all the monatomic ions: $C_p°(I^{z\pm}, ig) = (5/2)R = 20.79$ J·K⁻¹·mol⁻¹. Monatomic ions with unpaired electrons have a contribution from electronic spin to the heat capacity and entropy and poly-atomic ions have contributions from their rotational and vibrational modes. The standard molar volume of an isolated ion is a trivial quantity, being the same for all ions: $V°(I^{z\pm}, ig) = RT°/P° = 0.02479$ m³·mol⁻¹, where $R = 8.31451$ J·K⁻¹·mol⁻¹ is the gas constant.

The shape of isolated monatomic ions is spherical, but they may be deformed slightly by external forces (strong electrical fields). Ions that consist of several atoms may have any shape, but common ones are planar (NO_3^-, CO_3^{2-}), tetrahedral (NH_4^+, SO_4^{2-}), octahedral $(Fe(CN)_6^{4-})$, elongated (SCN^-), or more irregular $(CH_3CO_2^-, HCO_3^-)$. Tetrahedral and octahedral ions approximate spherical shape for many purposes and are termed globular.

The size of an isolated ion cannot be specified readily, because its outer electron shell extends indefinitely around the inner ones and the nucleus. In a series of isoelectronic monatomic species, the sizes diminish, for example, $O^{2-} > F^- > Ne > Na^+ > Mg^{2+}$, because of the increasing positive nuclear charge that pulls in the electrons. Nevertheless, radii for isolated monatomic ions, $r_I^g = r(I^{z\pm}, ig)$ have been specified with the quantum-mechanical scaling principle relative to those of the noble gases (of known collision diameters), for example, Ne in the above series.

An isolated ion ($I^{z\pm}$, ig) may be assigned a self-energy, due to its being charged. Per mole of isolated ions the self-energy is

$$E_{self}\left(I^{z\pm}, ig\right) = \frac{N_A z_I^2 e^2}{4\pi\varepsilon_0 r_I^g} \tag{2.2}$$

Where $\varepsilon_0 = 8.85419\times10^{-12}$ $C^2 \cdot J^{-1} \cdot m^{-1}$ is the permittivity of free space. The size of an isolated ion is an ill-defined quantity, as stated above, and so are its radius r_I^g and self-energy E_{self}.

Some other properties of isolated ions have been determined: the magnetic susceptibility, the polarizability, and the softness/hardness. Unless they have one or more unpaired electrons in their electronic shells, ions are diamagnetic, that is, they are repulsed out from a magnetic field. Their molar magnetic susceptibilities, χ_{Im}, are negative and range from a few to several tens of the unit $(-10^{-12} m^3 \cdot mol^{-1})$ with the dimension of a molar volume. Ions that have one or more unpaired electrons in their electronic shells are paramagnetic, are attracted into a magnetic field, and have positive molar susceptibilities. A paramagnetic ion at $T^\circ = 298.15\,K$ has $\chi_{Im} = +1.676n(n+2)\times10^{-9}\,m^3 \cdot mol^{-1}$ where n is the number of unpaired electrons. The values of χ_{Im} for many ions are shown in Table 2.4, some of the data having been determined in aqueous solutions as noted, but they should be valid for the isolated ions too, because χ_{Im} is rather insensitive to the environment in which the ion is situated.

The polarizability, α_I, of an ion is obtained indirectly from the molar refractivity at infinite frequency $R_{I\infty}$ that is proportional to the polarizability:

$$R_{I\infty} = \left(\frac{4\pi N_A}{3}\right)\alpha_I = 2.5227\times10^{24}\alpha_I \tag{2.3}$$

In lieu of the infinite frequency value R_∞, the molar refractivity R_D determined from the refractive index at the sodium D line (589 nm), n_D, can be used. It is given by the Lorenz–Lorentz expression:

$$R_D = \frac{V\left(n_D^2 - 1\right)}{n_D^2 + 2} \tag{2.4}$$

Where $V = (M/\rho)$ is the molar volume and M and ρ are the molar mass and the density. However, only for neutral species can the molar refractivity be obtained

TABLE 2.4 The Magnetic Susceptibility, χ_{lm} [6], the Molar Refraction at the Sodium D Line, R_D (normalized to the Value for Na⁺) [6], and the Softness Parameter [6], Modified by ±0.3 Units, $\sigma_{\pm0.3}$, of Isolated Ions

Cation	$-\chi_{lm}/10^{-12}\,m^3\cdot mol^{-1}$	$R_D/10^{-6}\,m^3\cdot mol^{-1}$	$\sigma_{\pm0.3}$	Anion	$-\chi_{lm}/10^{-12}\,m^3\cdot mol^{-1}$	$R_D/10^{-6}\,m^3\cdot mol^{-1}$	$\sigma_{\pm0.3}$
H⁺	6.6[a]	−0.3	−0.30	F⁻	13[a]	2.21	−0.36
Li⁺	3[a]	0.08	−1.32	Cl⁻	28[a]	8.63	+0.21
Na⁺	2.3[a]	0.65	−0.90	Br⁻	39[a]	12.24	+0.47
K⁺	11.2[a]	2.71	−0.88	I⁻	56.7[a]	18.95	+0.80
Rb⁺	20.1[a]	4.1	−0.83	OH⁻	12[a]	4.65	+0.30
Cs⁺	34[a]	6.89	−0.84	SH⁻		12.8	+0.95
Cu⁺	12	3.1	−0.52	CN⁻	18	7.9	+0.71
Ag⁺	24	5.1	−0.12	NCO⁻	21		+1.01
Au⁺	40		+0.14	SCN⁻	35	17	+1.15
Tl⁺	34	11.5	−0.10	N₃⁻		11	+1.06
H₃O⁺			−0.30	HF₂⁻			−1.54
NH₄⁺	11.5	4.7	−0.90	HO₂⁻			+0.73
C(NH₂)₃⁺		11.21 [18]		I₃⁻			+1.17
Me₄N⁺	65	22.9	+0.11	BO₂⁻	15[a]		−0.64
Et₄N⁺		43		NO₂⁻	23[a]	8.7	+0.45
Pr₄N⁺		61		NO₃⁻	32	10.43	+0.33
Bu₄N⁺		79		ClO₃⁻	40	12.1	+0.33
Ph₄As⁺	229	115.3	+6.61	BrO₃⁻	50	15.2	
Be²⁺	6		−0.93	IO₃⁻	34	18.85	
Mg²⁺	5 [24]	−0.7	−0.71	ClO₄⁻		12.77	0.00
Ca²⁺	8 [24]	1.59	−0.96	MnO₄⁻			+0.16
Sr²⁺	9[a]	2.65	−0.94	ReO₄⁻	60[c]		−0.10
Ba²⁺	21.5[a]	5.17	−0.96	BF₄⁻	39		0.00
Ra²⁺			−0.98	HCO₂⁻	21 [24]	9.43	−0.03
V²⁺	15[b,c]		−0.40	CH₃CO₂⁻	32.4	13.87	+0.08
Cr²⁺	−15[b,c]		−0.54	PhCO₂⁻		27.9	
Mn²⁺	−20.7[b]	2.2	−0.45	CF₃CO₂⁻	50		

Cation			
Fe^{2+}	-19.6[b]	2.1	-0.46
Co^{2+}	-18.5[b]	2.05	-0.41
Ni^{2+}	-17.5[b]	1.6	-0.41
Cu^{2+}	-16.4[b]	1.3	+0.08
Zn^{2+}	10	1.39	+0.05
Pd^{2+}	25[c]		+0.18
Cd^{2+}	22.5	3.22	+0.28
Sn^{2+}	20		-0.01
Sm^{2+}	-23[b,c]		-0.92
Eu^{2+}	-22[b,c]		-0.92
Yb^{2+}	20[c]		-0.96
Pt^{2+}	40[c]		+0.03
Hg^{2+}	37	6.13	+0.97
Pb^{2+}	28	11.9	+0.11
UO_2^{2+}	43[c]	14.1[a]	-0.67
Al^{3+}	3.1	-1.18	-0.61
Sc^{3+}	6[c]	1.6	-0.92
V^{3+}	-10[b,c]	3.5[a]	-0.59
Cr^{3+}	-16[b]		-0.40
Fe^{3+}	-15.6[b]	3.2	+0.03
Co^{3+}	-10[b,c]		+0.20
Ga^{3+}	8	5[24]	-0.01
Y^{3+}	12[c]	2.4	-0.99
In^{3+}	19	1.7[24]	+0.18
Sb^{3+}	14[c]	8.8[d]	+0.33
La^{3+}	20[c]	2.74	-1.05
Ce^{3+}	-20[b,c]	3.4[24]	-1.02
Pr^{3+}	-20[b,c]	3.3[24]	-0.93
Nd^{3+}	-20[b,c]	3.1[24]	-0.88

Anion			
BPh_4^-	215	108.7	+7.16
HCO_3^-		10.9	
HSO_4^-	37[24]		
$H_2PO_4^-$		14.6	
O^{2-}	12		
S^{2-}	38		+1.39
CO_3^{2-}	34[a]	11.45	-0.20
SO_3^{2-}	38	12.9	-0.04
SO_4^{2-}	40	13.79	-0.08
SeO_4^{2-}	51	16.4	
CrO_4^{2-}	51	27.5[d]	
MoO_4^{2-}	55	25.2[d]	
WO_4^{2-}	61	23.2[d]	
$S_2O_3^{2-}$	49	23.2	
SiF_6^{2-}		11.01[a]	
SO_3^{2-}	38	12.9	-0.04
SO_4^{2-}	40	13.79	-0.08
SeO_4^{2-}	51	16.4	
CrO_4^{2-}	51	27.5[d]	
PO_4^{3-}	50	15.1	
$Fe(CN)_6^{3-}$			-0.48
$Co(CN)_6^{3-}$		50.7[d]	+3.52
$Fe(CN)_6^{4-}$		46.7[d]	+3.93

(continued)

TABLE 2.4 (Continued)

Cation	$-\chi_{lm}/10^{-12}\,m^3\cdot mol^{-1}$	$R_D/10^{-6}\,m^3\cdot mol^{-1}$	$\sigma_{\pm 0.3}$	Anion	$-\chi_{lm}/10^{-12}\,m^3\cdot mol^{-1}$	$R_D/10^{-6}\,m^3\cdot mol^{-1}$	$\sigma_{\pm 0.3}$
Pm^{3+}	$-20^{b,c}$		-0.83				
Sm^{3+}	$-20^{b,c}$	2.9 [24]	-0.66				
Eu^{3+}	$-20^{b,c}$	2.7 [24]	-0.49				
Gd^{3+}	$-20^{b,c}$	2.6 [24]	-0.96				
Tb^{3+}	$-19^{b,c}$	2.5 [24]	-0.94				
Dy^{3+}	$-19^{b,c}$	2.4 [24]	-0.80				
Ho^{3+}	$-19^{b,c}$	2.2 [24]	-0.83				
Er^{3+}	$-18^{b,c}$	2.1 [24]	-0.87				
Tm^{3+}	$-18^{b,c}$	2.0 [24]	-0.73				
Yb^{3+}	$-18^{b,c}$	2.0 [24]	-0.57				
Lu^{3+}	17^c		-0.84				
Tl^{3+}	31	2.2 [24]	$+0.77$				
Bi^{3+}	25	8.1	$+0.52$				
Ac^{3+}			-0.97				
U^{3+}			-0.84				
Am^{3+}			-0.74				
Cm^{3+}			-1.01				
Zr^{4+}	12.5	1.0 [24]	-0.73				
Sn^{4+}	16	1.3 [24]	$+0.26$				
Ce^{4+}	21	1.9 [24]	-0.40				
Hf^{4+}	16^c	4.3^d	-0.91				
Th^{4+}	31.2	6.8^d	-0.97				
U^{4+}	35^c		-0.46				
Np^{4+}			-0.52				
Pu^{4+}		8.6^d	-0.51				

References are shown for data not selected in Ref. 6.

[a] In aqueous solutions.
[b] Paramagnetic.
[c] From Selwood [25].
[d] From Salzmann [26].

experimentally either in crystals or in dilute solutions. On the other hand, R_D is not very sensitive to the environment of the ions, and hence may be ascribed to the neutral combinations of the bare cations and anions. In order to ascribe a molar refractivity to an individual ion, the experimental R_D values must be split appropriately between the cations and the anions. There is no theoretically valid way to do this, so an empirical expedient is resorted to, namely using R_D (Na^+) = 0.65 cm$^3 \cdot$mol^{-1} at 25°C and the additivity of the stoichiometrically weighted values for the cations and anions, $R_D = \sum \nu_I R_{DI}$. The polarizability of an isolated ion $\alpha_I = \alpha(I^{z\pm}, ig)$ is equated with that obtained experimentally as $\alpha_I = 3R_{DI}/4\pi N_A$ (for R_{DI} in m$^3 \cdot$mol^{-1}) yielding values of the order of 10^{-30} m$^3 \cdot$particle^{-1}. The temperature coefficient of R_D is rather small, approximately +0.01 cm$^3 \cdot$mol$^{-1} \cdot$K^{-1}. Values of α_I are shown in Table 2.4 for many ions.

The "softness" that can be ascribed to an ion is loosely related to its polarizability. The ionic softness is obtained from the difference between the energetics of formation of the ion in the ideal gas phase from the neutral species on the one hand (loss or gain of electrons) and the transfer of the ions from there to an aqueous solution to produce the standard aqueous ions on the other. The gain or loss of pairs of electrons by coordination with the solvent in the hydration process neutralizes to some extent the charge on the ion that thus partly reverts to a neutral species. Normalized numerical values of the softness parameter are:

$$\sigma_+ = \frac{\left[\left\{ \sum I_p + \Delta_{hydr} H^\infty (I^+) \right\} / z_+ - \left\{ I_p(H^+) - \Delta_{hydr} H^\infty (H^+) \right\} \right]}{\left[I_p(H^+) - \Delta_{hydr} H^\infty (H^+) \right]} \tag{2.5}$$

for cations and

$$\sigma_- = \frac{\left[\left\{ -EA - \Delta_{hydr} H^\infty (I^-) \right\} / z_- + \left\{ -EA(OH) - \Delta_{hydr} H^\infty (OH^-) \right\} \right]}{\left[-EA(OH) - \Delta_{hydr} H^\infty (OH^-) \right]} \tag{2.6}$$

for anions.

The originally published [27] softness parameters σ_I are based on the arbitrary assignment of zero to the hydrogen ion for cations and to the hydroxide ion for anions, but a common scale for ions of both charge signs, $\sigma_{I\pm 0.3}$, is produced when 0.3 units are subtracted from these cation values and 0.3 is added to the anion values. Positive values of the softness parameter denote "soft" ions and negative values denote "hard" ions. The $\sigma_{I\pm 0.3}$ values of many ions are recorded in Table 2.4.

The magnetic susceptibility, the polarizability (molar refraction), and the softness parameter of isolated ions are portable and additive. This means that these properties of ions are not appreciably sensitive to the environment of the ions, whether they are isolated or in solution or in crystalline compounds or in molten salts. The property of a compound or an electrolyte is the sum of the stoichiometrically weighted properties of the cations and anions.

2.1.2 Ions in Clusters

Ions in the ideal gas phase may associate with solvent molecules to form clusters that are generally studied by means of mass spectroscopy. High pressure mass spectrometry and electrospray techniques, see, for example, Schroeder [28], have been employed and improved over the years to provide experimental information. Solvent molecules may be attached to an ion stagewise and the equilibrium constants are established experimentally, yielding the values of the molar Gibbs energy for the solvation of the ion in the ideal gas phase. The temperature dependence of the equilibrium constants yields the enthalpy and entropy for the cluster formation. Computer simulations, on the other hand, yield the geometry and bond distances of the species with minimal potential energy. High-level quantum-mechanical potential functions are employed for the molecular dynamics simulations.

The generalized equilibrium reaction in the gas phase between an ion $I^{z\pm}(g)$ and solvent molecules $S(g)$ is:

$$I^{z\pm}S_{n-1}(g) + S(g) \rightleftarrows I^{z\pm}S_n(g) \tag{2.7}$$

starting with $n=1$ and following n to as large values as could be achieved. The ion currents of $I^{z\pm}S_{n-1}(g)$ and $I^{z\pm}S_n(g)$, I_{n-1} and I_n, are measured in the mass spectrometer as a function of the solvent pressure P_S to yield the equilibrium constant $K_{n-1,n}$ from which the Gibbs energy of the solvation step $n-1$ to n is determined:

$$\Delta_{n-1,n}G° = -RT \ln K_{n-1,n} = -RT \ln\left[I_n/I_{n-1}(P_S/P°)\right] \tag{2.8}$$

A regularity was found in the stepwise Gibbs energies of the gaseous ion solvation—that of the second step is ~75% and that of the third step is ~50% of that for the first step [29]. For further steps, the values of $\Delta_{n-1,n}G°$ are approximately inversely proportional to $n-1$, the probability of the addition of another solvent molecule being statistical. The entropy change for each solvation step is rather indifferent to the solvation number n and in the case that the solvent S is water $T\Delta_{n-1,n}S° \sim 30\ kJ\cdot mol^{-1}$, making $\Delta_{n-1,n}H°$ more negative than $\Delta_{n-1,n}G°$ by this amount. Water has been studied extensively as the ion cluster solvent, but ammonia, methanol, and acetonitrile, among a few other solvents, have also been studied.

Alkali metal cations and halide anions have been studied in the earlier years, but alkaline earth metal and divalent transition- and post-transition metal cations have been studied more recently. However, for divalent ions, the second ionization potential ranging from 11.02 eV for Sr^{2+} through values for Ca^{2+}, Mg^{2+}, Pb^{2+}, Mn^{2+}, Cr^{2+}, Zn^{2+} to 20.27 eV for Cu^{2+} is larger than the ionization potential of the solvent molecules studied—water, ammonia, and methanol: 12.61, 10.16, and 10.85 eV respectively (except for Ca^{2+} and Sr^{2+} and water). Therefore, irreversible charge transfer $M^{2+}S_n(g) \rightarrow M^+S_{n-1}(g) + S^+(g)$ occurs below certain minimal n values that depend on the ion M^{2+} and solvent S. Only above n_{min}, the accumulated solvent molecules stabilize the divalent cation and therefore experimental values can be obtained only for the solvation steps with $n \geq n_{min}$ according to Chen and Stace [30]. In some cases, the stepwise solvation energies for multicharged metastable ions at $n < n_{min}$

have been obtained from theoretical calculations. The stepwise enthalpies of hydration of gaseous ions are shown in Table 2.5, and only few anions are included as the central ions of clusters. For several other gaseous anions, only the equilibrium constants of the stepwise hydration have been measured at a single temperature, leading to the Gibbs energies, Table 2.6, so that the enthalpies are not available.

Experimental and theoretically calculated values for the clustering energetics are not confined to the solvent $S = H_2O$, but for most solvents, only the addition of a single S molecule to an ion has been considered. Ammonia, acetonitrile, and methanol are exceptions to this situation and stepwise solvation of ions has been studied for these solvents (Table 2.7). Competition of such a solvent with water around an ion in a cluster is a readily available experimental method as suggested by Nielsen et al. [32].

Apart from the inherent interest in the gas phase ion clusters, the accumulation of solvent molecules around an ion should yield at the limit of very large values of n to a constant value of $\Delta_{n-1,n}H°(S,g)$. This would be the molar enthalpy of condensation of a solvent molecule into the bulk liquid solvent, because at this limit, the ion has no influence any more on the energetics of the process. Thus:

TABLE 2.5   The Standard Molar Enthalpies of Stepwise Clustering of Water Molecules Around Ions in the Ideal Gas Phase, $-\Delta_{n-1,n}H°/\text{kJ·mol}^{-1}$, at $T° = 298.15\,\text{K}$

Ion/n	1	2	3	4	5	6	Reference
Li$^+$	142	108	87	69	58	51	31
Na$^+$	100	83	66	58	51	45	31
K$^+$	75	67	55	49	45	42	31
Rb$^+$	67	57	51	47	44		31
Cs$^+$	57	52	47	44			31
Cu$^+$		69	70	61			32
Ag$^+$	139	106	63	62	57	56	32
H$_3$O$^+$	144	87	72	55	50		33
NH$_4^{+a}$	48	35	25	17	13		29
Mg^{2+}	342	301	237	184	117	103	34
Ca^{2+}	236	203	179	149	116	106	34
Sr^{2+}						95	35
Ba^{2+}					100	83	35
Zn^{2+}	431	368	233	179	105	101	34
F$^-$	97	69	57	56	55		36
Cl$^-$	55	53	49	46			36
Br$^-$	53	51	48	46			36
I$^-$	43	41	39	38			36
OH$^-$	105	69	63	59	59		37
NO$_2^-$	64	57	49	49			38
NO$_3^-$	61	60	58				38
HCO$_3^-$	66	62	57	56			39
SO$_4^{2-}$						61	40

^{a} $-\Delta_{n-1,n}G°/\text{kJ·mol}^{-1}$ values rather than the enthalpies.

TABLE 2.6 The Standard Molar Gibbs Energies of Stepwise Clustering of Water Molecules Around Anions in the Ideal Gas Phase, $-\Delta_{n-1,n}G°/\text{kJ·mol}^{-1}$, at $T° = 293.15\,K^a$

Anion/n	1	2	3
F[b]	76	52	35
Cl[a]	37	28	22
Br[b]	31	26	20
I[b]	24	18	13
NO_2^-	36	25	19
NO_3^-	30	22	16
ClO_2^-	38	26	20
ClO_3^-	26	20	
ClO_4^-	20		
BrO_3^-	27	21	
IO_3^-	27	21	18
HSO_4^-	25	20	
$CF_3SO_3^-$	19	16	
$H_2PO_4^-$	32	26	20
HCO_2^-	38	28	21
$CH_3CO_2^-$	39	28	22
$CF_3CO_2^-$	28	20	

[a] From Ref. 41.
[b] From Ref. 29.

TABLE 2.7 The Standard Molar Gibbs Energies of Stepwise Clustering of Acetonitrile, Methanol, and Ammonia Molecules Around Ions in the Ideal Gas Phase, $-\Delta_{n-1,n}G°/\text{kJ·mol}^{-1}$, at $T° = 298.15\,K^a$

Ion/n	1	2	3	4	5	6	7	8
Acetonitrile								
Na+	98	74	51	28	2			
K+	75	56	41	23	6			
Rb+	64	48	35	20	6			
Cs+	57	43	30	17	4			
F-	74	45	35	21	11	6		
Cl-	37	32	22	12	7	3	1	
Br-	39	28	18	12	8	5	4	
I-	29	21	14	9	6			
Methanol								
F-	66	52	34	23	16	11	9	7
Cl-	43	29	21	15	11	10	8	7
Br-	36	26	17	12	10	8	7	6
I-	26	18	13	10	7	6	4	3
Ammonia								
Li+	134	101	56	28	12	8		
Na+	90	64	41	26	8	3		
K+	55	40	28	17				
Rb+	48	34	23	14	5			
Cu+[b]			29	18	12			
Ag+[b]		114	31	17	11			
NH_4^+	72	39	26	15	1			

[a] From Ref. 29.
[b] From Ref. 32.

$$\lim(n \to \infty)\left[\Delta_{n-1,n}H^\circ\left(I^{\pm z},g\right) - \Delta_{n-1,n}H^\circ(S,g)\right] = 0 \tag{2.9}$$

For S = water $\Delta_{n-1,n}H^\circ(S,g)/kJ \cdot mol^{-1} = -44.01 + 42.7\left[n^{2/3} - (n-1)^{2/3}\right]$ at 298.15 K according to Coe [33], where $44.01\,kJ \cdot mol^{-1} = -\lim(n \to \infty)\Delta_{n-1,n}H^\circ(S,g)$ is the molar enthalpy of vaporization of water. The area between the curves $\Delta_{n-1,n}H^\circ(I^{\pm z},g) = f(n)$ and $\Delta_{n-1,n}H^\circ(S,g) = f'(n)$, taking n to be a continuous variable, is the molar enthalpy of solvation of the ion, that is, $\Delta_{solv}H^\circ(I^{\pm z},S)$, the molar enthalpy of transfer of the ion from its standard state in the ideal gas phase to its standard state in the solution, see Figure 2.1. More rapid convergence is achieved if the differences between the enthalpies of the stepwise solvation enthalpies of two ions I_α^z and I_β^z with the same charge number z are considered as functions of n.

$$\Delta_{solv}H^\circ\left(I_\alpha,S\right) - \Delta_{solv}H^\circ\left(I_\beta,S\right) = \lim(n \to \infty)\left[\Delta_{n-1,n}H^\circ\left(I_\alpha,S,g\right) - \Delta_{n-1,n}H^\circ\left(I_\beta,S,g\right)\right]$$
$$\tag{2.10}$$

Differences in the solvation enthalpies of ions with the same charge number z_I are well-defined thermodynamic quantities and can serve to test Equation 2.10 according to Coe [33].

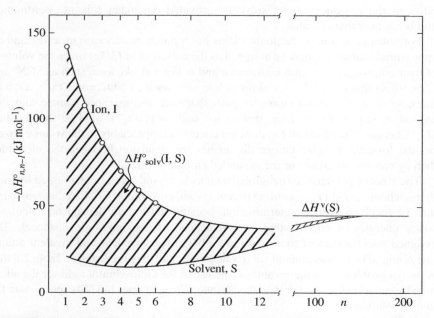

FIGURE 2.1 Schematic representation of the stepwise molar enthalpy $-\Delta_{n-1,n}H^\circ$ of the gas phase clustering of a solvent around an ion and a solvent molecule. The horizontal asymptote is the molar enthalpy of vaporization of the solvent and the area between the curves is the standard molar solvation enthalpy of the ion. (The circles are the experimental data for Li$^+$ in water.) From Ref. 42 by permission of the publisher, Wiley.

2.2 SIZES OF IONS

As mentioned earlier, the sizes of isolated ions are ill-defined, but in condensed phases, the ions can be assigned definite sizes, because of the strong repulsion of the contagious electronic shells. In crystals, the interionic distances can be measured by x-ray and neutron diffraction with an uncertainty of a fraction of a pm and individual ionic radii (at least for monatomic and globular ions) have been assigned. These radii, r_I, do depend on the coordination numbers of the ions and a set for the characteristic coordination in salt crystals that are usually met have been established by Shannon and Prewitt [43, 44].

In solutions, the distances between the centers of ions and of the nearest atoms of the surrounding solvent molecules can also be measured by x-ray and neutron diffraction, but with a somewhat larger uncertainty, ± 2 pm. In aqueous solutions, if the water molecule is assigned a constant radius $r_W = 138$ pm (one half of the experimental collision diameter), then the distances $d(I^{\pm z} - O_W)/\text{pm} = 138 + r_I/\text{pm}$ have been established by Marcus within the experimental uncertainty, with the same ionic radii r_I as in the crystals [45, 46]. These radii, as selected in Ref. 6 and annotated there, are listed in Table 2.8. The distances between the centers of ions in solutions in solvents other than water and of the nearest atoms of the solvents have also been determined in some cases reported by Ohtaki and Radnai [50] and confirm the portability of the r_I values among solvents, provided the mean solvent coordination number is near that in water.

For nonspherical ions, the ionic radius has a much more vague meaning and an approximate value of r_I may be assigned as the cube root of $(3/4\pi)$ times the volumes of their ellipsoids of rotation with axes a and b. For rod-like ions, such as SCN^- and I_3^-, $a > b$ so that $r_I \approx (a^2 b)^{1/3}$, and for oblate ions, such as NO_3^- and CO_3^{2-}, $a < b$ so that $r_I \approx (ab^2)^{1/3}$. Ions with a more irregular shape are assigned approximate values of the radius, noting, for instance, that an ion such as HSO_4^- should be smaller than SO_4^{2-}, because the added proton does not contribute appreciably to the volume (size), but the lowered negative charge diminishes the repulsion between the electrons, thereby causing shrinkage of the expanded anionic volume.

The sizes of polyatomic (nonglobular) ions in crystals are also expressed by their thermochemical radii r_{Ith} according to Jenkins and coworkers [47]. Circular reasoning may be involved in their determination, because these radii depend on calculated lattice energies of crystals that in turn depend on the interionic distances. The assigned uncertainties of these radii are ± 19 pm for univalent and divalent anions increasing to twice this amount for trivalent ones and they are listed in Table 2.8 too. A further problem with these values is the use of the Goldschmidt radii for the alkali metal counterions, r_+^G, rather than the Shannon–Prewitt ones [43, 44] appropriate for ions in solution.

A different approach to the sizes of ions (applicable to crystalline salts) is to consider their volumes rather than their radii as suggested by Jenkins et al. [48]. It is assumed that the volume of a formula unit of the salt M_pX_q is additive in the individual ionic volumes: $v(M_pX_q) = pv_+ + qv_-$. Again, the Goldschmidt radii of the alkali metal cations were used to define $v_+ = (4\pi/3)r_+^{G3}$, obtaining the anion volumes by difference. Individual ionic volumes calculated on this basis are also shown in Table 2.8. Optimized ion volumes have more recently been reported by Glasser and Jenkins [51] that differ considerably from previously reported values [24, 48] by the same authors. It has still to be demonstrated that these ionic volumes are pertinent also to the ions in solution.

A further measure of the sizes of ions, pertaining to ions in solution, is their intrinsic molar volumes, $V_{I \, intr}$. The molar volume of a "bare" unsolvated ion, $(4\pi N_A/3)r_I^3$, cannot represent the intrinsic volume of the ion in solution, because of the void spaces between the solvent molecules and the ion and among themselves. Mukerjee [52] proposed for aqueous alkali metal and halide ions at 25°C a factor of $k = 1.213$, producing:

$$V_{IIntr}^M = \left(4\pi N_A/3\right)\left(kr_I\right)^3 \tag{2.11}$$

This factor is near the value, 1.159, that is geometrically required for close-packed spheres of arbitrary but comparable sizes. These intrinsic volumes for the monatomic cations dealt with in this chapter and anions with radii <0.2 nm are shown as V_{IIntr}^M in Table 2.8.

Glueckauf [49] suggested that the void space between the water molecules surrounding an ion should be taken into account by the addition of an addend, a, to the radii of monatomic ions:

$$V_{Iintr}^G = 2522\sum_i \left(r_i + a\right)^3 \tag{2.12}$$

This addend is 0.055 nm at 25°C and is somewhat temperature dependent, and different values were suggested for polyatomic ions; some of the resulting intrinsic volumes are shown in Table 2.8.

Another method for the estimation of the intrinsic volumes of electrolytes, independent of values of the ionic radii, was proposed by Pedersen et al. [53], who employed the molar volume of the molten alkali metal halides, extrapolated to ambient temperatures, as a measure of their intrinsic volumes in aqueous solutions, but the extrapolation is quite long. A variant of this idea is to use the molar volumes of molten hydrated salts, proposed by Marcus [54], where the temperature extrapolation to 25°C is much shorter. It is then necessary to subtract the volume of the water of hydration, which is n times the molar volume of electrostricted water, 15.2 cm³·mol⁻¹ at 25°C [55], from the extrapolated molar volume of the undercooled molten hydrated salt containing n water molecules per formula unit of the salt. A cogent method, applicable to highly soluble salts, was proposed by Marcus [56]. The volumes considered, applied to aqueous solutions, are *intrinsic*, so they should be independent of the concentration c and to a certain extent also of the temperature T. The partial molar volume of an electrolyte, $V_E(c, T)$, describes the volume that it actually occupies in the solution and does not include the volume of the water. Therefore, a fairly short extrapolation of the linear $V_E(c, 25°C)$ from $c = 3$ M to such high concentrations at which all of the solvent is as closely packed as possible (completely electrostricted) is equivalent to considering the electrolyte as an undercooled molten hydrated salt

TABLE 2.8 Ionic Radii, r_I [6], Thermochemical Ionic Radii, r_{Ith} [47], Ionic Volumes, v_I [48], and Intrinsic Ionic Molar Volumes, $V_{I\,intr}^M$ (see the text)

Ion	r_I/pm	r_{Ith}/pm	v_I/nm^3	$V_{I\,intr}^M$/cm^3·mol^{-1}	$V_{I\,intr}^{aq}$/cm^3·mol^{-1}
H$^+$					0.0
Li$^+$	69		0.00199	1.5	5.1
Na$^+$	102		0.00394	4.8	5.8
K$^+$	138		0.00986	11.8	13.2
Rb$^+$	149		0.01386	14.9	
Cs$^+$	170		0.01882	22.1	24.7
Cu$^+$	96			4.0	
Ag$^+$	115			6.8	5.1
Au$^+$	137			11.6	
Tl$^+$	150			15.2	
H$_3$O$^+$	130				
NH$_4^+$	148		0.021		21.0
HONH$_3^+$	190	147	0.021		
H$_2$NNH$_3^+$	190	158	0.028		
C(NH$_2$)$_3^+$	210 [18]				
Me$_4$N$^+$	280	234	0.113		
Et$_4$N$^+$	337		0.199		
Pr$_4$N$^+$	379				
Bu$_4$N$^+$	413				
Pe$_4$N$^+$	443				
Ph$_4$P$^+$	424				
Ph$_4$As$^+$	425				
NO$^+$			0.010		
NO$_2^+$			0.022		
Be^{2+}	35		0.0002 [24]	0.2	
Mg^{2+}	72		0.00199	1.7	10.9
Ca^{2+}	100		0.00499	4.5	14.6
Sr^{2+}	113		0.00858	6.5	
Ba^{2+}	136		0.01225	11.3	
Ra^{2+}	143		0.0147 [24]	13.2	
V^{2+}	79		0.0016 [24]	2.2	
Cr^{2+}	82		0.0024 [24]	2.5	
Mn^{2+}	83		0.0032 [24]	2.6	8.6
Fe^{2+}	78		0.0022 [24]	2.1	
Co^{2+}	75		0.0022 [24]	1.9	11.4
Ni^{2+}	69		0.0020 [24]	1.5	11.7
Cu^{2+}	73			1.8	
Zn^{2+}	75		0.0024 [24]	1.9	11.1
Pd^{2+}	86			2.9	
Cd^{2+}	95		0.0046 [24]	3.9	19.8
Sn^{2+}	93			3.6	
Sm^{2+}	119			7.6	
Eu^{2+}	117		0.0080 [24]	7.2	
Yb^{2+}	105			5.2	
Pt^{2+}	80			2.3	
Hg^{2+}	102		0.0045 [24]	4.8	
Hg$_2^{2+}$	390				
Pb^{2+}	118		0.0069 [24]	7.4	

TABLE 2.8 (Continued)

Ion	r_I/pm	r_{Ith}/pm	v_I/nm^3	$V_{I\,intr}^{M}$/cm^3·mol^{-1}	$V_{I\,intr}^{aq}$/cm^3·mol^{-1}
UO_2^{2+}	280				
Al^{3+}	53		0.0008 [24]	0.7	10.0
Sc^{3+}	75		0.0024 [24]	1.9	
V^{3+}	64		0.0012 [24]	1.2	
Cr^{3+}	62		0.0011 [24]	1.1	10.5
Fe^{3+}	65		0.0013 [24]	1.2	12.7
Co^{3+}	65		0.0011 [24]	1.2	
Ga^{3+}	62		0.0010 [24]	1.1	
Y^{3+}	90		0.0031 [24]	3.3	
In^{3+}	79		0.0021 [24]	2.2	
Sb^{3+}	77		0.0019 [24]	2.1	
La^{3+}	105		0.0076 [24]	5.2	
Ce^{3+}	101		0.0069 [24]	4.6	
Pr^{3+}	100		0.0065 [24]	4.5	
Nd^{3+}	99		0.0064 [24]	4.4	
Pm^{3+}	97			4.1	
Sm^{3+}	96		0.0060 [24]	4.0	
Eu^{3+}	95		0.0060 [24]	3.9	
Gd^{3+}	94		0.0057 [24]	3.7	
Tb^{3+}	93		0.0054 [24]	3.6	
Dy^{3+}	91		0.0051 [24]	3.4	
Ho^{3+}	90		0.0049 [24]	3.3	
Er^{3+}	89		0.0047 [24]	3.2	
Tm^{3+}	88		0.0047 [24]	3.1	
Yb^{3+}	87		0.0042 [24]	3.0	
Lu^{3+}	86		0.0041 [24]	2.9	
Tl^{3+}	88		0.0048 [24]	3.1	
Bi^{3+}	102			4.8	
Ac^{3+}	118			7.4	
U^{3+}	104			5.1	
Am^{3+}	100			4.5	
Cm^{3+}	98			4.2	
Zr^{4+}	72		0.0028 [24]	1.7	
Sn^{4+}	69		0.0017 [24]	1.5	
Ce^{4+}	80		0.0045 [24]	2.3	
Hf^{4+}	71		0.0025 [24]	1.6	
Th^{4+}	100		0.0056 [24]	4.5	
U^{4+}	97		0.0049 [24]	4.1	
Np^{4+}	95			3.9	
Pu^{4+}	93			3.6	
H^-		148	0.033		
F^-	133	126	0.025	9.5	14.3
Cl^-	181	168	0.047	23.8	18.1
Br^-	196	190	0.056	30.3	27.8
I^-	220	211	0.072	26.9[a]	36.0
OH^-	133	152	0.032		17.6
SH^-	207	191	0.057		
ClO^-	158				

(continued)

TABLE 2.8 (Continued)

Ion	r_I/pm	r_{Ith}/pm	v_I/nm^3	$V_{I\ intr}^{M}$/cm^3·mol^{-1}	$V_{I\ intr}^{aq}$/cm^3·mol^{-1}
BrO$^-$	210				
IO$^-$	230				
CN$^-$	191	187	0.050		
NCO$^-$	203	193	0.054		
SCN$^-$	213	209	0.071	24.4[a]	46.6
N$_3^-$	195	180	0.058		
HF$_2^-$	172	172	0.047		
HO$_2^-$	180				
I$_3^-$	470	272	0.180 [24]		
BO$_2^-$	240				
ClO$_2^-$		195	0.063 [24]		
NO$_2^-$	192	187	0.055		29.3
NO$_3^-$	200	200	0.064	20.2[a]	29.0
ClO$_3^-$	200	208	0.073	32.2[a]	35.5
BrO$_3^-$	191	214	0.072		
IO$_3^-$	181	218	0.075		
ClO$_4^-$	240	225	0.082	34.9[a]	47.1
MnO$_4^-$	240	220	0.088		
TcO$_4^-$	250				
ReO$_4^-$	260	227	0.098 [24]		
BF$_4^-$	230	205	0.073		
HCO$_2^-$	204	200	0.056	34.1[a]	30.3
CH$_3$CO$_2^-$	232	194		50.2[a]	43.6
BPh$_4^-$	421				
HCO$_3^-$	156	207	0.064		
HSO$_4^-$	190	221	0.087 [24]		
H$_2$PO$_4^-$	200	213			
PF$_6^-$	245	242	0.109		
SbF$_6^-$	282	252	0.121		
O^{2-}	140	141	0.043		
S^{2-}	184	189	0.067		
CO$_3^{2-}$	178	189	0.061		
C$_2$O$_4^{2-}$	210				
SO$_3^{2-}$	200	204	0.071		
SO$_4^{2-}$	230	218	0.091	61.5[a]	29.3
SeO$_4^{2-}$	243	229	0.103		
CrO$_4^{2-}$	240	229	0.097		
MoO$_4^{2-}$	254	231	0.088		
WO$_4^{2-}$	270	237	0.088		
S$_2$O$_3^{2-}$	250	251	0.104		
SiF$_6^{2-}$	259	248	0.112		
Cr$_2$O$_7^{2-}$	320	292	0.167		
HPO$_4^{2-}$	200				
PO$_4^{3-}$	238	230	0.090		
AsO$_4^{3-}$	248	237	0.088		
Fe(CN)$_6^{3-}$	440	347	0.265		
Co(CN)$_6^{3-}$	430	349	0.263		
Fe(CN)$_6^{4-}$	450				

[a]According to Glueckauf's method [49].

but avoiding the temperature extrapolation. These molar volumes of the electrolytes $V_{\text{I intr}}^{\text{aq}}$ pertain to aqueous solutions at 25°C and probably also to solutions in other solvents and near ambient temperatures, the volumes being *intrinsic*. In order to obtain from them individual ionic molar volumes, it is necessary to split them appropriately among the constituent ions. This was done using the Mukerjee factor, Equation 2.11, as a guide, and the available data for Na^+, K^+, and Cs^+ salts. The resulting intrinsic ionic molar volumes are shown in Table 2.8 and have an uncertainty of $\pm 2.0\,\text{cm}^3\cdot\text{mol}^{-1}$.

2.3 IONS IN SOLUTION

The largest body of data for electrolyte solutions pertains to water as the solvent, and values for other solvents are best described in terms of the transfer functions of the ions from water as the source to the required solvent as the target (Section 4.3). This is because the transfer quantities are only a small fraction of the total and can be determined much more accurately than can the difference between the large values in the source and target solvents. Attention is, therefore, first directed toward aqueous solutions. After the relevant properties of the solvents that are involved in solutions of electrolytes have been dealt with in Chapter 3, the transfer of ions from aqueous solutions to solutions in these solvents, and eventually also to solutions in mixed solvents (Section 6.1) is presented and discussed.

The thermal movement of all the particles in the solution, the solvent molecules and the cations and anions making up an electrolyte, competes with the electrostatic interactions of the ions with their surroundings and with the hydrogen bonding and other interactions of the solvent molecules among themselves. At infinite dilution, specified for the standard state (at $T° = 298.15\,\text{K}$ and $P° = 0.1\,\text{MPa}$ in the neat solvent), an ion interacts only with the surrounding solvent and not with other ions. The overall interactions, involving ion solvation and effects of ions on the structure of the solvent, are quite complicated. In order to handle the resulting behavior of the system theoretically or by means of computer simulations, approximations have to be applied.

The *restricted primitive model* is the simplest approximation. It considers the ions as charged conducting particles dispersed uniformly in a continuum fluid made up of a compressible dielectric. The ions are characterized by their masses, charges (magnitudes), and sizes (radii), and are assumed to be spherical. The sign of the charge does not play a role in this model. The solvent is characterized by its permittivity, compressibility, and thermal expansibility. The standard state properties of the ions may then be estimated by the application of electrostatic theory and compared with the experimental values.

More sophisticated models allow for the molecular nature of the solvent and take into account the interactions between its molecules. It is then possible to ascribe concentric solvation shells to the ions made up with an average number of solvent molecules in the first and sometimes a second shell. This (first shell) solvation number becomes a definite integer if the solvent molecules form coordinate bonds with the ion. Such a model may still be treated by appropriate theoretical tools and used in computer simulations.

However, there are very few experimental determinations that can be applied unambiguously to individual ions in aqueous solution. These pertain to the mobility of an individual ion (diffusion coefficient and conductivity). Other determinations have to be conducted on entire electrolytes or pertain to differences between ions of the same sign and magnitude of charge. However, in a thought process, a single ion $I^{z\pm}$ may be transferred from the ideal gas phase into a neat solvent (Section 4.1). Such a process involves the passage of the ion through the gas–solvent interface and is connected with not well-defined consequences. Once an individual ion is in solution, its properties depend, in principle, on its location with respect to the surface and the walls of the vessel, due to its long-range electric field. It is assumed that when such a thought process is carried out simultaneously for ions of opposite charges, the effects of their positions and their passage through the gas–solvent interface cancel out, so that valid quantities can be derived from the process.

The more common process that can be carried out experimentally is to dissolve in the solvent an entire electrolyte, consisting of a matched number of cations and anions to produce a neutral species. Infinite dilution may be very well approximated as a limit of extrapolation from low, finite, and diminishing concentrations. This limit corresponds to the dissolution of a mole of electrolyte in a huge amount of solvent or of an infinitesimal amount of electrolyte in a finite amount of solvent. The molar quantities pertaining to the electrolyte at infinite dilution may then be dealt with. Some means to deduce from the measured quantities those pertaining to the individual ions must still be devised, in order to relate experimental values to those obtained from theory or computer simulations.

The individual ionic quantities contributing to the measured molar properties of the infinitely dilute electrolyte are additive, because then each ion is surrounded by solvent molecules only and is remote from other ions and does not interact with them. These quantities are weighted by their stoichiometric coefficients v_+ and v_- in the electrolyte $C_{v+}^{z+}A_{v-}^{z-}$. It follows that if the value for one ion is known, those of other ions (of opposite sign) can be derived by subtracting this value, appropriately weighted, from the values for electrolytes containing it and so forth.

An expedient for assigning a definite value to one ion is the use of so-called *conventional* values. The generally used convention for any additive property Y^∞ at infinite dilution is to assign the value zero to the hydrogen ion, mainly applied to the aqueous one $Y^\infty(H^+, aq) = 0$ at all temperatures. Sums of appropriately weighted conventional values of cations $v_+ Y^\infty(C^{z+})^{conv}$ and of anions $v_- Y^\infty(A^{z-})^{conv}$ represent the true values for electrolytes, even those not measured directly. Values of $Y^\infty(C^{z+})^{conv}$ of cations of the same magnitude of charge can be compared among themselves and be discussed, and similarly for anions among themselves. The conventional values may *not* be construed as representing the actual values of the properties that individual ions have.

If the properties Y^∞ of ions of opposite charge are to be compared, it is necessary to use the so-called *absolute* property values of individual ionic species. These are also needed for comparison with and validation of theoretical values of these properties and of those obtained by computer simulations. In the case of individual aqueous ions, a detailed discussion of the validity of methods for obtaining absolute property values was presented by Conway [57] and more recently by Marcus [58] and by Hünenberger and Reif [59]. These issues are treated in the following sections dealing with the various properties of aqueous ions.

The consequences of the electric charge on the ion in solutions depend on the huge size of the electric field at the boundary between the ion proper and its solvation shell. The electric field strength right near the surface of a potassium ion, at 0.138 nm from its center (its "bare" radius), is 102.3 GV·m^{-1}, and for a barium ion with a radius of 0.136 nm, it is nearly twice as large. On the other hand, electric fields achievable experimentally in the laboratory are of the order of 1 GV·m^{-1} only. Therefore, field effects related to the ions in solution are obtained from theory rather than from direct experiments according to Liszi et al. [60].

The permittivity of liquids at very high fields is given by the nonlinear dielectric effect:

$$\varepsilon(E) = \varepsilon(0) + \beta E^2 \tag{2.13}$$

For water, $\beta = -1.080 \times 10^{-15} \, \text{V}^{-2} \cdot \text{m}^2$ at ambient temperatures and is only moderately temperature dependent at very high fields. Values for other solvents were reported by Marcus and Hefter in Ref. 61. However, the field strength at a distance $r > r_I$ in the solvent depends in turn on the permittivity:

$$E(\varepsilon, r) = |z| e / (4\pi\varepsilon_0) \varepsilon(E, r) \cdot r^2 \tag{2.14}$$

so that iterative calculations between Equations 2.13.and 2.14 are required. One of the important consequences of the large fields under discussion is dielectric saturation in the solvation shell around an ion. It prevails at a short distance from the periphery of an ion: $\sim 0.08|z|^{1/2}$ nm. The relative permittivity diminishes to near the optical limit which is n_∞^2, the infinite frequency refractive index squared, ≈ 1.95 in water at 25°C. The dipoles of the solvent molecules can then no longer be oriented by external fields and the residual permittivity is due to the electronic polarization of these molecules. The permittivity grows as r is increased to eventually reach its bulk value; see Figure 2.2.

Another consequence of the ionic electric field is the large compressive pressure that it exerts on the solvent near the ion. Bockris and Saluja [62] calculated the effective pressure in the middle of the first hydration shell of aqueous ions, the numerical coefficient being valid at 25°C with the radii in nanometer:

$$P_{\text{eff}} / \text{GPa} = 0.18305 (r_I + r_w)^{-3} \tag{2.15}$$

For the aqueous potassium ion, the pressure is 8.7 GPa at this site ($r = 0.207$ nm), which is commensurate with the highest experimental pressures that can be applied to water or electrolyte solutions in the laboratory. At such large pressures the water in the hydration shell is highly compressed—it is strongly electrostricted.

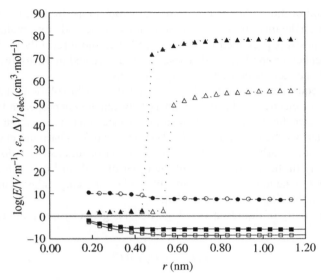

FIGURE 2.2   The electric field, $\log(E/V \cdot m^{-1})$ (circles and dashed line); the relative permittivity, ε_r (triangles and dotted lines); and the electrostriction, $\Delta V_{1\,els}/cm^3 \cdot mol^{-1}$ (squares and full lines) near aqueous fluoride anions at 25°C (filled symbols) and 100°C (empty symbols), as functions of the distance from the center of the ion. From Ref. 55 by permission of the publisher, ACS.

2.3.1   Thermodynamics of Ions in Aqueous Solutions

Ions in solutions are characterized by several thermodynamic quantities, including the standard molar heat capacities (at constant pressure) and entropies. Other important quantities are the standard molar enthalpy and Gibbs energy of formation of the ions in solution from the elements. As said earlier, in all these measures, it is possible to deal experimentally only with entire electrolytes or with such sums or differences of ions that are neutral. The assignment of *absolute* values to individual ions requires the splitting of the electrolyte values by some *extra thermodynamic* assumption that cannot be proved or disproved within the framework of thermodynamics.

2.3.1.1   Heat Capacities of Aqueous Ions   The difference between the specific heat of a dilute solution of an electrolyte and that of water can be obtained by flow microcalorimetry that requires also knowledge of the corresponding densities. Extrapolation to infinite dilution of these differences yields the standard partial molar (constant pressure) heat capacity of the electrolyte, C_{PE}^{∞}. Alternatively, the heat of solution of an electrolyte in water to form a dilute solution can be measured calorimetrically at several temperatures and extrapolation of the temperature coefficients of these heats of solution to infinite dilution yields the same quantity, but somewhat less accurately. A recent review of the experimental methods by Hakin and Bhuiyan [63] may be consulted for details. Determinations of C_{PE}^{∞} are accurate to ±1 to

± 3 J·K^{-1}·mol^{-1}, and compilations of values by Abraham and Marcus [21] and by Hepler and Hovey [64] are available.

As mentioned earlier, it is necessary to have a value for one ion in order to obtain the absolute standard molar ionic heat capacities, C_{PI}^{∞}, these values being additive at infinite dilution. As an expedient for obtaining absolute values, the TPTB assumption has been employed, equating the standard molar heat capacities of aqueous tetraphenylphosphonium and tetraphenylborate ions. These ions should have similar values due to their chemical similarity and similar sizes and to the charges of opposite sign being buried well inside the tetraphenyl structure [21]. The drawback of this expedient is that the C_{PI}^{∞} of these bulky ions are large, so that slight differences in the sizes and induced partial charges in the phenyl rings cause a large uncertainty in equating them. However, a more satisfactory method for splitting C_{PE}^{∞} into the C_{PI}^{∞} of the constituent ions has not been found. The resulting absolute value C_P^{∞} (H$^+$, aq) $= -71 \pm 14$ J·K^{-1}·mol^{-1} at 298.15 K has been suggested [21]. C_{PI}^{∞} values at 298.15 K for other aqueous ions, mainly from the most recent compilation for electrolytes [64] and some other sources, are shown in Table 2.9. Values for many electrolytes containing organic ions (such as carboxylate and alkylammonium ions), from which the C_{PI}^{∞} of these ions can be evaluated, are reported in Ref. 21. The values of C_{PI}^{∞} of polyatomic ions are more positive (or less negative) than those of monatomic ions of the same charge class and those of multicharged ions are seen to be large and negative.

The uncertainty in the absolute values of C_{PI}^{∞} is large (see above, ± 14 J·K^{-1}·mol^{-1}), but the uncertainties for the comparison of ions of the same sign of charge is much lower, being near those for the experimental C_{PI}^{∞}, 1–3 J·K^{-1}·mol^{-1}, for most ions except for the lanthanides, for which it is 20–30 J·K^{-1}·mol^{-1}, showing disagreement between the values in Refs. 21 and 64 for La^{3+} and Gd^{3+} (the only ones in the latter publication).

The partial molar heat capacities of electrolyte solutions at appreciable concentrations are obtained from electrolyte-specific parameters reported by Criss and Millero [74, 75]. The semiempirical Helgeson–Kirkham–Flowers [76] expression may be used for similar calculations according to Schock and Helgeson [65], particularly useful for elevated temperatures.

2.3.1.2 Entropies of Aqueous Ions

The temperature coefficients of the electromotive forces of galvanic cells are the most reliable experimental sources of the standard molar entropies of aqueous electrolytes, S_E^{∞}. Alternative sources are the temperature coefficients of the solubilities of sparingly soluble salts. The experimental uncertainties need not be larger than 0.5 J·K^{-1}·mol^{-1}. The values for individual ions are based on data from thermocells or from the potential of a mercury electrode at the point of zero charge. The assumption involved in the former method concerns the negligible entropy of transport across a boundary of similar solutions at different temperatures. That the temperature dependence of the surface potential of mercury in water is negligible is the assumption made for the latter method. Consistent values are obtained from both methods [13, 57] and the resulting reference value at 298.15 K is S^{∞}(H$^+$, aq) $= -22.2 \pm 1.4$ J·K^{-1}·mol^{-1}. The derived absolute values of the standard

TABLE 2.9 Standard Molar Heat Capacities [21] and Entropies [13] of Aqueous Ions and their Standard Molar Enthalpies and Gibbs Energies of formation (see the text)

Ion	C_{PI}^{∞}/J·K^{-1}·mol^{-1}	S_I^{∞}/J·K^{-1}·mol^{-1}	$\Delta_f H_I^{\infty}$/kJ·mol^{-1}	$\Delta_f G_I^{\infty}$/kJ·mol^{-1}
H$^+$	−71	−22.2	433.2	459.2
Li$^+$	−9	−8.8	154.7	165.9
Na$^+$	−29	36.8	193.1	197.3
K$^+$	−59	80.3	180.8	175.9
Rb$^+$	−80	99.3	182.0	175.2
Cs$^+$	−94	111.3	174.9	167.2
Cu$^+$	−14 [65]	18.4	504.9	509.2
Ag$^+$	−36	50.5	538.8	536.3
Au$^+$	−38 [65]		632.3	635 [66]
Tl$^+$	−89 [65]	103.3	438.6	426.8
H$_3$O$^+$		−22.2		
NH$_4^+$	−2	74.7	300.7	379.9
HONH$_3^+$		133		
H$_2$NNH$_3^+$		129		
C(NH$_2$)$_3^+$	103 [18]	201 [18]	292 [18]	
Me$_4$N$^+$	166 [21]	183 [67]	328.0 [68]	
Et$_4$N$^+$	443 [21]	255 [67]	218.1 [68]	
Pr$_4$N$^+$	852 [21]	308 [67]	97 [20]	
Bu$_4$N$^+$	1268 [21]	380 [67]	6.5 [68]	
Pe$_4$N$^+$	1603 [21]	506 [67]	483.6	538.7
Ph$_4$P$^+$	1141 [21]		399.6	463.6
Ph$_4$As$^+$	1168 [21]	348 [69]	323.6	364.8
NO$^+$		−125	320.6	358.9
NO$_2^+$		−115	328.8	357.6
Be^{2+}		−174.1	338.8	356.9
Mg^{2+}	−158	−182.5	640 [66]	700.4 [66]
Ca^{2+}	−169	−97.5	722.9	752.4 [66]
Sr^{2+}	−179	−77.0	645.7	690.3
Ba^{2+}	−190	−37.8	777.3	839.5
Ra^{2+}	−201 [65]	9.6	808.2	864.0
V^{2+}		−118.4	812.0	872.8
Cr^{2+}		−126.4	801.6	983.9
Mn^{2+}	−152	−118.0	712.5	771.3
Fe^{2+}	−168	−182.1	1015.4	1094.8
Co^{2+}	−170	−157.4	790.5	840.8
Ni^{2+}	−184	−173.3	857.6	891.2
Cu^{2+}	−162	−144.0	368.9	404.4
Zn^{2+}	−164	−156.5	339.4	378.2
Pd^{2+}	−168 [65]	−139	339 [66]	391.4
Cd^{2+}	−150	−117.6	1135.1[a]	1173.2
Sn^{2+}	−189 [65]	−76.6	695.3	1082.8
Sm^{2+}	−127 [65]	−92.6	694.0	1072.2 [66]
Eu^{2+}	−117 [21]	−74.6	864.7	894.0
Yb^{2+}	−139 [21]	−113.6	−153.2	−35.1
Pt^{2+}		−123.4	768.6	892.6
Hg^{2+}	−157	−76.6	685.4	791.0
Hg$_2^{2+}$	−70 [65]	21.1	1041 [66]	1135.6 [70]
Pb^{2+}	−197	−33.9	1078 [66]	1569.6 [66]

TABLE 2.9 (Continued)

Ion	C_{Pl}^{∞}/J·K^{-1}·mol^{-1}	S_I^{∞}/J·K^{-1}·mol^{-1}	$\Delta_f H_I^{\infty}$/kJ·mol^{-1}	$\Delta_f G_I^{\infty}$/kJ·mol^{-1}
UO_2^{2+}	−125	−142.7	1251.1	1362.2
Al^{3+}	−332	−388.3	1379	1508.6
Sc^{3+}	−219 [65]	−321.6	1087.9	1218.6
V^{3+}		−374	576.2	683.8
Cr^{3+}	−240	−335.6	1196	1279.6
Fe^{3+}		−369.9		
Co^{3+}	−206 [65]	−408 [70]	592.5	693.9
Ga^{3+}	−200 [65]	−397.6	603.4	705.6
Y^{3+}	−220 [65]	−317.6	595.0	698.5
In^{3+}	−217 [65]	−217.6	603.4	706.0
Sb^{3+}		−337	612 [66]	714.6
La^{3+}	−339 [21]	−284.2	611.6	711.0
Ce^{3+}		−271.6	693.6	803.5
Pr^{3+}	−378 [21]	−275.6	693.6	716.6
Nd^{3+}	−367 [21]	−273.3	616.8	725.7
Pm^{3+}		−276.6	601	712.6
Sm^{3+}	−362 [21]	−278.3	595	703.9
Eu^{3+}	−351 [21]	−288.6	594.2	708.5
Gd^{3+}	−351 [21]	−272.5	601.7	715.6
Tb^{3+}	−340 [21]	−292.6	625.1	733.6
Dy^{3+}	−334 [21]	−297.6	635	749.6
Ho^{3+}	−339 [21]	−293.4	1103.0	1592.2
Er^{3+}	−352 [21]	−310.9	1219	1460.4
Tm^{3+}	−336 [21]	−309.6	647 [66]	737.6 [66]
Yb^{3+}	−350 [21]	−304.6	810.5	902.2
Lu^{3+}	−340 [21]	−330.6	682.9 [66]	778.5 [66]
Tl^{3+}	−235 [65]	−258.6	685 [66]	781.8 [66]
Bi^{3+}		−218.4		1460
Ac^{3+}	−280 [71]	−251	1763.3	1839.3
U^{3+}	−264 [71]	−243.1	1195.6	1333.0
Am^{3+}	−253 [71]	−270.5		1281.8 [66]
Cm^{3+}	−247 [71]	−255.0	964	1131.7
Zr^{4+}	−342 [71]	−598.1	1141.6	1305.8
Sn^{4+}		−205.8	1176.7 [66]	1333.9 [66]
Ce^{4+}	−312 [71]	−389.8	1196.4 [66]	1355.2 [66]
Hf^{4+}		−554.5		
Th^{4+}	−360 [71]	−511.4	−765.8	738.0
U^{4+}	−336 [71]	−502.8	−600.4	590.4
Np^{4+}	−332 [71]	−477.8	−554.8	563.2
Pu^{4+}	−324 [71]	−477.8	−488.4	510.8
			−663.2	616.4
F^-	−45	8.4	−480.8	447.1
Cl^-	−55	78.7	−540.3	496.0
Br^-	−61	104.8	−527.3	492.6
I^-	−50	133.5	−540.7	497.7
OH^-	−69	11.5	−282.6	286.8
SH^-	−22	88.2	−579.2	556.6
ClO^-	−135 [65]	64	−356.8	366.5

(continued)

TABLE 2.9 (Continued)

Ion	C_{PI}^{∞}/J·K^{-1}·mol^{-1}	S_I^{∞}/J·K^{-1}·mol^{-1}	$\Delta_f H_I^{\infty}$/kJ·mol^{-1}	$\Delta_f G_I^{\infty}$/kJ·mol^{-1}
BrO$^-$	−135 [65]	64	−158.1	111.0
IO$^-$	−197 [65]	17.0	−1083.1	1037.4
CN$^-$	−65 [65]	116.3	−593.5	526.6
NCO$^-$		128.9	−484.7	510.6
SCN$^-$	45 [72]	166.5	−1205	1138.1
N$_3^-$		130.1	−499.7	442.0
HF$_2^-$	−68 [65]	114.7	−537.8	491.4
HO$_2^-$		46.0	−638	567.9
I$_3^-$	126 [65]	261.5	−537.2	467.2
BO$_2^-$	−101 [65]	−15.0	−500.3	440.6
ClO$_2^-$	−57 [65]	123.5	−654.5	587.2
NO$_2^-$	−23	145.2	−562.5	467.7
NO$_3^-$	0	168.8	−974.6	906.4
ClO$_3^-$	14	184.5	−474.5 [66]	495.2 [66]
BrO$_3^-$	−19	183.9	−1220.6	1153.7
IO$_3^-$	0	140.6	−2008.1	1946
ClO$_4^-$	45	206.2	−858.8	810
MnO$_4^-$	60	213.4	−919.2	828.5
TcO$_4^-$	57	219.7		
ReO$_4^-$	57	223.5	−1125.2	1046.0
BF$_4^-$	−48 [65]	202.2	−1320.6	1215.1
HCO$_2^-$	−14	114.2	−1729.5	1589.2
CH$_3$CO$_2^-$	97	108.8	−833.3	832.6
CF$_3$CO$_2^-$			−1543.5	1446.2
CF$_3$SO$_3^-$			−1691.5	1592.3
BPh$_4^-$	1141	347.8 [69]	−1501.9	1404.9
HCO$_3^-$	19	120.6	−1776.1	1662.9
HSO$_4^-$	53	154.0	−1465.5	1359.7
H$_2$PO$_4^-$	37	114.7	−1747.6	1646.2
PF$_6^-$	197 [73]		−1864.4	1754.7
O^{2-}		−41.6	−1942.1	1835.3 [21]
S^{2-}	−184 [71]	29.8	−1514.9	1440.9
CO$_3^{2-}$	−158	0.9	−3203.2 [66]	3056 [66]
C$_2$O$_4^{2-}$		90.0 [1]	−2356.7	2219
SO$_3^{2-}$	−120	15	−2158.5	2007
SO$_4^{2-}$	−134	63.2	−2318.3	2397
SeO$_4^{2-}$	−110 [65]	98.4	−2187.7	2026
CrO$_4^{2-}$	−126	94.6	−1861.5	648.2
MoO$_4^{2-}$	−65	71.6		1141.7
WO$_4^{2-}$	−52	85.0	−1277.2	
S$_2$O$_3^{2-}$	−106	111.4		
SiF$_6^{2-}$		166.6		
Cr$_2$O$_7^{2-}$	13	306.3		
HPO$_4^{2-}$	−111	10.9		
PO$_4^{3-}$	−282	−155.4		
AsO$_4^{3-}$		−102.3		
Fe(CN)$_6^{3-}$	−38	−336.9		
Co(CN)$_6^{3-}$		−299.2		
Fe(CN)$_6^{4-}$	−239	183.8		

molar ionic entropies for other aqueous ions, S_I^∞, at 298.15 K are shown in Table 2.9. The values increase with the masses of the ions but are small or negative for multicharged ions. The uncertainties in the absolute S_I^∞ values should also be not more than $\pm 2\,\mathrm{J\cdot K^{-1}\cdot mol^{-1}}$.

2.3.1.3 Enthalpies of Formation of Aqueous Ions

The standard molar enthalpy of formation of the infinitely dilute aqueous electrolyte, $\Delta_f H^\infty(E, aq)$, is obtained experimentally from the standard molar enthalpy of formation of the pure electrolyte (generally for the crystalline salt but also for electrolytes such as H_2SO_4 and HBr that are liquid and gaseous in their standard states), $\Delta_f H^\circ(E, c)$, and the heats of solution extrapolated to infinite dilution:

$$\Delta_f H^\infty(E, aq) = \Delta_f H^\circ(E, c) + \lim(c \to 0)\Delta_{sln} H_E \qquad (2.16)$$

The values of $\Delta_f H^\circ(E, c)$ and $\Delta_f H^\infty(E, aq)$ are critically compiled for many electrolytes by Wagman et al. [1] and are supplemented by more recent values for $\Delta_f H^\infty(E, aq)$. Conventional values, $\Delta_f H^\infty(I^{\pm z}, aq)^{conv}$, for individual ions are obtained from setting $\Delta_f H^\infty(H^+, aq) = 0$ at all temperatures and are also reported in Ref. 1.

As for the other thermodynamic quantities, a value must be estimated for one ion in order to split the experimentally available values of $\Delta_f H^\infty(E, aq)$ into the ionic contributions. This is done by assigning the value $433.2\,\mathrm{kJ\cdot mol^{-1}}$ to $\Delta_f H^\infty(H^+, aq)$ and using:

$$\Delta_f H^\infty\left(I^{\pm z}, aq\right) = \Delta_f H^\infty\left(I^{\pm z}, aq\right)^{conv} + 433.2 \cdot z_I \qquad (2.17)$$

with values in $\mathrm{kJ\cdot mol^{-1}}$ and taking the ionic charge number z_I algebraically (positive for cations, negative for anions). The numerical value for the hydrogen ion is based on the standard molar enthalpy of hydration $\Delta_{hyd} H^\infty$ (TPTB) of tetraphenylphosphonium tetraphenylborate (see Section 4.2.3) according to Marcus [77]. Contrary to the case of the heat capacities (Section 2.3.1.1), the values of the ionic enthalpy of hydration for these bulky ions are small compared to those of ordinary, small ions. Therefore, the uncertainty involved in setting $\Delta_{hyd} H^\infty\left(Ph_4P^+\right) = \Delta_{hyd} H^\infty\left(BPh_4^-\right)$ is also small. This leads to $\Delta_{hyd} H^\infty(H^+) = -1103 \pm 7\,\mathrm{kJ\cdot mol^{-1}}$ [77] and together with the well established $\Delta_f H^\infty(H^+, ig) = 1536.2\,\mathrm{kJ\cdot mol^{-1}}$ [1] to the value $433.2\,\mathrm{kJ\cdot mol^{-1}}$ for $\Delta_f H^\infty(H^+, aq)$ given above.

The values of $\Delta_f H^\infty(I^{\pm z}, aq)$ for many ions obtained from the conventional values according to Equation 2.17 are shown in Table 2.9 and are expected to be accurate to within $\pm 7 z_I\,\mathrm{kJ\cdot mol^{-1}}$. The values are all negative, as expected (heat is released). For singly charged ions, whether cations or anions, $\Delta_f H^\infty(I^{\pm z}, aq)$ values are of similar magnitude but become less negative with increasing sizes. For multicharged ions, they are considerably more negative than for singly charged ones by a factor of the order of z_I^2.

At finite concentrations of electrolytes, their molar enthalpies of hydration may be estimated by adding the relative partial molar heat content of the solute, L_E, to the sum of the stoichiometrically weighted cation and anion values of $\Delta_{hyd} H_I^\infty$. The value of L_E is equal and of opposite sign to the experimentally measurable enthalpy of dilution of the electrolyte, $\Delta_{dil}H_E$. At finite concentrations the heat content and the enthalpy of hydration may therefore be smaller or larger than at infinite dilution, depending on the enthalpies involved in the interactions between neighboring ions. These heat contents are obtainable from the temperature derivatives of the activity coefficients ($\gamma_{\pm}$ on the molal scale):

$$L_E = -\nu RT^2 \left(\frac{\partial \ln \gamma_{\pm}}{\partial T}\right)_{P,m} \tag{2.18}$$

where ν is the number of ions in a formula of the solute electrolyte. However, at finite concentrations, the ionic values are no longer additive and a value for an electrolyte must be obtained for each separate case.

2.3.1.4 Gibbs Energies of Formation of Aqueous Ions

The standard molar ionic Gibbs energies of formation of the aqueous ions, $\Delta_f G^{\infty}(I^{\pm z}, aq)$, are mostly obtained from the listed conventional values $\Delta_f G^{\infty conv}(I^{\pm z}, aq)$ [1] to which $459.2 z_I$ is added, with values in $kJ \cdot mol^{-1}$ and taking the ionic charge number z_I algebraically (positive for cations, negative for anions) (cf. Eq. 2.17). The numerical value is the sum of $\Delta_f G^{\circ}(H^+, ig) = 1523.2\ kJ \cdot mol^{-1}$ of the gaseous hydrogen ion, Table 2.3, and $\Delta_{hyd} G^{\infty}(H^+, aq) = -1064$, see Section 4.2.1. The values of $\Delta_f G^{\infty}(I^{\pm z}, aq)$ are also shown in Table 2.9.

2.3.1.5 Ionic Molar Volumes in Aqueous Solutions

The densities, ρ, of electrolyte solutions, as dependent on their concentrations at constant temperatures and pressures, are measurable with high accuracy. In a solution made up from n_w moles of water and n_E moles of electrolyte, the total volume of the solution is $V = M/\rho$, where $M = n_w M_w + n_E M_E$. The apparent molar volume of an electrolyte, ${}^{\varphi}V_E$, is the part of V remaining for the electrolyte per mole of it after subtraction of the volume assigned to the water, $n_w V_w^*$:

$$ {}^{\varphi}V_E = \frac{V - n_w V_w^*}{n_E} \tag{2.19}$$

Here V_w^* is the molar volume of pure water, disregarding any effect due to the ions. In an electrolyte solution of molality m_E or concentration c_E and density ρ, the apparent molar volume is:

$$ {}^{\varphi}V_E = M_E/\rho + 1000\left(\rho - \rho_w^*\right)/\rho\rho_w^* m_E \tag{2.20a}$$

$$ {}^{\varphi}V_E = M_E/\rho_w^* + 1000\left(\rho - \rho_w^*\right)/\rho_w^* c_E \tag{2.20b}$$

where M_E is the molar mass of the solute and ρ_w^* the density of pure water. However, the actual molar volume of the solute, V_E, differs from the apparent molar volume ${}^{\varphi}V_E$, because the ions of the electrolyte affect the volume of the water. The water near the ions is compressed, electrostricted, by the electrical fields of the ions. The partial molar volume of the electrolyte is the volume actually occupied by the ions in the solution. For a solution of molality m_E, it is obtained from:

$$V_E = {}^{\varphi}V_E + m_E \left(\frac{\partial^{\varphi}V_E}{\partial m_E} \right)_T = {}^{\varphi}V_E + m_E^{1/2} \left(\frac{\partial^{\varphi}V_E}{\partial m_E^{1/2}} \right)_T \qquad (2.21)$$

The second version of Equation 2.21 is preferred, because of the square root dependence of ${}^{\varphi}V_E$ on m_E in dilute solutions. Extrapolation of V_E or of ${}^{\varphi}V_E$ to infinite dilution yields the standard partial molar volume of the electrolyte: $V_E^{\infty} = {}^{\varphi}V_E^{\infty}$. It is noteworthy that molar volumes are not available for salts containing highly hydrolyzable cations, such as those of Bi^{3+}, Zr^{4+}, or tervalent and tetravalent actinides.

At infinite dilution, the standard molar volumes of the cations and anions are additive and conventional values V_I^{conv}, based on $V^{\infty}(H^+, aq)^{conv} = 0 \, cm^3 \cdot mol^{-1}$ at all temperatures, have been listed by Millero [78] at several temperatures (0, 25, 50, and 75°C). Some of the 25°C values have since been revised [55]. The absolute standard partial ionic molar volumes are $V^{\infty}(I^{\pm}, aq) = V_I^{conv} + z_I V^{\infty}(H^+, aq)$ and the temperature-dependent value for the absolute molar volume of the aqueous hydrogen ion is:

$$V^{\infty}\left(H^+, aq\right)/cm^3 \cdot mol^{-1} = -5.1 - 0.008(t/°C) - 1.7 \times 10^{-4}(t/°C)^2 \qquad (2.22)$$

valid to 200°C and resulting in $V^{\infty}(H^+, aq) = -5.4 \, cm^3 \cdot mol^{-1}$ at 25°C. The value for the aqueous hydronium ion, $V^{\infty}(H_3O^+, aq)$, is indistinguishable from that of $V^{\infty}(H^+, aq)$ [79]. Ionic standard partial molar volumes, $V^{\infty}(I^{\pm}, aq)$, at 25°C for many ions are shown in Table 2.10. The absolute ionic values have uncertainties of at least $\pm 0.2 z_I \, cm^3 \cdot mol^{-1}$, due to the steps that have led to Equation 2.22. The values of $V^{\infty}(I^{z\pm}, aq)$ are negative for some cations, and in particular for multivalent ones, because these ions cause a large electrostriction of the hydrating water.

Values of the ionic standard partial molar volumes of the hydrogen, alkali metal, alkaline earth metal, and ammonium cations and hydroxide, halide, nitrate, perchlorate, and sulfate anions from 0 to 200°C at 25°C intervals have been reported by Marcus in Refs. 79–81. Conventional values V_I^{conv} for these ions (except ClO_4^-) and also for HCO_3^- and HS^- are reported by Tanger and Helgeson [84] from 0 to 350°C at 25°C intervals.

At finite concentrations, interionic interactions cause the additivity of the individual ionic molar volumes V_I to break down. The apparent molar volumes of electrolytes can be expressed according to Redlich [85] as:

$$^{\varphi}V_E = {}^{\varphi}V_E^{\infty} + S_V c_E^{1/2} + b_E c_E \qquad (2.23)$$

where S_V is the theoretical slope of the square root term according to the Debye–Hückel theory (1.85 $dm^{3/2} \cdot mol^{-1/2}$ for aqueous solutions at 25°C), and b_E is an empirical parameter specific for each electrolyte. A linear relationship was found by Marcus [86] between the b_E values normalized with respect to the molar volume of the solvated ions and the B_{η} coefficients of the viscosities. Although the latter is additive and established for individual ions (Section 2.3.2.3), the linear relationships did *not* permit the estimation of individual ionic b_I values.

TABLE 2.10   Ionic Standard Partial Molar Volumes [6, 78], the Electrostriction Volumes Obtained from a Shell-by-shell Electrostatic Calculation [80], and the Ionic Standard Partial Molar Expansibilities and Compressibilities, all at 25°C

Ion	V_I^∞/cm³·mol⁻¹	$-V_{Ielec}^\infty$/cm³·mol⁻¹	E_I^∞/cm³·mol⁻¹·K⁻¹	K_I^∞/cm³·mol⁻¹·GPa⁻¹
H^+	−5.4	7.5 [79]	−0.017	16
Li^+	−6.4	12.9	−0.032	−26
Na^+	−6.7	8.6	0.048	−34
K^+	3.5	5.4	0.027	−29
Rb^+	8.6	5.3	0.031	−24
Cs^+	15.8	4.4	0.034	−21
Cu^+	−13.5			
Ag^+	−7.1 [55]		0.057	
Au^+	5.6			
Tl^+	5.1			
H_3O^+	−5.4 [79]	7.5 [79]	−0.016	
NH_4^+	12.4	4.5 [81]	0.018	
$C(NH_2)_3^+$	46.3 [18]	~0 [18]		
Me_4N^+	84.1		0.035	5
Et_4N^+	143.6		0.066	4
Pr_4N^+	208.9		0.110	2
Bu_4N^+	270.2		0.197	−5
Pe_4N^+	333.7			
Ph_4P^+	286.8			
Ph_4As^+	295.2		0.357	
Be^{2+}	−21.3 [55]			−7
Mg^{2+}	−32.2	52.5	−0.015	−64
Ca^{2+}	−28.9	38.5	−0.001	−61
Sr^{2+}	−28.8 [55]	33.9	0.028	−77
Ba^{2+}	−23.5	27.5	0.048	−76
Ra^{2+}	−21.6			
V^{2+}	−21.5			
Mn^{2+}	−28.5 [55]		−0.089	−62
Fe^{2+}	−33.6 [55]		−0.079	
Co^{2+}	−36.8 [55]		−0.054	−71
Ni^{2+}	−40.5 [55]		−0.084	−75
Cu^{2+}	−38.8		−0.038	−74
Zn^{2+}	−37.6 [55]		−0.049	−75
Pd^{2+}	−31.8			
Cd^{2+}	−24.0 [55]		−0.025	−49
Sn^{2+}	−26.5			
Sm^{2+}	−16.7			
Eu^{2+}	−13.2			
Yb^{2+}	−21.3			
Hg^{2+}	−25.4 [55]			
Hg_2^{2+}	3.4			
Pb^{2+}	−29.0			
UO_2^{2+}	−5.1			

TABLE 2.10 (Continued)

Ion	$V_I^\infty/cm^3\cdot mol^{-1}$	$-V_{Ielec}^\infty/cm^3\cdot mol^{-1}$	$E_I^\infty/cm^3\cdot mol^{-1}\cdot K^{-1}$	$K_I^\infty/cm^3\cdot mol^{-1}\cdot GPa^{-1}$
Al^{3+}	−61.5 [55]			
Sc^{3+}	−58.4			
Cr^{3+}	−53.2 [55]			
Fe^{3+}	−53.0 [55]			
Co^{3+}	−60.4			
Ga^{3+}	−61.4			
Y^{3+}	−57.1 [55]			
In^{3+}	−42.5			
La^{3+}	−55.6		−0.115	−115
Ce^{3+}	−56.3			−128
Pr^{3+}	−59.3 [55]			
Nd^{3+}	−59.8			
Pm^{3+}	−61.8			
Sm^{3+}	−58.8			
Eu^{3+}	−58.2 [55]			
Gd^{3+}	−56.4			
Tb^{3+}	−56.7			
Dy^{3+}	−57.3			
Ho^{3+}	−58.3			
Er^{3+}	−59.4			
Tm^{3+}	−59.7 [55]			
Yb^{3+}	−60.7			
Lu^{3+}	−62.4 [55]			
Tl^{3+}	−55.8			
Th^{4+}	−75.5			
F^-	4.3	6.2	0.033	−41
Cl^-	23.3	4.0	0.048	−16
Br^-	30.2	3.5	0.065	−9
I^-	41.7	2.8	0.096	3
OH^-	1.2 [79]	6.6 [79]	0.060	−48
SH^-	26.7			−18
ClO^-	15.5			
BrO^-	15.5			
IO^-	4.2			
CN^-	30.6			
NCO^-	31.6			
SCN^-	46.1 [55]		0.091	7
N_3^-	30.5			
HF_2^-	27.6			
HO_2^-	11.2			
I_3^-	62.7			
BO_2^-	−9.0 [65]			
ClO_2^-	29.7			
NO_2^-	31.7			

(continued)

TABLE 2.10 (Continued)

Ion	$V_I^\infty/cm^3 \cdot mol^{-1}$	$-V_{Ielec}^\infty/cm^3 \cdot mol^{-1}$	$E_I^\infty/cm^3 \cdot mol^{-1} \cdot K^{-1}$	$K_I^\infty/cm^3 \cdot mol^{-1} \cdot GPa^{-1}$
NO_3^-	34.5	2.7 [81]	0.054	−41
ClO_3^-	42.2			
BrO_3^-	40.8			
IO_3^-	30.8			
ClO_4^-	49.6	2.4	0.105	
MnO_4^-	48.0		0.177	
TcO_4^-				
ReO_4^-	54.6			
BF_4^-	50.6			13
HCO_2^-	31.6			
$CH_3CO_2^-$	46.2			−18
$CF_3CO_2^-$	63.8 [82]			
$CF_3SO_3^-$	81 [82]			
BPh_4^-	283.1		0.402	
HCO_3^-	28.96			
HSO_4^-	41.2			
$H_2PO_4^-$	34.6			
PF_6^-	38 [83]			
S^{2-}	2.8			
CO_3^{2-}	6.7		0.144	
$C_2O_4^{2-}$	27.0			−6
SO_3^{2-}	19.9			
SO_4^{2-}	25.0	13.8	0.132	−83
SeO_4^{2-}	32.0			
CrO_4^{2-}	30.7		0.156	−81
MoO_4^{2-}	39.9			
WO_4^{2-}	36.7			
$S_2O_3^{2-}$	38.2			
SiF_6^{2-}	53.6			
$Cr_2O_7^{2-}$	84.0			
HPO_4^{2-}	16.3 [55]			
PO_4^{3-}	−9.1 [55]			
AsO_4^{3-}	0.9			
$Fe(CN)_6^{3-}$	137.3			
$Co(CN)_6^{3-}$	131.9			
$Fe(CN)_6^{4-}$	96.0			

As mentioned earlier, the actual volume to be assigned to an ion in the solution at infinite dilution is its standard partial molar volume V_I^∞. It may be negative, in particular for small highly charged ions, because the electrostriction (volume diminution), V_{Iels}^∞, such ions cause in the water surrounding the ion may be numerically larger than the intrinsic volume of the ion, V_{Iintr} [80]. Ionic intrinsic volumes that are independent of the concentration are discussed in Section 2.2 and are shown for many ions in Table 2.8. The electrostriction that the ion causes in the surrounding water at infinite dilution may then be calculated as $V_{\text{Iels}}^{\infty} = V_{\text{I}}^{\infty} - V_{\text{Iintr}}$.

An alternate way to estimate the electrostriction caused by an ion is based on the electrostatic effects that the very high electric field of an ion has on the water surrounding it. This field compresses the water strongly and decreases its permittivity down to dielectric saturation, Figure 2.2. A calculation of the volume diminution in consecutive concentric shells around the ion numbered from $j = 1$ up to such a value that the incremental change in V_{Iels}^{∞} is negligible yields the desired quantity according to Marcus and Hefter [87]:

$$V_{\text{Ielec}}^{\infty} = -\left(8\pi^2 N_A \varepsilon_0\right) \sum_j \left[r(j)^3 - r(j-1)^3\right]\left\{\varepsilon_{\text{W}}(j)\left[\left(\frac{\partial \ln \varepsilon_{\text{W}}}{\partial P}\right)_T - \kappa_{\text{TW}}\right] + \kappa_{\text{T}}\right\}E(j)^2$$

(2.24)

An iterative calculation is required due to the mutual dependence of the field $E(j)$ and the permittivity $\varepsilon_{\text{W}}(j)$ in each shell as discussed earlier (Eqs. 2.13 and 2.14). The resulting values for 25°C determined by Marcus [79–81] are shown in Table 2.10, but values for rounded temperatures from 0 to 200°C are also available. Note that discrepancies may exist between the ionic intrinsic molar volumes calculated from $V_{\text{Iintr}} = V_{\text{I}}^{\infty} - V_{\text{Ielec}}^{\infty}$ and those listed in Table 2.8, due to different modes of consideration of the void volumes between the particles in the solution.

The temperature and pressure dependencies of the ionic molar volumes are also of interest. The ionic standard partial molar isobaric expansibilities, E_{I}^{∞}, at 25°C are obtained from the V_{I}^{∞} data at 0 and 50°C as $E_{\text{I}}^{\infty} = [V_{\text{I}}^{\infty}(50\,^{\circ}C) - V_{\text{I}}^{\infty}(0\,^{\circ}C)]/50$, using the recently established $E^{\infty}(\text{H}^+, \text{aq}) = -0.017\,\text{cm}^3\cdot\text{mol}^{-1}\cdot\text{K}^{-1}$ by Marcus [79]. The values listed in the older compilation [6] based mainly on data in the Millero review [78] and on $E^{\infty}(\text{H}^+, \text{aq}) = 0.064\,\text{cm}^3\cdot\text{mol}^{-1}\cdot\text{K}^{-1}$ must, therefore, be corrected by subtracting $0.081z_{\text{I}}\,\text{cm}^3\cdot\text{mol}^{-1}\cdot\text{K}^{-1}$. The available data are shown in Table 2.10. The ionic standard partial molar isothermal compressibilities at 25°C are based on the estimate by Mathieson and Conway [88] $K_T^{\infty}(\text{K}^+, \text{aq}) = -16.0 \pm 1.5\,\text{cm}^3\cdot\text{mol}^{-1}\cdot\text{GPa}^{-1}$. The available values [6] are also shown in Table 2.10.

2.3.2 Other Properties of Aqueous Ions

2.3.2.1 Ionic Conductivities in Aqueous Solutions
The thermodynamic quantities for ions in solution dealt with in the previous sections could be measured only for complete electrolytes (or for charge balanced differences between ions of the same sign) but not for individual ions. On the contrary, this is not the case for ionic conductivities (and diffusion coefficients, see Section 2.3.2.2). These can be determined experimentally for individual ions from the electrolyte conductivities and the transference numbers. The conductivity of an electrolyte solution is accurately measured with an alternating external electric field at a rate of ~1 kHz imposed on the solution with a high impedance instrument in a virtually open circuit (zero current). The molar conductivity, Λ_{E}, can then be determined per unit concentration. Ion–ion interactions cause the conductivities of electrolytes to diminish as the concentration

increases, and the standard molar conductivity, Λ_E^∞, is obtained on extrapolation to infinite dilution. An appropriate theoretical expression is required for this extrapolation that takes into account the indirect ion–ion and ion–solvent effects. The molar conductivity of a completely dissociated electrolyte at a concentration c_E molar is according to Fernandez-Prini [89]:

$$\Lambda_E = \Lambda_E^\infty - S c_E^{1/2} + E c_E \ln c_E + J'(R') c_E - J''(R'') c_E^{3/2} \tag{2.25}$$

The coefficients S, E, J', and J'' are explicit expressions, containing contributions from relaxation and electrophoretic effects, the latter two coefficients depending also on ion–ion distance parameters R. The commonly used units of the molar ionic conductivities are $S \cdot cm^2 \cdot mol^{-1}$ ($S = \Omega^{-1}$).

The limiting molar ionic conductivities, λ_I^∞, are obtained by application of the experimentally measured (and extrapolated to infinite dilution) transference numbers, t_+^∞ and $t_-^\infty = 1 - t_+^\infty$. Thus $\lambda_+^\infty = t_+^\infty \cdot \Lambda_E^\infty / v_+$ and $\lambda_-^\infty = t_-^\infty \cdot \Lambda_E^\infty / v_-$, so that $\Lambda_E^\infty = v_+ \lambda_+^\infty + v_- \lambda_-^\infty$, the $v_\pm$ being the stoichiometric coefficients of the electrolyte. The limiting (standard) molar ionic conductivities λ_I^∞ for many ions in water at 25°C are shown in Table 2.11 with uncertainties not larger than $\pm 0.01\ S \cdot cm^2 \cdot mol^{-1}$. Between 0 and 100°C Λ_E^∞ increase about fivefold, mainly because the viscosity of the solvent diminishes in this direction by a similar factor. The transference numbers t_+ and t_- are temperature dependent too, but only mildly.

The rates of movement of ions in an electric field are expressed by their mobilities u_I, measuring their speed at unit field. The mobilities at infinite dilution, u_I^∞, are directly proportional to the limiting ionic molar conductivities:

$$u_I^\infty = \frac{\lambda_I^\infty}{|z| F} \tag{2.26}$$

their units being $m^2 \cdot s^{-1} \cdot V^{-1}$. The mobility of an ion, hence its electric conductivity, depends on its size and on the viscosity of the solvent (η_W^* for aqueous solutions). A so-called Stokes radius may be assigned to an aqueous ion according to Nernst, Stokes, and Einstein:

$$r_{ISt} = \left(\frac{F^2}{6 \pi N_A} \right) |z_I| / \eta_W^* \lambda_I^\infty \tag{2.27}$$

Although the r_{ISt} can be calculated formally by Equation 2.27, they have no physical significance and their use ought to be discouraged [117].

2.3.2.2 Ionic Self-Diffusion in Aqueous Solutions   The inherent movement of ions in the absence of an external field is their self-diffusion, but a directional concentration gradient at finite concentration (a gradient of chemical potentials) causes directional diffusion of ions. The limiting ionic self-diffusion coefficients are generally obtained from the limiting conductivities

$$D_I^\infty = \frac{RT \lambda_I^\infty}{z^2 F^2} \tag{2.28}$$

The rate of self-diffusion of individual ions can also be measured directly by using isotopically labeled ions in capillaries or diaphragm cells or by spin-echo NMR measurements of suitable nuclei. When isotopically labeled ions are employed, it is assumed that the slight mass difference between ions that differ only by their isotopic composition does not affect their rate of diffusion. If the diffusing solvated ions carry their solvation shells along, the validity of this assumption is enhanced, because then their mass includes the mass of the cotransported solvent. The labeled ion diffuses in a solution comprising nonlabeled ions of the same elemental species at a given concentration. A diaphragm cell may be employed with equal concentrations of the electrolyte in the two stirred compartments, in one of which ions of one kind are labeled by an isotopic tracer, the concentration of which can be monitored. This method is more appropriate for fairly concentrated solutions according to Mills [118], whereas capillary methods, in which the labeled ion diffuses out of a capillary into a nonlabeled solution, can also be used at very low concentrations as suggested by Passiniemi [119]. However, such measurements take a long time (several days) to perform.

Spin-echo NMR measurements with nonlabeled isotopes (elements at their natural isotopic composition) but involving suitable nuclei have also been used. The interplay between the signal relaxation and the diffusion requires nuclei with large gyromagnetic ratios and long relaxation times. Also, fairly concentrated solutions, of the order of 1 M, are required. Nuclei such as ^7Li, ^{19}F, and ^{133}Cs, present in the natural elements, could be used according to Braun and Weingärtner [120]. The advantage of this method is the short duration of the experiment, though its accuracy is somewhat less that that using isotopically labeled ions.

The values of the limiting diffusion coefficient, D_I^∞, of aqueous ions of the order of $10^{-9}\,m^2 \cdot s^{-1}$, at 25°C are shown in Table 2.11. The self-diffusion coefficients increase with increasing temperatures and a fivefold increase in aqueous solutions from 0 to 100°C has been noted. This is mainly because the viscosity of the solvent diminishes in this direction (Table 3.7).

The rate of self-diffusion of ions in aqueous solutions is inversely related to the bulk of their hydration shell. Notable exceptions to this rule are the hydrogen and hydroxide ions that do not diffuse while carrying their hydration shells. These ions, being essentially a part of the solvent, migrate by the Grotthuss mechanism, according to which the charge (i.e., a missing or an extra electron) hops from one water molecule in the hydrogen-bonded network to the next, which is much faster than massive movement of an ion.

2.3.2.3 *Ionic Effects on the Viscosity*

Ions affect the dynamic viscosity, η, of aqueous solutions: some electrolytes enhance it (e.g., lithium acetate), whereas others diminish it (e.g., cesium iodide) as demonstrated by Marcus [121]. Up to fairly concentrated solutions, this effect is described by the Jones–Dole expression [122]:

$$\left(\frac{\eta}{\eta_w^*} \right) = 1 + A_\eta c_E^{1/2} + B_\eta c_E + \cdots \tag{2.29}$$

The coefficients of the square root of the concentration term, A_η, can be calculated theoretically from the conductivities according to Falkenhagen [123] but the B_η

TABLE 2.11 Ionic Transport and Dynamic Properties in Aqueous Solutions: Limiting Molar Conductivities, λ_I^∞ [90], Limiting Diffusion Coefficients, D_I^∞ [91], Viscosity B-coefficients, $B_{\eta I}$ [92], and NMR B_{nmrI} Values [93] at 25°C

Ion	λ_I^∞/S·cm²·mol⁻¹	D_I^∞/10⁻⁹m²·s⁻¹	$B_{\eta I}$/dm³·mol⁻¹	B_{nmrI}/dm³·mol⁻¹
H^+	349.8	9.31	0.068	0.06
D^+	249.9 [94] [a]	6.66		
Li^+	38.7	1.03	0.146	0.14
Na^+	50.1	1.33	0.085	0.06
K^+	73.5	1.96	−0.009	−0.01
Rb^+	77.8	2.07	−0.033	−0.04
Cs^+	77.3	2.06	−0.047	−0.05
Cu^+	44 [95]	1.2 [95]		
Ag^+	61.9	1.65	0.090	0.06
Au^+	67.4 [95]	1.8 [95]		
Tl^+	74.3	1.99	−0.036	
H_3O^+	349.8		0.068	
NH_4^+	73.6	1.96	−0.008	
$H_2NNH_3^+$	59 [95]	1.57		
$C(NH_2)_3^+$	39.6 [18]	1.06 [18]	0.058 [18]	
Me_4N^+	44.9	1.20	0.123	0.18 [96]
Et_4N^+	32.7	0.87	0.385	0.44 [96]
Pr_4N^+	23.4	0.63	0.916	0.89 [96]
Bu_4N^+	19.5	0.52	1.275	1.33 [96]
Pe_4N^+	17.9	0.47		
Ph_4P^+	20.2 [97]		1.073	0.83 [96]
Ph_4As^+	19.7 [97]		1.073	
Be^{2+}	90	0.60	0.45	
Mg^{2+}	106.1	0.71	0.385	0.50
Ca^{2+}	118.9	0.79	0.298	0.27
Sr^{2+}	118.9	0.79	0.272	0.23
Ba^{2+}	127.3	0.85	0.229	0.18
Ra^{2+}	133.6 [98]	0.89		
Cr^{2+}	123.3 [99]			
Mn^{2+}	100 [98]	0.71	0.39	
Fe^{2+}	107 [98]	0.72	0.42	
Co^{2+}	100	0.73	0.376	
Ni^{2+}	98 [98]	0.66	0.375	
Cu^{2+}	107.2	0.71	0.368	
Zn^{2+}	105.6	0.70	0.361	
Pd^{2+}	139.8 [100] [a]	1.99 [100]		
Cd^{2+}	108	0.72	0.36	
Sn^{2+}	150.9 [95]	0.9 [95]		
Hg^{2+}	137.2 [91]	0.85		
Hg_2^{2+}	127.2 [91]	0.91		
Pb^{2+}	139	0.95	0.233	
UO_2^{2+}	114 [101]	0.43		
Al^{3+}	183 [98]	0.54	0.67	
Sc^{3+}	194.1 [98]	0.57		

TABLE 2.11   (Continued)

Ion	λ_I^∞/S·cm²·mol⁻¹	D_I^∞/10⁻⁹m²·s⁻¹	$B_{\eta I}$/dm³·mol⁻¹	B_{nmrI}/dm³·mol⁻¹
Cr^{3+}	201 [91]	0.60	0.737	
Fe^{3+}	204 [91]	0.60	0.69	
Y^{3+}	186 [98]	0.55		
In^{3+}	186.3 [99]			
La^{3+}	209.1	0.62	0.576	
Ce^{3+}	209.41	0.62	0.57	
Pr^{3+}	208.8	0.62	0.581	
Nd^{3+}	208.2	0.62	0.576	
Sm^{3+}	205.5	0.61	0.599	
Eu^{3+}	203.4	0.60	0.618	
Gd^{3+}	201.9	0.60	0.64	
Tb^{3+}	197.3		0.647	
Dy^{3+}	196.8	0.58	0.656	
Ho^{3+}	198.9	0.59	0.667	
Er^{3+}	197.7	0.59	0.657	
Tm^{3+}	196.2	0.58	0.672	
Yb^{3+}	196.8	0.58	0.665	
Lu^{3+}	196.4		0.675	
Th^{4+}	288 [102]		0.726 [103]	
U^{4+}	272 [104]			
F^-	55.4	1.48	0.127	0.14
Cl^-	76.4	2.03	−0.005	−0.01
Br^-	78.1	2.08	−0.033	−0.04
I^-	76.4	2.05	−0.073	−0.08
OH^-	198.3	5.27	0.122	0.18
SH^-	65 [91]	1.73	0.025 [105]	
CN^-	82 [98]	2.08	−0.024	−0.04
NCO^-	64.6 [91]	1.72	−0.032 [105]	
SCN^-	66 [98]	1.76	−0.032 [106]	−0.07
N_3^-	69	1.84	−0.018	0.00
HF_2^-	75 [91]	2.00	0.058	
HO_2^-		0.5 [107]		
I_3^-	42.5 [108]	1.13 [99]		
BO_2^-	30.6 [109]	0.81 [109]		
ClO_2^-	52 [91]	1.39	0.067 [105]	
NO_2^-	71.8 [91]	1.91	−0.024 [105]	−0.05
NO_3^-	71.5	1.90	−0.045	−0.05
ClO_3^-	64.6	1.72	−0.022	−0.08
BrO_3^-	55.7	1.48	0.009	−0.06
IO_3^-	40.5	1.08	0.14	0.02
ClO_4^-	67.4	1.79	−0.058	−0.085
MnO_4^-	61 [98]	1.63	−0.059	
ReO_4^-	54.9 [91]	1.46	−0.055	−0.03
BF_4^-	37.7 [96]^{d,e}			

(continued)

TABLE 2.11   (Continued)

Ion	λ_I^∞/S·cm^2·mol^{-1}	D_I^∞/10^{-9}m^2·s^{-1}	$B_{\eta I}$/dm^3·mol^{-1}	$B_{nmrI/}$dm^3·mol^{-1}
HCO$_2^-$	54.6	1.45	−0.052 [105]	
CH$_3$CO$_2^-$	41.4 [98]	1.09	0.246	
CF$_3$CO$_2^-$				
CF$_3$SO$_3^-$	44.5 [109]			
BPh$_4^-$	19.9 [97]	0.56	1.114	
HCO$_3^-$	44.5	1.19	0.13 [110]	
HSO$_4^-$	50 [91]	1.33	0.127 [110]	
H$_2$PO$_4^-$	33 [91]	0.88	0.34 [110]	
PF$_6^-$	56.9 [91]^e	1.52^e	−0.21 [111]	
S^{2-}	191 [112]	2.65 [113]^b		
CO$_3^{2-}$	138.6	0.92	0.294	0.25
C$_2$O$_4^{2-}$	148.3	0.99	0.174	
SO$_3^{2-}$	159.8 [101]	1.06	0.282 [105]	0.22
SO$_4^{2-}$	160	1.07	0.206	0.12
SeO$_4^{2-}$	151.4 [101]	1.01	0.165 [105]	
CrO$_4^{2-}$	166 [98]	1.13	0.165	
MoO$_4^{2-}$	149 [91]	0.98^c	0.22 [105]	
WO$_4^{2-}$	138 [91]	0.92		
S$_2$O$_3^{2-}$	169.8 [98]	1.13		
SiF$_6^{2-}$			0.374 [105]	
Cr$_2$O$_7^{2-}$	120.7 [114]		0.084 [110]	
HPO$_4^{2-}$	66 [91]	0.44	0.382 [115]	
PO$_4^{3-}$	207 [91]	0.61	0.495 [110]	
AsO$_4^{3-}$			0.52 [110]	
Fe(CN)$_6^{3-}$	325.9 [116]	0.90	0.114 [116]	
Co(CN)$_6^{3-}$	304.5 [116]	0.88	0.152 [116]	
Fe(CN)$_6^{4-}$	406 [116]	0.741	0.35 [116]	

^{a} Calculated from the diffusion coefficients, cf. Equation 2.29.

^{b} The value appears to be excessively large.

^{c} Correction of the value given in Ref. 6.

^{d} Extrapolated from data for LiBF$_4$ in aqueous 2-methoxyethanol to zero content of the cosolvent.

^{e} Ref. 107 yields different values for BF$_4^-$ (λ_6^∞ = 69.8 and D_I^∞ = 1.86) and PF$_6^-$ (λ_I^∞ = 29.0 and D_I^∞ = 0.77).

coefficients are empirical. They represent the limiting slopes of plots of $\left[\left(\eta/\eta_W^*\right)-1\right]c_E^{-1/2}$ vs. $c_E^{1/2}$. The B_η coefficients are additive and the electrolyte values can be split into the ionic contributions in a manner related to the mobilities of the ions: $B_{\eta+}/B_{\eta-} \approx u_+/u_-$. Over a fairly wide temperature range, the ratio u_+/u_- for aqueous RbBr is constant, and hence B_η (Rb$^+$, aq) = B_η (Br$^-$, aq) was adopted as the splitting assumption by Jenkins and Marcus [92]. Viscosity B_η coefficients of many ions at 25°C are listed in Table 2.11. They are negative for large univalent ions and positive for small and multivalent ions. These algebraic signs constitute the classification of ions into water-structure makers ($B_{\eta I}>0$) and water-structure breakers ($B_{\eta I}<0$) as suggested by Gurney [124] and recently reviewed by Marcus [121, 125]. Negative B_η values become less negative as the temperature is raised and may change sign at a characteristic temperature. This is explained by the diminishing extent of hydrogen-bonded structure in the water as the temperature is raised, so that structure-breaking ions have less structure to break.

2.3.2.4 Ionic Effects on the Relaxation of NMR Signals

In analogy with the viscosity B-coefficients, the longitudinal relaxation time T_1 of the proton NMR signals of aqueous solutions of electrolytes obeys the Engel and Hertz expression [93]:

$$\left[\left(1/T_{1E}\right)/\left(1/T_{1W}\right)-1\right]=B_{nmr}c_E+\cdots \tag{2.30}$$

A good correspondence exists between the ionic $B_{\eta I}$ values and the ionic B_{nmrI}, the latter calculated according to the convention that $B_{nmr}(K^+)=B_{nmr}(Cl^-)$ [93, 126]. Values of B_{nmrI} for many ions at 25°C are shown in Table 2.11, supplemented by data from the ^{17}O NMR spin-lattice relaxation of D_2O molecules in aqueous salt solutions, again split according to $B_{nmr}(K^+)=B_{nmr}(Cl^-)$. These values [127] agree in sign and generally in magnitude with the proton NMR B_{nmrI} and the viscosity B_{η} values and are also shown in Table 2.11.

Ions were classified [93] as being "positively hydrated" and "negatively hydrated" according to the signs of their $\Delta_{exch}G^{\neq}$, the Gibbs energy of activation for the rate of exchange of hydration shell water molecules with the bulk water as proposed by Samoilov [128]. These epithets are no longer in common use, having been replaced by water-structure-breaking ions for those that have $B_{nmrI}<0$ and water-structure-making ions for those that have $B_{nmrI}>0$.

Only aqueous diamagnetic ions can be studied by means of proton NMR measurements of longitudinal relaxation times, T_1. Therefore, for paramagnetic ions, for example, transition metal cations, other kinds of NMR measurements are required, but no such data have been found.

The relative rotational correlation times τ_r of the water molecules near ions are $\tau_r=\tau_{rI}/\tau_{rW}=1+(55.51/h_1)B_{nmr}$, where the h_1 are the hydration numbers of the ions. The relative water molecule reorientation times τ_{rI}/τ_{rW} at 22°C are less than 1 for Br^-, I^-, NH_4^+, NO_3^-, and N_3^-, ~1.0 for K^+, and >1 for Li^+, Na^+, Mg^{2+}, Ca^{2+}, Sr^{2+}, Ba^{2+}, F^-, Cl^-, H_3O^+, SO_4^{2-}, and CO_3^{2-} according to Chizhik [129]. The signs obtained for the τ_{rI}/τ_{rW} values at relatively large concentrations agree more or less with the signs of the B_{nmr} in dilute solutions.

2.3.2.5 Ionic Dielectric Decrements

Ions in dilute aqueous solutions diminish the permittivity of the solution, in a manner proportional to the concentration, an effect called the dielectric decrement. The permittivity of electrolyte solutions is measured as a function of both the concentration c and the frequency of the applied electric field ω and extrapolated to zero values of both, hence obtaining the static decrement $\delta_{0E}=\lim(c\rightarrow0,\omega\rightarrow0)d\varepsilon/dc$. The infinite dilution electrolyte values at 25°C are additive in the ionic contributions and Marcus [130] proposed to split them into the latter, δ_{0I}, on the assumption adopted for the viscosity B-coefficients (Section 2.3.2.3), namely $\delta_0(Rb^+)=\delta_0(Br^-)$, with results shown in Table 2.12. The uncertainties of the values are $\pm2\,M^{-1}$. The values of δ_{0I} are approximately linearly related to the corresponding $B_{\eta I}$ values (correlation coefficient of 0.85). Ionic values for divalent anions, such as sulfate could not be established, because ion pairing takes place at the concentrations at which measurements could be made, so that the electrolyte values could not be extrapolated to infinite dilution to obtain additive ionic values.

2.3.2.6 Ionic Effects on the Surface Tension   The surface layer of the aqueous solutions, of thickness ~1 nm, has air (or dilute water vapors) on the one side and bulk water on the other. Ions may be positively or negatively sorbed in this layer, depending on whether they decrease or increase the surface tension, σ, of water. According to the Gibbs adsorption law:

$$\Gamma_E = -\left(\frac{a_E}{RT}\right)\left(\frac{\partial \sigma}{\partial a_E}\right)_{T,P} \approx -\left(\frac{c_E}{RT}\right)\left(\frac{\partial \sigma}{\partial c_E}\right)_{T,P} \tag{2.31}$$

where Γ_E denotes the number of moles of electrolyte sorbed per unit increase of the surface energy and a_E is the activity of the electrolyte. At low molar concentrations c_E, these approximate the activities a_E. The molar surface tension increment, $k_E = (\partial \sigma / \partial c_E)_{T,P}$, is a key quantity in dealing with the role of ions at surfaces. Over considerable ranges of composition, the derivatives of the surface tension increment $\Delta\sigma = \sigma - \sigma_W^*$ with the composition are linear up to at least 1 M as shown by Marcus [131]. The additivity of the ionic contributions $k_I = d\sigma/dc_i$ to k_E holds within $\pm 0.2 \, mN \cdot m^{-1} \cdot mol^{-1} \cdot dm^3$. Ionic values k_I are derived on the *arbitrary* but plausible basis that $d\sigma/dc_i$ in $mN \cdot m^{-1} \cdot mol^{-1} \cdot dm^3$ are 0.90 for Na^+ and 1.20 for Cl^- and are shown in Table 2.12.

The values of k_I exhibit some clear trends that are independent of the arbitrary assumption of the values for Na^+ (and/or Cl^-) since they pertain to sequences among cations separately from those among anions. One feature is obvious: negative values of k_I are rare, meaning that most (small) ions are desorbed from the surface layer and their concentration in it is lower than in the bulk solution ($\Gamma_I < 0$ for $k_I > 0$). Outstanding cases of negative values are H^+ and $(CH_3)_4N^+$ among the cations and large singly charged anions, such as ClO_4^-. Large hydrophobic groups in ions favor their sorption into the surface layer: $k_I/mN \cdot m^{-1} \cdot mol^{-1} \cdot dm^3 = -2.95$ for propanoate, -6.45, for butanoate, -18 for butylammonium, -49 for dibutylammonium, and -136 for tri- or tetrabutylammonium as chlorides. Another trend that emerges from the data is that the positive values of $d\sigma/dc_i$ tend to increase with the ionic charge, whether positive or negative. The higher the charge, the larger their centrally symmetric hydration spheres and the less well are they accommodated in the nonisotropic hydrogen-bonded structure of the surface layer [131].

As mentioned, cations with $k_i > 0$ are repelled from the surface ($\Gamma_I < 0$), and if $k_+ > k_-$, they are repelled more than the anions. This causes a charge imbalance in the surface layer leading to the establishment of an electric double layer. The surface potential of electrolyte solutions over that of pure water (with respect to vacuum/air/dilute water vapor), $\Delta\Delta\chi$, was measured as a function of the electrolyte concentration. The available $\Delta\Delta\chi$ values at 1 M MX, for $M = H^+$, Na^+, K^+, and NH_4^+ with a variety

TABLE 2.12 Ionic Static Permittivity Decrements, $\delta_{OI}/dm^3 \cdot mol^{-1}$ [130] and Surface Tension Increments, $d\sigma/dc_i/mN \cdot m^{-1} \cdot mol^{-1} \cdot dm^3$ year, in Aqueous Solutions at 20–30°C

Cation	δ_{OI}	$d\sigma/dc_i$	Anion	δ_{OI}	$d\sigma/dc_i$
H^+		−1.35	OH^-	8	1.35
Li^+	8	0.65	F^-	4	1.10
Na^+	7	0.90	Cl^-	5	1.20
K^+	6	0.80	Br^-	6	0.95
Rb^+	6	(0.65)	I^-	7	0.35
Cs^+	5	0.50	SCN^-		0.20
NH_4^+	4	0.40	NO_3^-	5	0.45
$(CH_3)_4N^+$		−0.40	ClO_3^-		0.00
$C(NH_2)_3 H^+$	5	−0.24	ClO_4^-	1	−0.50
Ag^+	1	0.40	BF_4^-	1	
Tl^+		0.30	HCO_2^-	4	0.15
Mg^{2+}	24	1.65	$CH_3CO_2^-$		0.05
Ca^{2+}	17	1.50	$C_2H_5CO_2^-$		−2.95
Sr^{2+}	14	1.20	$C_3H_7CO_2^-$		−6.45
Ba^2	10	0.90	KH_2PO_4		1.25
Mn^{2+}		0.75	CO_3^{2-}		0.95
Co^{2+}		1.05	SO_4^{2-}		1.15
Ni^{2+}		1.10	CrO_4^{2-}		1.45
Cd^{2+}	20		$S_2O_3^{2-}$		1.45
Pb^{2+}		1.80	PO_4^{3-}		2.00
UO_2^{2+}		1.40	KH_2PO_4		1.25
Al^{3+}		1.75	CO_3^{2-}		0.95
Y^{3+}	22				
La^{3+}	25	2.30			

of anions X^- were compared by Marcus [131]. In the case of the acids, the hydrogen ions are attracted to the surface rather than repelled from it as are the other cations, hence 30 mV had to be added to the $\Delta\Delta\chi$ values for HX, with results shown in Figure 2.3. The scatter in the plot of $\Delta\Delta\chi$ vs. k_E is appreciable, but the plot is linear:

$$\Delta\Delta\chi/mV = (55 \pm 2) - (29 \pm 1)\left(k_E/mN \cdot m^{-1} \cdot M^{-1}\right) \quad (2.32)$$

The right-hand panel of Figure 2.3 shows the plot of the $\Delta\Delta\chi$ values against $k_+ - k_-$, the *difference* between the individual ionic values. The linear plot passes essentially through the origin for the KX salts and does so for the other salts when −80, +20, and −15 mV are added to the HX, NaX, and NH_4X electrolyte data:

$$\Delta\Delta\chi/mV = (-3 \pm 2) + (30 \pm 2)\left[(k_+ - k_-)/(mN \cdot m^{-1} \cdot M^{-1})\right] \quad (2.33)$$

This means that the surface potential is dominated by the ion sorption/desorption at the surface and shows the practical value of the splitting of the electrolyte molar k_E into the individual ionic values.

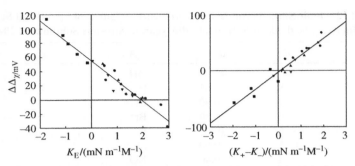

FIGURE 2.3   Surface potential increment of aqueous 1:1 electrolyte solutions at 1 M over that of pure water, $\Delta\Delta\chi$. Left-hand panel: plotted against their molar surface tension increments, k_E; right-hand panel: plotted against the difference between the cation and anion molar surface tension increments, $k_+ - k_-$. Squares are for acids HX, circles are for NaX, upright triangles are for KX, and downward triangles are for NH$_4$X, with X being the anion. From Ref. 131 with permission of the publisher, ACS.

The individual ionic k_I values normalized with respect to the ionic charge are also roughly linear with increasing ionic sizes:

$$\left(k_I / z_I\right) / \left(\text{mN} \cdot \text{m}^{-1} \cdot \text{M}^{-1}\right) = (1.51 \pm 0.12) - (5.8 \pm 0.7)\left(r_I / \text{nm}\right) \qquad (2.34)$$

for 37 ions of both signs, univalent as well as multivalent, and monatomic as well as polyatomic ones [131].

Further linear correlations of the k_I values with independent properties of the ions, such as their molar refractivity (polarizability), their softness, and their effects on the structure of the water in their vicinity, have also been reported by Marcus [131]. Water-structure-breaking ions (see Chapter 5) have small positive or even negative k_I values and structure-making ions are desorbed from the water surface with positive k_I values. On the whole, however, hardly any correlation exists when all the ions for which both k_I values and also $\Delta_{hyd}G_I^\infty$ values are known are considered. Thus, ion dehydration appears not to play a *major* role in the desorption of ions from the surface layer.

REFERENCES

[1] D. D. Wagman, W. H. Evans, V. B. Parker, R. H. Schumm, I. Halow, S. M. Bailey, K. L. Churney, R. L. Nuttall, J. Phys. Chem. Ref. Data **11**, Suppl. 2, (1982).

[2] S. G. Lias, J. E. Bartmess, J. F. Liebman, J. L. Holmes, R. D. Levin, W. G. Mallard, J. Phys. Chem. Ref. Data **17**, Suppl. 1, (1988).

[3] R. W. Taft, F. Bordwell, Acc. Chem. Res. **21**, 463 (1988).

[4] A. A. Viggiano, M. J. Henchman, F. Dale, C. A. Deakyne, J. F. Paulson, J. Am. Chem. Soc. **114**, 4299 (1992).

[5] D. R. Lide, ed., Handbook of Chemistry and Physics, CRC Press, Boca Raton, FL, 82nd ed. (2001/2002).

[6] Y. Marcus, Ion Properties, Dekker, New York (1997) and references therein.

[7] G. L. Gutsev, A. I. Boldyrev, J. Phys. Chem. **94**, 2256 (1990).

[8] X.-B. Wang, L.-S. Wang, J. Chem. Phys. **113**, 10928 (2000).

[9] H. Wei, G.-L. Hou, W. Huang, N. Govind, J. Chem. Phys. **135**, 184309 (2011).

[10] R. Yamdagni, P. Kebarle, Ber. Bunsenges. Phys. Chem. **78**, 181 (1974).

[11] Y. Marcus, J. Inorg. Nucl. Chem. **37**, 493 (1975).

[12] S. B. Bratsch, J. L. Lagowski, J. Phys. Chem. **90**, 307 (1986); J. Solution Chem. **16**, 583 (1987).

[13] Y. Marcus, A. Loewenschuss, Annu. Rep. **81C**, 81 (1985).

[14] A. Loewenschuss, Y. Marcus, J. Phys. Chem. Ref. Data **16**, 61 (1987).

[15] M. W. Chase Jr., J. L. Curnutt, J. R. Downey, J. Phys. Chem. Ref. Data **11**, 695 (1982).

[16] S. Olivella, F. Urpi, J. Vilarrasa, J. Comput. Chem. **5**, 230 (1984).

[17] D. A. Dixon, J. S. Francisco, Y. Alexeev, J. Phys. Chem. A **110**, 185 (2006).

[18] Y. Marcus, J. Chem. Thermodyn. **48**, 70 (2012).

[19] Y. Marcus, A. Loewenschuss, J. Phys. Chem. Ref. Data **25**, 1495 (1996).

[20] Y. Marcus, J. Chem. Thermodyn. **71**, 196 (2014).

[21] M. H. Abraham, Y. Marcus, J. Chem. Soc. Faraday Trans. **1**, 82, 3255 (1986).

[22] W. E. Dasent, Inorganic Enedrgetics, Cambridge University Press, Cambridge, UK, 2nd ed. (1982).

[23] G. Caldwell, R. Renneboog, P. Kebarle, Can. J. Chem. **67**, 611 (1989).

[24] Y. Marcus, H. D. B. Jenkins, L. Glasser, *J. Chem. Soc. Dalton Trans.* 3795 (2002).

[25] P. W. Selwood, Magnetochemistry, Interscience, New York, 2nd ed. 403 (1956).

[26] J. J. Salzmann, C. K. Jørgensen, Helv. Chim. Acta **51**, 1276 (1968).

[27] Y. Marcus, Thermochim. Acta **104**, 389 (1987).

[28] D. Schroeder, Phys. Chem. Chem. Phys. **14**, 6375 (2012).

[29] J. Rais, T. Okada, Anal. Sci. **22**, 533 (2006).

[30] X. Chen, A. J. Stace, J. Phys. Chem. A **117**, 5015 (2013).

[31] L. Dzidic, P. Kebarle, J. Phys. Chem. **74**, 1466 (1970).

[32] S. B. Nielsen, M. Masella, P. Kebarle, J. Phys. Chem. A **103**, 9891 (1999).

[33] J. V. Coe, Chem. Phys. Lett. **229**, 161 (1994).

[34] M. Peschke, A. T. Blades, P. Kebarle, J. Am. Chem. Soc. **122**, 10440 (2000).

[35] M. Peschke, A. T. Blades, P. Kebarle, J. Phys. Chem. A **102**, 9978 (1998).

[36] M. Arshadi, R. Yamdagni, P. Kebarle, J. Phys. Chem. **74**, 1475 (1970).

[37] W. R. Davidson, P. Kebarle, J. Am. Chem. Soc. **98**, 6125 (1976).

[38] N. Lee, R. G. Keese, A. W. Castleman, J. Chem. Phys. **72**, 1089 (1980).

[39] R. G. Keese, N. Lee, A. W. Castleman, J. Am. Chem. Soc. **101**, 2599 (1979).

[40] A. T. Blades, P. Kebarle, J. Phys. Chem. A **109**, 8293 (2005).

[41] A. T. Blades, J. S. Klassen, P. Kebarle, J. Am. Chem. Soc. **117**, 10563 (1995).

[42] Y. Marcus, Ion Solvation, Wiley, Chichester (1985).

[43] R. D. Shannon, C. T. Prewitt, Acta Cryst. **B25**, 925 (1969); **B26**, 1046 (1969).

[44] R. D. Shannon, Acta Cryst. **A32**, 751 (1976).

[45] Y. Marcus, J. Solution Chem. **12**, 271 (1983).

[46] Y. Marcus, Chem. Rev. **88**, 1475 (1988).

[47] H. K. Roobottom, H. D. B. Jenkins, J. Passmore, L. Glasser, J. Chem. Educ. **76**, 1570 (1999).

[48] H. D. B. Jenkins, H. K. Roobottom, J. Passmore, L. Glasser, Inorg. Chem. **38**, 3609 (1999).

[49] E. Glueckauf, Trans. Faraday Soc. **61**, 914 (1965).

[50] H. Ohtaki, T. Radnai, Chem. Rev. **93**, 1167 (1993).

[51] L. Glasser, H. D. B. Jenkins, Inorg. Chem. **47**, 6195 (2008).

[52] P. Mukerjee, J. Phys. Chem. **65**, 740 (1961).

[53] T. G. Pedersen, C. Dethlefsen, A. Hvidt, Carlsberg Res. Commun. **49**, 445 (1984).

[54] Y. Marcus, J. Chem. Eng. Data **58**, 488 (2013).

[55] Y. Marcus, Chem. Rev. **111**, 2761 (2011).

[56] Y. Marcus, J. Solution Chem. **39**, 1031 (2010). 488.

[57] B. E. Conway, J. Solution Chem. **7**, 721 (1978).

[58] Y. Marcus, in D. D. Bostrelli, ed., Solution Chemistry Research Progress, Nova Science Publishers, Hauppauge, NY, 51 (2008).

[59] P. Hünenberger, M. Reif, Single-Ion Solvation: Experimental and Theoretical Approaches to Elusive Thermodynamic Quantities, Royal Society of Chemistry, Cambridge (2010).

[60] J. Liszi, L. Mészáros, I. Ruff, I. J. Chem. Phys. **74**, 6896 (1981).

[61] Y. Marcus, G. Hefter, J. Solution Chem. **28**, 579 (1999).

[62] J. O'M. Bockris, P. P. S. Saluja, J. Phys. Chem. **76**, 2140 (1972).

[63] A. W. Hakin, M. M. Bhuiyan, in W. Wilhelm, T. Letcher, eds., Heat Capacities: Liquids, Solutions, and Vapours, The Royal Society of Chemistry, London, 132 (2010).

[64] L. G. Hepler, J. K. Hovey, Can. J. Chem. **74**, 639 (1996).

[65] E. L. Schock, H. C. Helgeson, Geochim. Cosmochim. Acta **52**, 2009 (1989); correction: ibid. **53**, 215 (1989).

[66] A. J. Bard, R. Parsons, J. Jordan, eds., Standard Potentials in Aqueous Solutions, Dekker, New York (1985).

[67] Y. Marcus, J. Solution Chem. **37**, 1071 (2008).

[68] Y. Nagano, H. Mizuno, M. Sakiyama, T. Fujiwara, Y. Kondo, J. Phys. Chem. **95**, 2536 (1991).

[69] Y. Marcus, A. Loewenschuss, J. Chem. Soc. Faraday Trans. 1, **82**, 993 (1986).

[70] D. A. Johnson, P. G. Nelson, J. Chem. Soc. Dalton Trans., **1** (1990).

[71] G. A. Krestov, Thermodynamics of Solvation, Ellis Horwood, Chichester (1991).

[72] H. L. Friedman, C. V. Krishnan, in F. Franks, ed., Water: A Comprehensive Treatise, Plenum, New York, Vol. **3**, 54 (1973).

[73] O. D. Bonner, P. J. Cerutti, J. Chem. Thermodyn. **8**, 105 (1976).

[74] C. M. Criss, F. J. Millero, J. Phys. Chem. **100**, 1288 (1996).

[75] C. M. Criss, F. J. Millero, J. Solution Chem. **28**, 849 (1999).

[76] H. C. Helgeson, D. H. Kirkham, G. C. Flowers, Am. J. Sci. **281**, 1249 (1981).

[77] Y. Marcus, J. Chem. Soc. Faraday Trans. **83**, 339 (1987).

[78] F. J. Millero, Chem. Rev. **71**, 147 (1971).

[79] Y. Marcus, J. Chem. Phys. **137**, 154501 (2012).

[80] Y. Marcus, J. Phys. Chem. B **113**, 10285 (2009).

[81] Y. Marcus, J. Phys. Chem. B **116**, 7232 (2012).

[82] P. Lepori, L. Lepori, J. Solution Chem. **25**, 1 (1996).

[83] D. Tran, J. P. Hunt, S. Wherland, Inorg. Chem. **31**, 2460 (1992).

[84] J. C. Tanger IV, H. C. Helgeson, Am. J. Sci. **288**, 19 (1988).

[85] O. Redlich, J. Phys. Chem. **44**, 619 (1940).

[86] Y. Marcus, J. Solution Chem. **35**, 1271 (2006).

[87] Y. Marcus, G. Hefter, J. Solution Chem. **28**, 575 (1999).

[88] J. G. Mathieson, B. E. Conway, J. Solution Chem. **3**, 435 (1974).

[89] R. Fernandez-Prini, Trans. Faraday Soc. **65**, 3311 (1969).

[90] R. A. Robinson, R. H. Stokes, Electrolyte Solutions, Butterworths, London, 2nd ed. (1965).

[91] P. Vanysek, in D. R. Lide, ed., Handbook of Chemistry and Physics, CRC Press, Boca Raton, FL, 82nd ed. (2001/2002).

[92] H. D. B. Jenkins, Y. Marcus, Chem. Rev. **95**, 2695 (1995).

[93] G. Engel, H. G. Hertz, Ber. Bunsengesell. Phys. Chem. **72**, 808 (1968).

[94] C. Biondi, L. Bellugi, Chem. Phys. **62**, 145 (1981).

[95] E. H. Olkers, H. C. Helgeson, J. Solution Chem. **18**, 691 (1989).

[96] B. Das, D. K. Hazra, J. Solution Chem. **27**, 1021 (1998).

[97] Y. Marcus, J. Chem. Soc. Faraday Trans. **1**, 83, 2985 (1987).

[98] H. Falkenhagen, Theorie der Elektrolyte, S. Hirzel Verlag, Leipzig (1971).

[99] A. N. Campbell, Can. J. Chem. **51**, 3006 (1973); **54**, 3732 (1976).

[100] B. P. Yurev. S. P. Shkuryakova, Y. V. Borisoglebskii, Zh. Prokl. Khim. (Leningrad) **56**, 1892 (1983).

[101] P. Vanysek, in D. R. Lide, ed. CRC Handbook of Chemistry and Physics, CRC Press, Boca Raton, FL, 75th ed., 5–90 (1994).

[102] A. Apelblat, D. Azoulay, A. Sahar, J. Chem. Soc. Faraday Trans. **1**, 69, 1618 (1973).

[103] B. Sinha, P. K. Roy, B. K. Sarkar, D. Brahman, M. N. Roy, J. Chem. Thermodyn. **42**, 380 (2010).

[104] F. Arndt, G. Marx, Z. Naturf. A **36**, 1019 (1981).

[105] G. A. Krestov, Termodyn. Ionikh Protsesov v Rastvorov, Khimiya, Leningrad, 1984.

[106] Y. Marcus, J. Chem. Eng. Data **57**, 617 (2012).

[107] E. L. Littauer, K. C. Tsai, Electrochim. Acta **24**, 681 (1979).

[108] M. Spiro, A. M. Creeth, J. Chem. Soc. Faraday Trans. **86**, 3572 (1990).

[109] S. E. Okan, D. C. Champeney, J. Solution Chem. **26**, 405 (1997).

[110] A. N. Soriano, A. M. Agapito, L. J. L. I. Lagumbay, A. R. Caparangaq, M.-H. Li, J. Taiwan Chem. Eng. **42**, 258 (2011).

[111] E. R. Nightingale, in B. E. Conway, R. G. Barradas, eds., Chemical Physics of Electrolyte Solutions, Wiley, New York (1966).

[112] J. L. Frahn, J. Chromatogr. **37**, 279 (1968).

[113] K. Shimizu, R. A. Osteryoung, Anal. Chem. **53**, 2350 (1981).

[114] J. A. Swamy, B. Sethuran, T. N., Rao, Indian J. Chem. **A15**, 449 (1977).

[115] E. C. Bingham, J. Phys. Chem. **45**, 885 (1941).

[116] J. G. Mathieson, G. Curthoys, Austr. J. Chem. **28**, 975 (1975).

[117] Y. Marcus, J. Solution Chem. **41**, 2082 (2012).

[118] R. Mills, Rev. Pure Appl. Chem. (Melbourne) **11**, 78 (1961).

[119] P. Passiniemi, J. Solution Chem. **12**, 801 (1983).

[120] B. M. Braun, H. Weingärtner, J. Phys. Chem. **92**, 1342 (1988).

[121] Y. Marcus, Pure Appl. Chem. **82**, 1889 (2010).

[122] G. Jones, M. Dole, J. Am. Chem. Soc. **51**, 2950 (1929).

[123] H. Falkenhagen, M. Dole, Phys. Z. **30**, 611 (1929).

[124] R. W. Gurney, Ionic Processes in Solution, McGraw-Hill, New York (1953).

[125] Y. Marcus, Chem. Rev. **109**, 1346 (2009).

[126] M. H. Abraham, J. Liszi, E. Papp, J. Chem. Soc. Faraday Trans. **78**, 197 (1982).

[127] K. Yoshida, K. Ibuki, M. Ueno, J. Solution Chem. **25**, 435 (1996).

[128] O. Ya. Samoilov, Disc. Faraday Soc. **24**, 141 (1957).

[129] V. I. Chizhik, Mol. Phys. **90**, 653 (1997).

[130] Y. Marcus, J. Solution Chem. **42**, 2354 (2013).

[131] Y. Marcus, Langmuir **29**, 2881 (2013).

3

SOLVENTS FOR IONS

3.1 SOLVENT PROPERTIES THAT SUIT ION DISSOLUTION

Liquids that are to act as solvents for electrolytes and permit their dissociation into ions must have certain properties. Their molecules should have an ability to interact with the ions sufficiently strongly in energetic terms in order to overcome the binding energy of the ions in the pure electrolyte: lattice energy in crystalline ionic substances or bond energies in molecular ionogenic substances. In other words, the molecules of the solvent should be able to solvate the ions of the electrolyte. The solvent should have a sufficiently high permittivity in order to prevent the ions from sticking together nearly completely by electrostatic forces. They should be chemically stable and have relatively low vapor pressures at the temperatures at which they are to be employed. Other properties, desirable but not indispensable, include a large enough potential window to permit electrochemical studies of the ions, transparency to visible and near ultraviolet light to permit spectrometric studies of the ions, and similar properties. Most studies of ions in solution are carried out at ambient conditions (near the standard temperature of $T° = 298.15$ K and the standard pressure of $P° = 0.1$ MPa), but, of course, measurements of the temperature and pressure derivatives of the properties of the ions in solution require departures from these standard conditions.

In view of the requirements noted here, hydrocarbons are unsuitable as solvents for ions, although aromatic hydrocarbons do interact with electrolytes that themselves have aromatic or large aliphatic groups attached to an ionized atom or group

Ions in Solution and Their Solvation, First Edition. Yizhak Marcus.
© 2015 John Wiley & Sons, Inc. Published 2015 by John Wiley & Sons, Inc.

of atoms and permit their dissolution and some ionic dissociation. Ethers and aliphatic amines, although containing an atom with good electron donor properties (respectively oxygen and nitrogen), generally have low permittivities and are rarely used as solvents for ions. There is, therefore, a rather limited, but not exclusive, list of solvents that are widely employed for studying ions in solution. Table 3.1 provides the names of the solvents, their common abbreviations, their molecular formulas, and their CAS registry numbers. The CAS registry numbers help to find further information concerning the solvents.

In the following, the physical and chemical properties of such solvents that make them suitable for dissolving electrolytes and permit their ionic dissociation are presented and discussed. Properties of liquids not included in this limited list can be found in such compilations as [1–3]. Some of the solvents that are listed have low permittivities and are not particularly suited for the dissolution of electrolytes and their ionic dissociation, but have been employed for special purposes, such as in mixtures with water (e.g., 1,4-dioxane) or because of special donor properties (such as tetrahydrothiophene). Another solvent listed, ammonia, is not liquid at ambient conditions, but becomes so at low temperatures and moderately high pressures, and its ion solvation properties have been extensively studied.

There are many physical properties of solvents for electrolytes that have been compiled ([1–3] and elsewhere) but are of less interest in the context of the solvation of the ions and the properties of electrolyte solutions in the solvents. Such properties include the critical temperature, pressure, and density on the one hand, when the solvent is heated, and the glass transition temperature, when the solvent is cooled and forms a glass on the other hand. The glass is homogeneous and isotropic and has a viscosity $\geq 10^{10}$ Pa·s, but is not in internal equilibrium.

3.2 PHYSICAL PROPERTIES OF SOLVENTS

The liquid range of solvents for electrolytes is of prime importance for their use. The freezing point of the liquid solvents (or the melting point of the frozen solvent, t_m) and its normal boiling point (t_b at 1 atm = 0.101325 MPa, note that this differs slightly from $P° = 0.1$ MPa) are listed in Table 3.2. Two solvents on the list have freezing points >25°C: t-butanol (2-methyl-2-propanol) and sulfolane (tetramethylenesulfone) $t_m = 28.5°C$. If not otherwise noted, their physical properties at 25°C listed in the following tables are for the super-cooled liquid. One solvent, ammonia, has t_b below ambient (−33.5°C) and the data pertaining to it are at t_b. Also included in this table are the molar masses, M, of the solvents. These data are taken mostly from the compilations by Riddick et al. and by Marcus [1, 3].

3.2.1 Volumetric Properties

The volumetric properties of solvents for electrolytes: the densities, ρ, the molar volumes, $V = M/\rho$, the isobaric expansibilities, $\alpha_P = V^{-1}(\partial V/\partial T)_P$, and isothermal compressibilities, $\kappa_T = -V^{-1}(\partial V/\partial P)_T$, all at 25°C, are shown in Table 3.3. Most

TABLE 3.1 Solvents, their Common Abbreviations, Formulas, and CAS Registry Numbers

Solvent	Abbreviation	Formula	CAS registry
Water	W	H_2O	7732-18-5
Heavy water		D_2O	11105-15-0
Methanol	MeOH	CH_3OH	67-56-1
Ethanol	EtOH	C_2H_5OH	64-17-5
n-Propanol	n-PrOH	C_3H_7OH	71-23-8
i-Propanol	i-PrOH	$(CH_3)_2CHOH$	67-63-0
t-Butanol	t-BuOH	$(CH_3)_3COH$	76-65-0
Trifluoroethanol	TFE	CF_3CH_2OH	75-89-8
Ethylene glycol	EG	HOC_2H_4OH	107-21-1
Tetrahydrofuran	THF	$c\text{-}(CH_2)_4O$	109-99-9
1,4-Dioxane	Diox	$c\text{-}(C_2H_4O\text{-})_2$	123-91-1
Acetone	Me_2CO	$(CH_3)_2CO$	67-64-1
Propylene carbonate	PC	^{a}	108-32-7
γ-Butyrolactone	gBulac	$c\text{-}(CH_2)_3C(O)O$	96-48-0
Formamide	FA	$HC(O)NH_2$	75-12-7
N-Methylformamide	NMF	$HC(O)NHCH_3$	123-39-7
N,N-Dimethylformamide	DMF	$HC(O)N(CH_3)_2$	68-12-2
N,N-Dimethylacetamide	DMA	$CH_3C(O)N(CH_3)_2$	127-19-5
Tetramethylurea	TMU	$[(CH_3)_2N]_2C(O)$	632-22-4
N-Methylpyrrolidinone	NMPy	^{b}	872-50-4
Ammonia		NH_3	7664-41-7
Pyridine	Py	$c\text{-}(CH)_5N$	110-86-1
Acetonitrile	MeCN	CH_3CN	75-05-8
Benzonitrile	PhCN	C_6H_5CN	100-47-0
Nitromethane	$MeNO_2$	CH_3NO_2	75-52-5
Nitrobenzene	$PhNO_2$	$C_6H_5NO_2$	98-95-3
Chloroform		$CHCl_3$	67-66-3
1.1-Dichloroethane	1,1-DClE	$CHCl_2CH_3$	75-34-3
1,2-Dichloroethane	1,2-DClE	CH_2ClCH_2Cl	107-06-2
Dimethylsulfoxide	DMSO	$(CH_3)_2SO$	67-68-5
Tetramethylenesulfone^c	TMS	$c\text{-}(CH_2)_4SO_2$	126-33-0
Tetrahydrothiophene	THTh	$c\text{-}(CH_2)_4S$	110-01-0
Trimethylphosphate	TMP	$(CH_3O)_3PO$	512-56-1
Hexamethyl phosphoric triamide	HMPT	$[(CH_3)_2N]_3PO$	680-31-8

$^a c\text{-}CH(CH_3)CH_2OC(O)O.$
$^b c\text{-}(CH_2)_3C(O)N(CH_3).$
^{c} Also called sulfolane.

of the data are from the compilation by Marcus [3], generally derived from that of Riddick et al. [1] with some additional values. The densities, ρ, are generally available for solvents over a range of temperatures, and hence the molar volume and isobaric expansibilities are also generally available.

TABLE 3.2   The Freezing/Melting Points, t_m, Normal Boiling Points, t_b, of Solvents, and their Molar Masses

Solvent	$t_m/°C$	$t_b/°C$	$M/kg \cdot mol^{-1}$
Water	0.0	100.0	0.01802
Heavy water	3.8	101.4	0.02002
Methanol	−97.7	64.7	0.03204
Ethanol	−114.1	78.2	0.04607
n-Propanol	−126.2	97.2	0.06010
i-Propanol	−88.0	88.0	0.06010
t-Butanol	25.8	82.4	0.07412
Trifluoroethanol	−43.5	73.8	0.10004
Ethylene glycol	−13.0	197.5	0.06207
Tetrahydrofuran	−108.5	66.0	0.07211
1,4-Dioxane	11.8	74.3	0.08811
Acetone	−94.7	56.3	0.05808
Propylene carbonate	−54.5	242	0.10208
γ-Butyrolactone	−43.5	204.0	0.08609
Formamide	2.6	210.5	0.04504
N-Methylformamide	−3.8	182.4	0.05907
N,N-Dimethylformamide	−60.4	153.0	0.07310
N,N-Dimethylacetamide	−20.0	166.1	0.08712
Tetramethylurea	−1.2	175.2	0.11616
N-Methylpyrrolidinone	−24.4	202	0.09913
Ammonia	−77.8	−33.5	0.01703
Pyridine	−41.6	115.3	0.07910
Acetonitrile	−43.6	81.6	0.04105
Benzonitrile	−12.8	191.1	0.10313
Nitromethane	−28.6	101.2	0.06104
Nitrobenzene	5.8	210.8	0.12311
Chloroform	−63.6	61.2	0.11938
1.1-Dichloroethane	−97.0	57.3	0.09896
1,2-Dichloroethane	−53.7	83.5	0.09896
Dimethylsulfoxide	18.5	189.0	0.07813
Tetramethylenesulfone	28.5	287.3	0.12017
Tetrahydrothiophene	−96.2	120.9	0.08817
Trimethylphosphate	−46.0	197.2	0.14000
Hexamethyl phosphoric triamide	7.2	233	0.17920

The molar volume of a solvent at a temperature other than 25°C is readily calculated from:

$$V(T) = V(298.15)\left[1 + \alpha_P\left\{(T\,K^{-1}) - 298.15\right\}\right] \tag{3.1}$$

on the assumption that α_P is not appreciably temperature dependent.

TABLE 3.3 Volumetric Properties of Solvents at 25°C: Densities, ρ, Molar Volumes, V, Isobaric Expansibilities, α_p, Isothermal Compressibilities, κ_T, and Intrinsic Volumes, V_X

Solvent	ρ/kg·m^{-3}	V/cm^3·mol^{-1}	$10^3\alpha_p$/K^{-1}	κ_T/GPa^{-1}	V_X/cm^3·mol^{-1}
Water	997.045	18.07	0.257	0.457	16.7
Heavy water	1104.48	18.13	0.218	0.468	
Methanol	787.2	40.7	1.19	1.248	30.8
Ethanol	784.8	58.7	1.09	1.153	44.9
n-Propanol	800.3	75.1	1.09	1.025	59.0
i-Propanol	781.5	76.9	1.08	1.332	59.0
t-Butanol	781.0	94.9	1.26	0.989	73.1
Trifluoroethanol	1381.8	72.4	1.19	1.220	41.5
Ethylene glycol	1110.4	55.9	0.62	0.392	50.8
Tetrahydrofuran	883.7	81.6	1.29	0.953	62.2
1,4-Dioxane	1028.1	85.7	1.10	0.738	68.1
Acetone	784.9	74.0	1.45	1.324	54.7
Propylene carbonate	1198.1	85.2	0.95	0.590	69.7
γ-Butyrolactone	1125.4	76.5	0.86	0.610	63.8
Formamide	1128.8	39.9	0.75	0.399	36.5
N-Methylformamide	999.5	59.1	0.88	0.560	50.6
N,N-Dimethylformamide	943.3	77.4	1.00	0.642	58.1
N,N-Dimethylacetamide	936.8	93.0	0.98	0.630	78.8
Tetramethylurea	965.6	120.3	1.41	0.910	102.8
N-Methylpyrrolidinone	1028.3	96.4	0.86	0.620	82.0
Ammonia	6812	25.0	1.85		20.8
Pyridine	977.8	80.9	1.02	0.715	67.5
Acetonitrile	776.0	52.9	1.38	1.070	40.4
Benzonitrile	1000.3	103.1	0.83	0.621	87.1
Nitromethane	1044.1	71.9	1.14	0.790	42.4
Nitrobenzene	1198.7	102.7	0.85	0.508	89.1
Chloroform	1479.3	80.7	1.29	1.033	61.7
1.1-Dichloroethane	1168.4	84.7	1.33	1.148	63.5
1,2-Dichloroethane	1246.3	79.4	1.15	0.816	63.5
Dimethylsulfoxide	1095.8	71.3	0.91	0.524	61.3
Tetramethylenesulfone	1261.0	95.3	0.62	0.430	84.5
Tetrahydrothiophene	994.0	88.7	0.95	0.660^a	72.7
Trimethylphosphate	1069.6	115.3	0.96	1.490	114.5
Hexamethyl phosphoric triamide	1019.9	175.7	0.86	0.790	126.8

^{a}Calculated from the adiabatic compressibility κ_S as $\kappa_T = \kappa_S + \alpha_p^2 VT/C_p$.

The Tait expression may be employed in this context for the calculation of the isothermal compressibility, recast in the form:

$$\kappa_T = \frac{(\ln 10)C}{(B+P^\circ)} \tag{3.2}$$

where the empirical constants C and B are generally temperature dependent. For a solvent not on the list, but for which the adiabatic compressibility $\kappa_S = \rho^{-1} u^{-2}$, where u is the speed of ultrasound, is known, the isothermal compressibility may be calculated from:

$$\kappa_T = \kappa_S + \alpha_P^2 V T / C_P \tag{3.3}$$

where C_P is the constant pressure molar heat capacity; see also the paper by Marcus and Hefter [4] for further possibilities to obtain κ_T.

The volumetric properties of solvents include also their so-called intrinsic volumes, $V_X = \sum n_j V_{Xj} + \sum n_k V_{bond}$, with n_j and n_k being the numbers of the atoms of kind j and of bonds irrespective of whether single or double, as proposed by Abraham and McGowan [5]. The atomic volume contributions $V_{Xj}/\text{cm}^3 \cdot \text{mol}^{-1}$ are C=16.35, H=8.71, O=12.43, N=14.39, F=10.48, Cl=20.95, S=22.91, and P=24.87, and $V_{bond}/\text{cm}^3 \cdot \text{mol}^{-1} = -6.56$. The intrinsic volumes, V_X, do not distinguish between isomers. The molar van der Waals volumes of the solvent molecules are linear with V_X (within $1 \, \text{cm}^3 \cdot \text{mol}^{-1}$) [6]:

$$V_{vdW} = 1.8 + 0.674 \, V_X / \text{cm}^3 \cdot \text{mol}^{-1} \tag{3.4}$$

Values of the intrinsic volumes, V_X, of solvents are listed in Table 3.3. They are independent of the temperature and pressure. The difference $V - V_X$ expresses the compressible "free volume" in the solvent and plays also a role in their transport properties (Section 3.2.4).

The diameters of solvents play a role in theoretical considerations, such as the application of the scaled particle theory. For gaseous solvent molecules, the collision diameter, σ, is related to the Lennard–Jones pair potential energy, u^{LJ}:

$$u^{LJ}(r) = 4\varepsilon \left[\left(r_0 / r \right)^{12} - \left(r_0 / r \right)^6 \right] \tag{3.5}$$

where $r_0 = 2^{1/6} \sigma$ is the equilibrium distance between the pair of colliding molecules, where $u^{LJ}(\sigma) = 0$, and $u^{LJ}(r_0) = -\varepsilon$ is the depth of the potential well. When these collision diameters are assigned also to the molecules in the liquid solvents, they bear a linear relationship to the intrinsic volumes:

$$\sigma^{LJ} \, \text{nm}^{-1} = 0.336 + 0.00285 \left(V_X \, \text{cm}^{-3} \cdot \text{mol}^{-1} \right) \tag{3.6}$$

These diameters range from 0.343 nm for water molecules to 0.702 nm for those of hexamethyl phosphoric triamide (HMPT), but most of the values for the solvents listed here cluster between 0.4 and 0.6 nm.

Another estimate of the diameters of solvent molecules is related to the molar volumes at 25°C according to Kim [7]:

$$\sigma^K \, \text{nm}^{-1} = -0.085 + 0.1363 \left(V \, \text{cm}^{-3} \cdot \text{mol}^{-1} \right)^{1/3} \tag{3.7}$$

The σ^K diameters agree well with the collision diameters σ^{LJ} (except for water), but are on the average 0.021 ± 0.026 nm smaller.

3.2.2 Thermodynamic Properties

The thermodynamic properties of liquids that could be of interest for solvents of electrolytes are the vapor pressure, p, the molar enthalpy of vaporization, ΔH^{v}, the molar constant pressure heat capacity, C_p, and the surface tension, σ, that are listed for 25°C in Table 3.4.

TABLE 3.4 Some Thermodynamic Properties of Solvents for Electrolytes: The Vapor Pressure, p, the Molar Enthalpy of Vaporization, ΔH^{v}, the Molar Constant Pressure Heat Capacity, C_p, the Surface Tension σ, and the Hildebrand Solubility Parameter, at 25°C

Solvent	p/kPa	$\Delta_{\mathrm{V}} H$/kJ·mol^{-1}	C_p/J·K^{-1}·mol^{-1}	σ/mN·m^{-1}	δ_{H}/MPa$^{1/2}$
Water	3.17	43.87	75.38	71.96	47.9
Heavy water	2.74	46.38	84.52	71.85	50.6
Methanol	16.9	37.3	81.47	22.3	29.3
Ethanol	7.89	42.32	112.3	21.9	26.0
n-Propanol	2.73	47.45	143.8	23.1	24.4
i-Propanol	6.03	45.39	154.6	21.2	23.7
t-Butanol	5.60	46.69	220.3	20.1	21.6
Trifluoroethanol	10.1	44.00	155.8	20.7 [8]	23.9
Ethylene glycol	0.0117	61.1	150.5	48.0	32.4
Tetrahydrofuran	23.5	31.8	123.9	26.4	19.0
1,4-Dioxane	4.95	35.70	150.6	32.8	19.7
Acetone	30.8	30.99	124.9	22.7	22.1
Propylene carbonate	0.0062	65.28	160.2	45.0 [9]	21.8
γ-Butyrolactone	0.430	52.20	141.4	38.5	25.5
Formamide	0.0088	60.57	107.6	58.2	39.6
N-Methylformamide	0.0338	56.25	123.8	39.5	31.1
N,N-Dimethylformamide	0.530	47.57	148.3	36.4	24.1
N,N-Dimethylacetamide	0.240	50.23	176.0	31.7	18.3
Tetramethylurea	0.0261	63.61	345.7	34.6 [10]	18.9
N-Methylpyrrolidinone	0.0510	53.96	166.1	40.7	23.6
Ammonia		19.8	82.26	21.1	29.2
Pyridine	2.77	40.15	135.6	36.3	21.7
Acetonitrile	12.2	33.23	91.8	28.3	24.1
Benzonitrile	0.0864	54.45	190.3	38.5	22.7
Nitromethane	4.89	38.62	105.8	36.3	25.7
Nitrobenzene	0.0350	55.01	177.2	42.4	22.1
Chloroform	26.3	31.28	117.0	26.5	19.5
1.1-Dichloroethane	30.3	30.62	126.2	24.2	18.3
1,2-Dichloroethane	10.6	35.16	128.9	31.5	20.0
Dimethylsulfoxide	0.0770	52.89	153.1	43.0	26.6
Tetramethylenesulfone	0.0009	79.5	180.0	35.5	27.2
Tetrahydrothiophene	2.45	38.62	139.8	35.0	20.5
Trimethylphosphate	0.117	47.30		36.9	21.1
Hexamethyl phosphoric triamide	0.0061	61.1	321.30	33.8	19.1

The temperature dependence of the vapor pressure is generally reported in terms of the three-parameter Antoine expression:

$$\log p = A - \frac{B}{(T-C)} \tag{3.8}$$

The values of the parameters A, B, and C are listed in [1] for some of the solvents on the present list. On the other hand, the temperature dependence of the vapor pressure may be estimated from:

$$\left(\frac{\partial \ln p}{\partial T}\right)_{\sigma} \approx \frac{\Delta H^{V}}{RT^{2}} \tag{3.9}$$

Here the subscript σ means that the derivative is taken along the saturation line, where the liquid is in equilibrium with the vapor. On the contrary, if the Antoine parameters are known, the molar heat of vaporization can be estimated from:

$$\Delta H^{V} \approx \frac{-R(\ln 10) BT^{2}}{(T-C)^{2}} \tag{3.10}$$

Although in principle, ΔH^{V} is temperature-dependent, over short ranges of temperature near ambient, it may be approximated as being constant.

If the constant volume (isochoric) heat capacity of the solvent, C_{V}, is needed, it can be obtained from the thermodynamic relationship with the isobaric expansibility and isothermal compressibility (see Section 3.2.2):

$$C_{V} = C_{P} - \alpha_{P}^{2} VT / \kappa_{T} \tag{3.11}$$

The surface tension of solvents, σ, represents the work that has to be applied to the solvent in order to increase its surface area by one unit, but the units of σ are generally given as the force acting normal to a unit length, that is, $mN\,m^{-1}$. The phase on the other side of the liquid surface is implied to be the vapor, but air at atmospheric pressure is usually used with no appreciable difference. There are several methods used to measure the surface tension, such as the force applied to a ring or plate touching the surface or the capillary rise of the solvent. The values of σ at 25°C for the solvents dealt with here are listed in Table 3.4, and their temperature dependence is negative but rather small.

The Hildebrand solubility parameter $\delta_{H} = \left[(\Delta H^{V} - RT)/V\right]^{1/2}$ is readily calculated from the listed molar enthalpies of vaporization and the volumes and is also listed in Table 3.4. It varies from $\delta_{H}/MPa^{-1/2} = 47.9$ for water through 39.6 for formamide, 32.4 for ethylene glycol, 31.1 for N-methylformamide, and 29.3 for methanol to values from 28 down to 19 for the other solvents listed here. In contrast to solutions of nonelectrolytes, the δ_{H} of solvents does not play a very important role in solutions of ions.

3.2.3 Electrical, Optical, and Magnetic Properties

The relative permittivities ε (at 25°C) of the liquid solvents and their temperature derivatives, $(\partial\varepsilon/\partial T)_{P}$, taken from [3], are listed in Table 3.5.

TABLE 3.5 The Relative Permittivities of Solvents at 25°C and Ambient Pressure, ε, and their Temperature and Pressure Derivatives $(\partial\varepsilon/\partial T)_P$ and $(\partial\varepsilon/\partial P)_T$, their Refractive Indexes, n_D, their Polarizabilities, α, and their Dipole Moments, μ

Solvent	ε	$-(\partial\varepsilon/\partial T)_P/\mathrm{K}^{-1}$	$(\partial\varepsilon/\partial P)_T/\mathrm{MPa}^{-1}$	n_D	$10^{30}\alpha/\mathrm{m}^3$	μ/D^a
Water	78.46	0.360	0.0370	1.3325	1.456	1.834
Heavy water	78.06	0.362		1.3284	1.536	1.84
Methanol	32.70	0.197	0.0392	1.3265	3.3	2.87
Ethanol	24.55	0.147	0.0255	1.3594	5.1	1.66
n-Propanol	20.33	0.142	0.0197	1.3837	7.0	3.09
i-Propanol	19.92	0.131	0.0207^b	1.3752	7.0	1.66
t-Butanol	12.47	0.175	0.058^c	1.3852	8.8	1.66
Trifluoroethanol	26.67	0.193		1.2907	5.2	2.52
Ethylene glycol	40.70	0.194	0.0233^d	1.4306	5.7	2.31
Tetrahydrofuran	7.58	0.030		1.4050	7.9	1.75
1,4-Dioxane	2.21	0.004		1.4203	8.6	0.45
Acetone	20.70	0.098	0.0230^c	1.3560	6.4	2.69
Propylene carbonate	64.92	0.236	0.0325	1.419	8.6	4.94
γ-Butyrolactone	41.68	0.129^e	0.0175	1.434	7.9	4.12
Formamide	109.5	1.653	0.0493	1.446	4.2	3.37
N-Methylformamide	182.4	1.620		1.430	6.1	3.86
N,N-Dimethylformamide	36.71	0.178		1.428	7.8	3.82
N,N-Dimethylacetamide	37.78	0.230		1.455	9.6	3.72
Tetramethylurea	23.60			1.449	12.8	3.47
N-Methylpyrrolidinone	32.20	0.122^f	0.0232	1.583	10.6	4.09
Ammonia	22.38	0.074		1.325	2.0	1.47
Pyridine	12.91	0.063	0.0107^g	1.507	9.6	2.37
Acetonitrile	35.94	0.150	0.0385	1.341	4.4	3.92
Benzonitrile	25.20	0.091	0.0159^c	1.525	12.5	4.18
Nitromethane	35.87	0.156	0.0319^c	1.379	5.0	3.56
Nitrobenzene	34.82	0.180	0.0202^c	1.550	13.0	4.22
Chloroform	4.90	0.0177	0.0055	1.442	8.5	1.15
1.1-Dichloroethane	10.00	0.048	0.0163^c	1.413	8.4	1.82
1,2-Dichloroethane	10.36	0.056	0.0189^c	1.442	8.3	1.83
Dimethylsulfoxide	46.68	0.106		1.477	8.0	4.06
Tetramethylenesulfone	43.26^a			1.481	10.8	4.81
Tetrahydrothiophene	8.61			1.502	10.4	1.90
Trimethylphosphate	16.39			1.395	10.9	3.12
Hexamethyl phosphoric triamide	29.30	0.604		1.457	18.9	5.54

^{a} 1 D = 3.33564 × 10⁻³⁰ C·m.
^{b} From Ref. 14.
^{c} At 30°C.
^{d} From Ref. 13.
^{e} From Ref. 11.
^{f} From Ref. 12.
^{g} At 15°C.

Less widely available is the pressure derivative of the permittivity $(\partial \varepsilon / \partial P)_T$ and is provided in Table 3.5 where known [4]. For solvents that have no entries in the $(\partial \varepsilon / \partial P)_T$ column, the values may be estimated from:

$$\left(\frac{\partial \varepsilon}{\partial P}\right)_T = 1.057[\varepsilon(P^\circ) - 1]\kappa_T \qquad (3.12)$$

where κ_T is the isothermal compressibility from Table 3.3. This expression has a standard deviation of $1.5 \times 10^{-4}\varepsilon(P^\circ)\,\mathrm{MPa}^{-1}$ for the 50 solvents for which experimental data had been available at the time [4]. Most of the solvents for electrolytes should have $\varepsilon \geq 10$, if the ionic dissociation is to occur to an appreciable extent. There are four solvents on the list with lower permittivities: tetrahydrofuran, $\varepsilon = 7.58$, 1,4-dioxane, $\varepsilon = 2.21$, chloroform, $\varepsilon = 4.90$, and tetrahydrothiophene, $\varepsilon = 8.61$. These have been used in connection with electrolyte solutions for some special purposes.

The refractive index (at the sodium D-line, 589 nm), n_D, the range of which for the solvents listed is fairly narrow, from 1.3265 for methanol to 1.550 for nitrobenzene, is also listed in Table 3.5. From the refractive index is derived the molar refraction by means of the Lorenz–Lorentz expressions:

$$R_D = \frac{V(n_D^2 - 1)}{(n_D^2 + 2)} \qquad (3.13)$$

These values cover a much larger range because of the dependence on the molar volume, namely from $R_D/\mathrm{cm}^3 \cdot \mathrm{mol}^{-1} = 3.8$ for water to 47.7 for HMPT. The polarizability, α, of the molecules of the solvent is proportional to the molar refraction:

$$\alpha \approx \left(\frac{3}{4\pi N_A}\right) R_D \qquad (3.14)$$

The approximation here is the use of R_D instead of the infinite frequency value, R_∞. The range of the α values is, of course, as large as for the molar refractions from $\alpha/10^{-30} \mathrm{~m}^3 = 1.5$ for water to 18.9 for HMPT.

The commonly used solvents for electrolytes are all dipolar and the dipole moments, μ, of the molecules of these solvents range from 1.66 D (1 D (Debye unit) $= 3.33564 \times 10^{-30}\,\mathrm{Cm}$) for ethanol and the two isomeric propanols to 5.54 D for HMPT. Several of the solvents listed in Table 3.5 are very polar having dipole moments >4 D: propylene carbonate, γ-butyrolactone, N-methylpyrrolidinone, benzonitrile, nitrobenzene, dimethyl-sulfoxide, sulfolane, and HMPT. The polarizability α and the polarity (dipole moment) μ together with some chemical properties dealt with in Section 3.3 bear on the ability of the solvents to solvate the ions in electrolyte solutions.

A quantity that is known for only a few of the relevant solvents is the surface potential at the vapor/liquid interface, $\Delta\chi$. This has been obtained from Volta potential measurements and reported in the publications of Trasatti and of Krishtalik et al. [15, 16] and more recently in those of Parfenyuk [17, 18]. The values are based on the $\Delta\chi$ of water, +0.10V established by Parfenyuk [19], probably known to be no better than ±0.1 V. The change in $\Delta\chi$ on going from water to aqueous and co-solvent mixtures is noted. Extrapolation of the latter quantity to zero water content leads to the $\Delta\chi$ values of the neat organic solvents, Table 3.6.

TABLE 3.6 Some Properties of Solvents at 25°C: their Surface Potential $\Delta\chi$/V [18], their Electric Field Dependence of the Permittivity, β/GV^{-2}·m^2 [4], their Limiting Slope of the Electrolyte Molar Volumes, S_V/cm^3·dm$^{3/2}$·mol$^{-1/2}$ [4], their Molar Diamagnetic Susceptibility, χ/10^{-6} cm^3·mol^{-1} [3, 20], and their Mean Molar Electrostriction, $\Delta_{els}V_S$/cm^3·mol^{-1} [21]

Solvent	$\Delta\chi$	$-\beta$	S_V	$-\chi$	$\Delta_{els}V_S$
Water	0.10	1080	1.85	12.9	2.91[a]
Methanol	−0.18	660	18.57	21.4	5.99
Ethanol	−0.26	385	26.17	33.5	
n-Propanol	−0.27	330	27.42	45.2	14.41
i-Propanol	−0.28			45.7	26.67
t-Butanol		−80		57.4	
Trifluoroethanol					
Ethylene Glycol	−0.13[b]		5.42	38.9	9.46
Tetrahydrofuran					
1,4-Dioxane	0.26[c]			51.1	
Acetone	−0.34	84	29.08	34.0	25.50
Propylene Carbonate			2.35	54.5	7.12
γ-Butyrolactone	−0.16[b]		3.77		9.26
Formamide			1.12	23.1	
N-Methylformamide			0.66	34.3	
N,N-Dimethylformamide	−0.44		8.13	38.8	
N,N-Dimethylacetamide				56.1	
Tetramethylurea				75.7	
N-Methylpyrrolidinone			11.47	61.7	21.40
Ammonia				16.3	
Pyridine			51.47	48.5	
Acetonitrile	−0.11		13.40	27.6	4.93
Benzonitrile			13.54	65.2	
Nitromethane			12.24	20.9	
Nitrobenzene	−0.14[b]	315	8.09	61.8	
Chloroform		1.6		59.3	
1.1-Dichloroethane		28			
1,2-Dichloroethane	0.02[b]	34			
Dimethylsulfoxide	−0.24		4.65	43.9	4.98
Tetramethylenesulfone					
Tetrahydrothiophene				63.5	
Trimethylphosphate					
Hexamethyl phosphoric triamide			13.46	118.0	

[a] The value may be revised in view of more accurate pressure derivatives of the permittivity to 3.47 [24].
[b] From Ref. 22, using the value +0.10 V for water.
[c] From Ref. 23, extrapolated to pure dioxane from data on aqueous solutions.

The diamagnetic susceptibility of solvents is relevant to NMR studies of electrolyte solutions, where appropriate corrections to the observed chemical shifts related to this property are applied. The values for the solvents on our list are shown in Table 3.6. The negative molar (volume) magnetic susceptibilities are proportional to the molar refraction for most solvents at ambient temperatures:

$$-\chi = 2.46 \times 10^{-5} R_D \tag{3.15}$$

where both variables are expressed in the same units ($cm^3 mol^{-1}$).

Another quantity that should be considered in this context is the electric field dependence of the permittivity of the solvent, noting the huge electrical fields exerted by ions in their vicinity. This topic was already briefly brought up in Section 2.3, but deserves fuller attention. The following expression [25] relates the permittivity of the solvent to the electric field strength in terms of the nonlinear dielectric effect:

$$\varepsilon(E) = \frac{\varepsilon_\infty + \left[\varepsilon(0) - \varepsilon_\infty\right]}{(1 + bE^2)} \tag{3.16}$$

Here ε_∞ is the permittivity at infinite frequency of the alternating electric field; it expresses the inability of the molecular dipoles of the solvent to orient themselves in the direction of the field. Then only the electronic orientation within the atoms remains dependent on the electric field, which is expressed by the refractive index squared, n_∞^2, approximated as $1.1 n_D^2$. Equation 3.16 can be simplified by expressing it as a power series in E, truncated after the second term, resulting in Equation 2.13: $\varepsilon(E) = \varepsilon(0) + \beta E^2$. Values of the coefficient β have been compiled by Marcus and Hefter [4] but are known for only some of the solvents listed in the tables of this chapter.

The molar volumes of ions in solution depend on their concentrations according to Equation 2.25 as discussed in Section 2.3.1.5 for the apparent molar volumes. This expression involves the limiting slope of the square root of the concentration, S_V, which is according to the Debye–Hückel theory [4, 26]:

$$S_V = \left[N_A^2 e^3 \left(8\pi^2 \varepsilon_o^3 R\right)^{-1/2}\right] T^{-1/2} \varepsilon^{-3/2} \left[\left(\frac{\partial \ln \varepsilon}{\partial P}\right)_T - \kappa_T/3\right] \tag{3.17a}$$

At 25°C this becomes:

$$S_V / cm^{-3} \cdot dm^{3/2} \cdot mol^{-3/2} = 4046.4 \varepsilon^{-3/2} \left[\frac{(\partial \ln \varepsilon/\partial P)_T - \kappa_T}{3}\right] \tag{3.17b}$$

A further quantity that is related to the pressure dependence of the permittivity, but also to its electric field dependence, is the mean molar electrostricted volume of a solvent in the presence of ions. The mean molar electrostriction of the solvent S in the presence of the ionic electric field was given by Marcus [21] as:

$$\Delta_{els} V_S = \left\{2X / \left[X^2 - 2\left(\frac{\partial X}{\partial P}\right)_T\right]\right\} \kappa_{TS}^* V_S^* \tag{3.18}$$

where the auxiliary function $X = 3(\partial \ln \varepsilon / \partial P)_T - \kappa_{TS}^*$ and κ_{TS}^* and V_S^* are the isothermal compressibility and molar volume of the pure solvent. The mean volume of the solvent in the electrostricted solvation shell of an ion is then $V_S^* - \Delta_{els} V_S$.

Table 3.6 lists, as far as could be gleaned from the literature, the values of the vapor/liquid surface potentials $\Delta \chi$, the field dependence of the permittivity coefficient β (Eq. 2.13), the theoretical limiting slopes of the apparent molar volumes, S_V, the molar diamagnetic susceptibility, χ, and the mean electrostriction volumes, $\Delta_{els} V_S$, of solvents at ambient temperatures.

3.2.4 Transport Properties

The conductivity of well-purified solvents is very low and generally can be ignored. Values of the specific conductance, κ, have been reported by Riddick et al. [1] for many of the solvents listed here, Table 3.7, but those that are $>1\,S\,m^1$ are suspect, in that they are possibly due to the presence of impurities, not least of which is CO_2 absorbed from the air. The true conductivity of a solvent is proportional to the number of charge carriers, that is, ions resulting from self-ionization (autoprotolysis for protic solvents, see Section 3.3.3) and to its fluidity (the reciprocal of its viscosity, see below).

Of more consequence is the self-diffusion ability of the solvent molecules, D, obtained from isotopically labeled solvent molecules or from the band widths of NMR signals. The values thus determined are shown in Table 3.7. The self diffusion coefficient follows an Arrhenius-type expression with regard to the temperature: $D = A_D \exp(\Delta E_D / RT)$.

The viscosity of solvents is obtained either in the kinematic mode, ν, measured in a flow viscosimeter, or in the dynamic mode, $\eta = \nu / \rho$, the latter being most frequently quoted and applied. A common unit for the dynamic viscosity is the centipoise, cP, a non-SI unit, which is equivalent to mPa·s in SI units. The fluidity of a solvent, Φ, is the reciprocal of its dynamic viscosity. The viscosity of solvents diminishes strongly with increasing temperatures, again according to an Arrhenius-type expression, $\eta = A_\eta \exp(-\Delta E_\eta / RT)$. The activation energy for viscous flow is $\Delta E_\eta = -RT^2 (\partial \ln \eta / \partial T)_P$ and varies from 6 to 20 kJ mol^{-1}. Values of the dynamic viscosity at 25°C and of its temperature dependence (on a logarithmic scale) $(\partial \ln \eta / \partial T)_P$ are shown in Table 3.7.

A relationship has been found by Marcus [28] between the viscosities of solvents and their relative free molar volumes, $(V - V_X)/V = (1 - V_X/V)$, on the one hand, and their molar enthalpies of vaporization, ΔH^V, on the other. The former describes the space available for the movement of the solvent molecules and the latter measures the tightness of the mutual binding of the molecules. These values have to be modified by the number of hydroxyl groups, n_{OH}, pertinent to the hydrogen bonding in solvents containing such groups:

$$\log(\eta/\text{mPa·s}) = -3.44(1 - V_X/V) + 0.0352(\Delta H^V/RT) + 0.46 n_{OH} \quad (3.19)$$

TABLE 3.7   Transport Properties of Solvents at 25°C: their Specific Conductance κ [1], their Self-diffusion Coefficient D [27], their Dynamic Viscosity η, and its Temperature Derivative $(\partial\eta/\partial T)_P$

Solvent	$10^6\kappa/\text{S}\cdot\text{m}^{-1}$	$10^9 D/\text{m}^2\cdot\text{s}^{-1}$	$\eta/\text{mPa}\cdot\text{s}$	$100(\partial\ln\eta/\partial T)_P/\text{K}^{-1}$
Water	5.89	2.27	0.8903	2.21
Heavy water		1.87	1.121	2.71
Methanol	0.15	2.32	0.551	1.32
Ethanol	0.135	1.01	1.083	1.91
n-Propanol	0.917	0.65	1.943	2.37
i-Propanol	5.8	0.65	2.044	2.92
t-Butanol	2.66	0.51	4.438	2.80
Trifluoroethanol			1.755	2.57
Ethylene glycol	116		16.34	4.60
Tetrahydrofuran	9.3		0.462	1.04
1,4-Dioxane	0.000	1.01	1.19	1.77
Acetone	0.49	4.77	0.303	0.95
Propylene Carbonate	2		2.53	2.22
γ-Butyrolactone			1.717	1.91
Formamide	<20		3.302	2.62
N-Methylformamide	80	0.85	1.65	1.58
N,N-Dimethylformamide	6	1.61	0.802	1.22
N,N-Dimethylacetamide			0.927	1.19
Tetramethylurea	<6		1.395	
N-Methylpyrrolidinone	2	0.78	1.666	1.88
Ammonia	0.001	5.71	0.131	0.99
Pyridine	4.0	1.49	1.53	1.49
Acetonitrile	0.06	4.85	0.96	4.85
Benzonitrile	5		1.237	1.51
Nitromethane	50	2.11	0.614	1.17
Nitrobenzene	0.0205		1.784	1.80
Chloroform	<0.01		0.536	1.00
1.1-Dichloroethane	0.20		0.505	1.07
1.071,2-Dichloroethane	0.004		0.779	1.27
Dimethylsulfoxide	0.2	0.76	1.991	1.93
Tetramethylenesulfone	<2		10.286	2.27
Tetrahydrothiophene			0.971	1.31
Trimethylphosphate			2.03	1.75
Hexamethyl phosphoric Triamide	30		3.11	2.39

There is a relationship between the viscosity and the diffusion coefficient for those solvents, the molecules of which have a more or less globular shape:

$$D = \frac{k_B T}{S_D \pi \eta (\sigma/2)} \tag{3.20}$$

where S_D is a numerical coefficient equaling 6 for "slip" and 4 for "stick" conditions for the mutual movement of the solvent molecules and σ is their diameter.

Another transport property of liquids is the heat conductance, λ, which is partly dependent on the viscosity, but can be expressed better by the molar mass and the heat capacity per unit volume according to Marcus [28]:

$$\lambda / W \cdot m^{-1} \cdot K^{-1} = 0.00205 \left(M/kg \cdot mol^{-1} \right) + 0.0695 \left[\left(C_P / V \right) / J \cdot K^{-1} \cdot cm^{-3} \right] \quad (3.21)$$

The rotational orientation times and ultrasound absorbances are further transport properties of liquids that may be found in [28].

3.3　CHEMICAL PROPERTIES OF SOLVENTS

The chemical properties of solvents that are relevant to their dissolution abilities for electrolytes and the ionic dissociation of the latter include their structuredness or self-association and their donor (electron pair donation, basicity) and acceptor (hydrogen bonding ability, acidity) properties as well as their softness. The mutual solubility with other solvents, in particular water, is also of importance as are the windows for making spectroscopic and electrochemical measurements on solutions of ions in the solvents.

3.3.1　Structuredness

The structuredness of solvents is closely related to their self-association by means of hydrogen bonds or dipole–dipole interactions as described by Marcus [27, 29]. A measure of the structuredness is the "stiffness" of the solvents, measured by the difference in the cohesive energy density, U/V, where U is the molar internal energy, and the internal pressure, P_{int}. Solvents for which this difference is $>50\,J\,cm^{-3}$ are considered to be stiff, as are the majority of the solvents dealt with here. The cohesive energy density, $U/V = (\Delta H^V - RT)/V = \delta_H^2$, is available from the data in Table 3.4. The internal pressure is defined as the first of the following equalities and is shown for many liquids in the review by Marcus [30]:

$$P_{int} = \left(\frac{\partial U}{\partial V} \right)_T = T \left(\frac{\partial P}{\partial T} \right)_V - P = T \alpha_P / \kappa_T - P \quad (3.22)$$

Here $(\partial P/\partial T)_V$ is the isochoric thermal pressure coefficient that is seldom measured directly and P_{int} is generally obtained by the last equality in (3.22). The magnitude of P_{int} is >100 MPa, so that at ambient conditions and saturation vapor pressures, the last term, $-P$, in Equation 3.22 can be neglected. The isobaric expansibility, α_P, and isothermal compressibility, κ_T, are available in Table 3.3. The differences $U/V - P_{int}$ at 25°C for the solvent listed here are shown in Table 3.8, with "non-stiff" solvents marked by *italics* font. The value of $U/V - P_{int}$ for water is by far larger than for other structured solvents, but it diminishes with increasing temperatures [30] to become commensurate with P_{int} of other solvents above ~250°C.

TABLE 3.8   Measures of the Structuredness of Solvents At 25°C: Stiffness $= U/V - P_{int}$, order $= \Delta\Delta S^V(T, P°)/R$ and $\Delta C_r/V(l)$ [27], and the Kirkwood Dipole Orientation Correlation Parameter g [29]

Solvent	$(U/V - P_{int})$/J·cm^{-3}	$\Delta\Delta S^V(T, P°)/R$	$\Delta C_r/V(l)$/J·K^{-1}·cm^{-3}	g
Water	2143	7.82	2.31	2.57
Heavy water (D$_2$O)	2282	9.0	2.77	2.56
Methanol	570	6.26	0.92	2.82
Ethanol	383	7.45	0.80	2.90
n-Propanol	315	6.67	0.76	2.99
i-Propanol	316	6.28	0.86	3.08
t-Butanol	128	5.00	1.16	2.22
Trifluoroethanol	282	5.15[a]	0.93[b]	2.38
Ethylene glycol	548	21.20	0.92	2.08
Tetrahydrofuran	−45	0.73[a]	0.58	1.07
1,4-Dioxane	−56		0.60	
Acetone	157	2.36	0.68	1.05
Propylene carbonate	−69	5.57[a]		1.23
γ-Butyrolactone	227	3.57		1.07
Formamide	1014	7.58	1.56	1.67
N-Methylformamide	441	6.10	0.88	3.97
N,N-Dimethylformamide	101	4.00	0.74	1.03
N,N-Dimethylacetamide	135	3.34	0.72	1.26
Tetramethylurea	46	8.29[a]		1.16
N-Methylpyrrolidinone	120	3.36	0.64	0.92
Ammonia			1.8	1.68
Pyridine	41	1.95[a]	0.69	0.93
Acetonitrile	186	4.38	0.74	0.74
Benzonitrile	106	1.45[a]	0.79	0.66
Nitromethane	161	3.36	0.90	0.92
Nitrobenzene	12	0.39[a]	0.49	0.88
Chloroform	−8	2.01	0.64	1.30
1.1-Dichloroethane	−13	1.84	0.59	0.78
1,2-Dichloroethane	−19	2.44	0.65	0.45
Dimethylsulfoxide	187	5.07	0.89	1.04
Tetramethylenesulfone	378	2.23[a]	0.41	0.92
Tetrahydrothiophene	−22	1.11[a]	0.53[c]	
Trimethylphosphate	206	0.09[a]		1.00
Hexamethyl phosphoric triamide	38	0.00[a]		1.39

[a] Not corrected for vapor phase association.
[b] C_p(i.g.) from [31] extrapolated to 298.15 K.
[c] C_p(i.g.) from [32].

The "openness" of solvents is measured by their relative free volumes, $(V V_X)/V$. The fluidity, $\Phi = 1/\eta$, of a liquid was shown by Hildebrand and Lamoreaux [33] to be a linear function of $(V - V_X)/V$ for various kinds of liquids as the temperature changes, but when this notion is applied to a comparison of solvents at a given temperature, this relationship is not valid. A proportionality does exist between the isothermal compressibilities, κ_T, of solvents with their openness, $\kappa_T/\text{GPa}^{-1} \approx 4.61(1 - V_X/V)$, if the values for TFE, DMF, MeNO$_2$, and HMPT are excluded as outliers. Because of these outliers with no apparent reason, this proportionality should not be used to predict unknown solvent compressibilities.

The "ordering" of solvents, as a measure of their structuredness, can be represented by two quantities that place solvents in categories of "ordered" and "unordered" in good agreement with each other. The one is the entropy deficit of the solvent relative to its vapor and compared with the relevant values for a completely unordered liquid, the alkane homologue. The alkane homologue is chosen to have the same skeleton, with non-carbon atoms exchanged for –CH$_3$, –CH$_2$–, or >CH– as the case may be. According to Marcus [27]:

$$\frac{\Delta\Delta S^V (T, P^\circ)}{R} = \left[\frac{\Delta S_S^V (T, P^\circ) - \Delta S_{\text{alk}}^V (T, P^\circ)}{R} \right] + (P^\circ/R) \frac{d(B_S - B_{\text{alk}})}{dT} \tag{3.23}$$

In this expression, ΔS^V is the molar entropy of vaporization and B is the second virial coefficient of the solvent ($_S$) and the homologous alkane ($_{\text{alk}}$). The temperature derivative of B compensates for possible association of the solvent molecules in the vapor phase. The dimensionless values of $\Delta\Delta S^V(T, P^\circ)/R$ from [27] are shown in Table 3.8 for the solvents dealt with here, but for some of them, the values of dB_S/dT are not known; these solvents are not expected to associate in the vapor phase anyway. For solvents for which the value of $\Delta\Delta S^V(T, P^\circ)/R$ has not been previously determined, the values of $\Delta S_S^V (T, P^\circ)$ and $\Delta S_{\text{alk}}^V (T, P^\circ)$ are obtained for 25°C from:

$$\Delta S^V (T, P^\circ) = \Delta H^V/T + \ln (p/P^\circ) \tag{3.24}$$

with the molar enthalpy of vaporization and the vapor pressure take from Table 3.4. The criterion for a solvent at 25°C to being "ordered" is $\Delta\Delta S^V (T, P^\circ)/R \geq 2$, so that practically all the solvents on the list for which there are data are "ordered" in some manner or other.

Another measure of the "ordering" in a solvent is the amount of thermal energy that has to be applied to it per unit volume and relative to the corresponding amount for the (ideal gas) solvent vapor to raise its temperature and thus increase its thermal disordering. The quantity that measures the "ordering" is then $\Delta C_P/V = [C_P(\text{l}) - C_P(\text{i.g.})]/V(\text{l})$, where (l) denotes the liquid solvent. The values of the heat capacity density in excess of the ideal gas values at 25°C are shown in Table 3.8 [27]. The criterion for solvents to be considered as ordered is $\Delta C_P/V \geq 0.6\,\text{J}\cdot\text{K}^{-1}\cdot\text{cm}^{-3}$. There exists a general agreement of the assignment of solvents to the categories of "ordered" and "non-ordered" (the latter being marked by values in *italics*) between the applications of the criteria for $\Delta\Delta S^V(T, P^\circ)/R$ and $\Delta C_P/V$, but there are some disagreements.

A further quantity that describes the structuredness and self-association of solvents is the Kirkwood dipole angular correlation parameter. This parameter is $g = 1 + Z |\cos\theta|$, where Z is the number of nearest neighbors a solvent molecule has and θ is the average angle between the dipole vectors of adjacent solvent molecules. This parameter is obtained from the semi-empirical expression:

$$g = \left(\frac{9k_B\varepsilon_0}{4\pi N_A}\right) VT\mu^{-2}\frac{\left(2\varepsilon - 1.1n_D^2\right)}{\varepsilon\left(2 + 1.1n_D^2\right)^2} \tag{3.25}$$

The factor 1.1 multiplying the square of the refractive index, n_D^2, should compensate for the use of n_D^2 instead of n_∞^2 to represent the infinite frequency permittivity of the solvent, ε_∞. The values of g at 25°C are shown in Table 3.8. Solvents having $g \geq 1.7$ were considered by Marcus [29] to be structured, in the sense that their molecules display a considerable extent of order in the liquid state, whereas solvents having $g \leq 1.3$ are considered as being unstructured. According to these criteria, mainly such solvents, the molecules of which are hydrogen-bonded to each other, are structured. Solvents with large dipole moments μ also display some order, although they fall short of the $g \geq 1.7$ criterion. However, these solvents are considered structured (displaying molecular ordering) according to other criteria, such as their entropy deficit and their heat capacity density relative to their vapors at ambient temperatures [27].

In the cases of solvents with a hydrogen-bonded tri-dimensional network, foremost water (both H_2O and D_2O) but also ethylene glycol, glycerol, and formamide, to name a few, the average number of hydrogen bonds existing in the liquid solvent per solvent molecule is an excellent measure of their structuredness. It depends on external conditions, such as the temperature and pressure, and on the presence of solutes that affect this hydrogen-bonded structure.

In the case of water, the average number of hydrogen bonds per water molecule diminishes with increasing temperatures according to several criteria used to define the presence of intact hydrogen bonds. Representative values of this average number are 3.50 at 25°C diminishing to 2.45 at 100°C as suggested by Marcus [34] (compared with 4.0 for ice). Dilute solutions of some cosolvents enhance the structure of the water, as measured by its excess partial molar volume or heat capacity as shown by Marcus [35]. The effects of ions on the structure of water are dealt with in Chapter 5.

3.3.2 Solvent Properties Related to Their Ion Solvating Ability

The molecules of solvents have several features that enable them to solvate ions effectively. These include the polarity, measured by the dipole moment, μ, and the molecular polarizability, α, of individual molecules, but have other manifestations in the bulk liquid solvent. The solvents generally include donor atoms that have one or more pairs of free electrons and some are protic and can donate a hydrogen atom toward the formation of hydrogen bonds and can accept a pair of electrons from a donor atom. Even aprotic solvents may be protogenic, in that under suitable circumstances, their electronic structure can be rearranged so as to make a hydrogen atom available for hydrogen bonding. The properties measured in bulk liquid solvents, mainly by the solvatochromic method, are described in the following sections.

The solvatochromic method employs an indicator substance as a probe at a low concentration, which is used as a stand-in for the ions or other solute molecules to be solvated. In order to serve as an indicator, the substance in question must have a large solvent-dependent energy gap between its ground state and an excited state resulting from light absorption (preferably in the visible range). Two reference solvent are then employed to utilize quantitatively the resulting energy gap for any desired solvent—one with the minimal gap to act as a baseline and one with the maximal observed gap to normalize the values.

It is, of course, a gross approximation that any one probe can act as a stand-in for any desired solute. Resort may be taken to the use of several probes having different functional groups that would be used in the solvation of a solute, provided that their use leads to convergent values of the desired property. The resulting parameter, the average of the values thus produced, suffers from "fuzziness," that is, has a small range of acceptable values, but should still in general represent the property of solvents better than the value obtained with a single probe, when applied to the solvation of very diverse solutes, as argued by Marcus [36].

A critical compilation of scales of solvent parameters pertaining to pure, non-hydrogen-bond donating solvents by Abboud and Notario is available in [37]. In addition to the solvent scales described in the following (with values listed in Table 3.9), many other scales are dealt with in [37] and elsewhere, but those shown here appear to be the most popularly used ones.

A large number of solvent effects on solutes reported by Reichardt [38] can be described by a general linear solvation energy relationship (LSER) according to Kamlet and Taft [39]. The quantities measured include solubility, partition coefficient, light absorption peak, NMR chemical shift, toxicity, etc. They depend linearly on a sum of terms, each of which is the product of a solute property and a corresponding solvent property. These terms include a measure of the solute volume and the solvent solubility parameter squared (describing the energy required for the formation of a cavity in the solvent to accommodate the solute) and terms involving the polarity, electron pair donicity (basicity), hydrogen bond donation ability (acidity), and eventually the softness of the solvent. These solvent properties are described in the following sections.

3.3.2.1 Polarity The solvents considered here are all polar in that their molecules have nonzero dipole moments (Table 3.5). However, the characterization of the polarity of solvents acting as bulk liquids is best done by the solvatochromic method. Popular in this respect is the Dimroth–Reichardt betaine indicator 2,6-diphenyl-4-(2,4,6-triphenyl-1-pyridino)-phen-oxide [40] that happened to be the 30th indicator studied at the time, and hence the name $E_T(30)$ for the transition energy between the ground and excited states. The range of the hypsochromic (blue shift) effect of solvents of increasing polarity for this probe is one of the largest among useful indicators. The literature [38] contains entries of the non-normalized parameter $E_T(30) = 28590/(\lambda/nm)$, where λ is the wavelength of the lowest energy (largest wavelength) peak of light absorption. The numerical coefficient arises from the values being expressed in kcal·mol^{-1} units ($1\,cal = 4.184\,J$) with an accuracy of ± 0.1 unit. The temperature dependence of $E_T(30)$ near ambient is minimal and the indicator is sufficiently soluble in most solvents without chemical reaction.

TABLE 3.9   Indexes of Solvent Solvation Ability: Polarity E_T^N and π^*, Electron Pair Donicity, DN and β, Hydrogen Bond Donicity, AN and α, and the Softness Parameter, μ

Solvent	E_T^N	π^*	DN	β	AN	α	μ
Water	1.000	1.09	18.0	0.47	54.8	1.17	0.00
Heavy water	0.991						0.0014
Methanol	0.762	0.60	30.0	0.66	41.5	0.98	0.02
Ethanol	0.654	0.54	32.0	0.75	37.1	0.86	0.08
n-Propanol	0.617	0.52	30.0	0.90	33.7	0.84	0.16
i-Propanol	0.545	0.48	36.0	0.84	33.5	0.76	
t-Butanol	0.389	0.41	38.0	0.93	27.1	0.42	
Trifluoroethanol	0.898	0.73		0.00	53.8	1.51	−0.12
Ethylene glycol	0.790	0.92	20.0	0.52	43.4	0.90	−0.03
Tetrahydrofuran	0.207	0.55	20.0	0.55	8.0	0.00	0.00
1,4-Dioxane	0.164	0.49	14.8	0.37	10.8	0.00	0.07
Acetone	0.355	0.62	17.0	0.48	12.5	0.08	0.03
Propylene carbonate	0.472	0.83	15.1	0.40	18.3	0.00	−0.09
γ-Butyrolactone	0.420	0.85	18.0	0.49	17.3	0.00	0.02
Formamide	0.775	0.97	24.0	0.48	39.8	0.71	0.09
N-Methylformamide	0.722	0.90	27.0	0.80	32.1	0.62	0.17
N,N-Dimethylformamide	0.386	0.88	26.6	0.69	16.0	0.00	0.11
N,N-Dimethylacetamide	0.377	0.85	27.8	0.76	13.6	0.00	0.17
Tetramethylurea	0.315	0.79	29.6	0.71	9.2	0.00	0.14
N-Methylpyrrolidinone	0.355	0.92	27.3	0.77	13.3	0.00	0.13
Ammonia	0.272		59.0				
Pyridine	0.302	0.87	33.1	0.64	14.2	0.00	0.64
Acetonitrile	0.460	0.66	32.0	0.37	18.9	0.19	0.34
Benzonitrile	0.333	0.88	11.9	0.37	15.5	0.00	0.34
Nitromethane	0.481	0.75	2.7	0.06	20.5	0.22	0.03
Nitrobenzene	0.324	0.86	4.4	0.30	14.8	0.00	0.23
Chloroform	0.259	0.58	4.0	0.10	23.1	0.20	
1.1-Dichloroethane	0.269	0.48		0.10	16.2	0.10	0.07
1,2-Dichloroethane	0.327	0.73	0.0	0.10	16.7	0.00	0.03
Dimethylsulfoxide	0.444	1.00	29.8	0.76	19.3	0.00	0.22
Tetramethylenesulfone	0.410	0.90	14.8	0.39	19.2	0.00	0.00
Tetrahydrothiophene	0.185	0.60		0.44		0.00	0.80
Trimethylphosphate	0.398	0.73	23.0	0.77	16.3	0.00	0.02
Hexamethyl phosphoric triamide	0.315	0.87	38.8	1.00	9.8	0.00	0.29

The normalized Dimroth–Reichardt polarity index is preferably used, because its values are dimensionless and with values between zero and unity for the sake of comparison with other solvent property indexes used in LSER expressions:

$$E_T^N = \frac{\left(\lambda_S^{-1} - \lambda_{TMS}^{-1}\right)}{\left(\lambda_W^{-1} - \lambda_{TMS}^{-1}\right)} \tag{3.26}$$

where the λs pertain to the studied solvent $(_\text{s})$, to tetramethylsilane $(_\text{TMS})$, and to water $(_\text{w})$, yielding $E_T^\text{N} = \left[E_T(30)_\text{s} - 30.7\right]/32.4$. The values of E_T^N according to Equation 3.26 for the solvents dealt with here are shown in Table 3.9. Highly acidic solvents (Section 3.3.3) protonate the betaine, and hence their E_T^N values are obtained indirectly according to Hormadaly and Marcus [41] from a correlation expression with the Kosower Z values obtained with a different indicator, 4-cyano-1-ethylpyridinium iodide: $E_T(30) = 0.752Z - 7.87$. Some other indirect determinations of $E_T(30)$ have been described more recently by Ceron-Carrasco et al. [42]. It turns out that E_T^N does not depend solely on the polarity of the solvent but has an important contribution from the ability of the molecules of the solvent to form hydrogen bonds with solutes (or with themselves); see the following text.

Another popular measure of solvent polarity is the Kamlet-Taft [43, 44] π^*, obtained as the average $\pi \rightarrow \pi^*$ transition energies from several probes: 4-nitro-N,N-diethylaniline, 3-nitro-N,N-diethylaniline, 4-nitroanisole, 4-nitro-1-ethylbenzene, and 4-(2-nitroethenyl)-anisole. These probes can be augmented by using 2,4-dinitro-N,N-diethylaniline, 4-cyano-N,N-dimethylaniline, 4-acetyl-N,N-dimethylaniline, and 4-carbomethoxy-N,N-dimethylaniline as suggested by Nicolet and Laurence [45]. These average values, normalized by setting $\pi^*(\text{cyclohexane}) = 0$ and $\pi^*(\text{dimethylsulfoxide}) = 1$, are listed in Table 3.9. Again, it is known that the π^* of a solvent involves a notable contribution from its polarizability, so that the net polarity is given by $\pi^*(1 - d\delta)$ where $d = 0.4$ for most applications and $\delta = 0.5$ for polychlorinated hydrocarbons (chloroform, 1,1- and 1.2-dichloroethane on the present list), $\delta = 1.0$ for aromatic solvents (benzonitrile and nitrobenzene), and 0.0 for all the others. If π^* values are sought for polar solvents for which it has not been determined by the solvatochromic method, it may be estimated from the dipole moments (in Debye units) $\pi^* \approx 0.03 + 0.23(\mu/\text{D})$ for aliphatic solvents and $\pi^* \approx 0.56 + 0.11(\mu/\text{D})$ for aromatic ones.

3.3.2.2 Electron Pair Donicity and Ability to Accept a Hydrogen Bond The
ability of solvent molecules to donate a free electron pair from their donor atoms (O, N, or S) to coordinate with acceptor atoms of solutes is a measure of the solvent donicity. It can also be construed as its basicity in the Lewis and the Brönsted senses, because it also describes the ability of the solvent molecules to accept a proton from a Brönsted acid to be protonated or to form a hydrogen bond.

This property is expressed in terms of the Gutmann donor number DN [46], which is the negative of the standard molar enthalpy of reaction of the solvent with the Lewis acid antimony pentachloride, both in dilute solution in the inert solvent 1,2-dichloroethane (in kcal·mol^{-1} units). It is assumed that this quantity, pertaining to solvent molecules in dilute solution, represents also the property of the bulk solvent. The scale was expanded by good correlations with suitable solvatochromic probes: acetylacetonato-N,N,N',N'-tetra-methyethylenediamino copper(II) perchlorate (ATMECu) and diacetylacetonatooxo-vanadium(IV) and by the stretching frequencies of the O–D bond of CH_3OD methanol and the O–H bond of phenol as probes as demonstrated by Marcus [47]. The combined values, taken from [3], are shown in Table 3.9.

Another approach is the use of averaged quantities obtained from protic solvato-chromic probes relative to structurally similar but aprotic probes, leading to the Kamlet-Taft β-scale [48]. The nonspecific effects of the solvent on the protic probe are assumed to be the same as those on the aprotic probe, expressed by the π^* of the latter (Section 3.3.2.2). The pairs of probes originally used were 4-nitrophenol and 4-nitroanisole, 4-nitroaniline, and 4-nitro-N,N-diethylaniline, and the expression used was:

$$\beta = b(\nu_0 - \nu_S) + s\pi^*(1 - d\delta) \tag{3.27}$$

where b and s are solvent-independent but probe-specific coefficients and the ν are the wavenumbers of peak absorbance of the probe in cyclohexane (ν_0) and in the solvent studied (ν_S). Use of the ATMECu probe mentioned above has the advantage that for it $s = 0$, the resulting β being independent of the solvent polarity expressed as π^*. It turns out that there are systematic differences between the β values obtained using 4-nitrophenol and 4-nitroaniline for solvents not having oxygen as their donor atom. The averaged β values of the latter solvents have, therefore, an uncertainty of ± 0.4 units. The β values are normalized by using $\beta = 0$ for cyclohexane and $\beta = 1$ for HMPT. The resulting values are shown in Table 3.9.

The two measures of solvent donicity are linearly correlated [37]:

$$DN = 0.5 + 38.2\beta \tag{3.28}$$

Hence the one donicity measure can be estimated for a solvent when the other is known, but outliers do occur. It was recommended [38] to abandon the averaged β values (listed in Table 3.9) in favor of those obtained by the 4-nitrophenol probe alone. It is also to be noted that for solvents associated by hydrogen bond networks (water and alkanols), their molecules in dilute solution in inert solvents have consider-ably lower values than those for the bulk solvents, in which the cooperative effect of the hydrogen bonding enhances the donicity.

3.3.2.3 Hydrogen Bond Donicity and Electron Pair Acceptance

The electron pair acceptance propensity of the molecules of a solvent are closely related to its ability to provide protons for hydrogen bonding, and hence is confined to protic or at least protogenic solvents among those on our list. The latter solvents include chloro-form and solvent having a methyl group adjacent to a C=O, C≡N, or NO_2 group.

The Gutmann–Mayer acceptor number scale AN [49] uses the NMR chemical shift δ of the ^{31}P atom of triethylphosphine oxide as the electron-pair donor probe in dilute solution in the solvent to be studied relative to the shift in n-hexane. The AN values are normalized to make $AN = 2.348(\delta/ppm)$, the δ values being corrected for the diamag-netic susceptibility χ of the solvent. The AN scale does include solvent polarity effects besides its ability to donate hydrogen bonds, as is seen in its being proportional to the E_T^N values: $AN = 56.8 E_T^N$ [37]. In fact, both scales are sensitive to both the solvent polarity and its acidity. The acceptor numbers of solvents are listed in Table 3.9.

The Kamlet-Taft α scale [48], on the other hand, was designed to show the net hydrogen bond donation or electron pair acceptance ability of a solvent, being averaged over the values obtained from several probes, but the values require the use of general polarity and eventually also solvent donicity corrections. Subsequently, a probe was introduced by Schneider et al. [50] to the original scale, with which such complications were absent. The differences in the ^{13}C NMR chemical shifts δ in the C(2) vs. C(4) or C(3) vs C(4) positions in pyridine-N-oxide provide the net hydrogen bonding ability of the solvent:

$$\alpha = 2.43 - 0.162[\delta C(2) - \delta C(4)]/ppm = 0.40 - 0.174[\delta C(3) - \delta C(4)]/ppm \quad (3.29)$$

For other probes, namely the AN and E_T^N scales, the values need corrections for the solvent polarity:

$$\alpha = 0.0337AN - 0.10 - 0.47\pi^* = 2.13E_T^N - 0.03 - 0.76\pi^* \quad (3.30)$$

The α values obtained in this manner have an uncertainty of ±0.08 and are listed in Table 3.9 adopted from [3]. Water has a large value of $\alpha = 1.17$, but certain phenols and halogen-substituted alkanols and carboxylic acids are considerably more acidic: hexafluoro-i-propanol has $a = 1.96$, dichloroacetic acid has $\alpha = 2.24$, and trifluoro-methanesulfonic acid has $AN = 131.7$, much higher than the value for water, 54.8, but these are rarely used as solvents for electrolytes and therefore are not dealt with further here.

3.3.2.4 Softness A further property of solvents that is relevant to their solvation of ions is their softness. The general rule that soft solvents preferably solvate soft ions (Section 2.1.1, Table 2.4) and hard solvents do so for hard ions is valid. Among the solvents on our list, the majority is hard (they have oxygen donor atoms) but a few are soft—pyridine, acetonitrile, tetrahydrothiophene, and these indeed solvate soft cations more strongly than expected from other properties that they have. On the other hand, the solvation of soft anions by the soft solvents is not particularly enhanced.

The softness of solvents is best expressed according to Marcus [51] by the difference of $\Delta_{tr}G^\infty$, the standard molar Gibbs energy of transfer from water as the source solvent to the target solvent under study, of silver ions on the one hand and the mean of the values for sodium and potassium on the other:

$$\mu = \{\Delta_{tr}G^\infty(Ag^+) - 0.5[\Delta_{tr}G^\infty(Na^+) + \Delta_{tr}G^\infty(K^+)]\}/(100 \text{ kJ} \cdot mol^{-1}) \quad (3.31)$$

Silver ions are soft whereas sodium and potassium ions are hard, and their mean is used in order to counteract the effect of the ionic size on the standard molar Gibbs energy of transfer, $\Delta_{tr}G^\infty$. The mean radius for the two alkali metal cations, $0.5[r(Na^+) + r(K^+)] = 0.5(0.102 + 0.138) = 0.120$ nm, is close enough to $r(Ag^+) = 0.115$ nm for this purpose. The uncertainties of $\Delta_{tr}G^\infty$ are ±6 kJ mol^{-1} so that the uncertainties of μ assigned to solvents is ±0.08 units.

Other measures of solvent softness have also been established by Marcus et al. [52], using the Raman spectrum of the (soft) mercury bromide or the infrared spectrum of iodine cyanide. These measures correlate with the μ scale and are used to supplement the values if the transfer Gibbs energies are lacking. The resulting values of μ, taken from [3], are shown in Table 3.9 and solvents with $\mu \geq 0.25$ are to be considered as soft, whereas solvents with negative μ values (trifluoroethanol and ethylene glycol) are very hard.

Solvents in which an oxygen donor atom has been replaced by a sulfur atom switch from hard to soft solvents. Tetrahydrofuran can thus be compared with tetrahydrothiophene: $\mu = 0.00$ and 0.80, respectively. Some other such solvent pairs are N,N-dimethyl-thioformamide ($\mu = 1.35$) compared to DMF ($\mu = 0.11$), N-methyl-thiopyrrolidinone ($\mu = 1.35$) compared to NMPy ($\mu = 0.13$), diethylsulfide ($\mu = 0.68$) compared to diethylether ($\mu = 0.00$), and hexamethyl thiophosporamide ($\mu = 1.57$) compared to HMPT ($\mu = 0.29$). Solvents with nitrogen and phosphorus donor atoms are also soft, for example, pyrrole ($\mu = 0.81$) and aniline ($\mu = 0.75$) [3].

3.3.3 Solvents as Acids and Bases

The electron pair donicity and acceptance ability of solvent molecules are reflected in their bulk properties dealt with in Sections 3.3.2.2 and 3.3.2.3. They are, of course, related to their reactions as acids and bases pertaining specifically to the deprotonation or protonation in the gas phase, on the one hand, and these reactions in dilute solutions in a reference solvent (water) on the other, as well as their autoprotolysis as pure liquids. These topics are dealt with in the present section.

The deprotonation of gaseous molecules of protic or protogenic solvents is measured by the standard molar Gibbs energy, ΔG_A, of the reaction $SH(g) \rightarrow S^-(g) + H^+(g)$ where SH stands for a protic solvent molecule. It is generally compared with that of hydrogen chloride in a competitive reaction

$$SH(g) + Cl^-(g) \rightleftarrows HCl(g) + S^- \qquad (3.32)$$

using the well established $\Delta G_A = 1535.1$ kJ·mol^{-1} for HCl(g). Values of the standard reaction Gibbs energy for 25°C of equilibrium (3.32), $\Delta\Delta G_A$, rather than the absolute values of ΔG_A, compiled by Lias et al. in [53], are shown in Table 3.10 to facilitate the ranking of the solvent molecule acidities.

The protonation of gaseous solvent molecules is measured by the standard molar enthalpy, called proton affinity, PA, of the reaction $S(g) + H^+(g) \rightarrow SH^+(g)$, where S stands for a solvent molecule. The proton affinity is generally compared with that of ammonia in a competitive reaction:

$$S(g) + NH_4^+(g) \rightleftarrows NH_3(g) + SH^+ \qquad (3.33)$$

using the well established $PA = 854.0$ kJ·mol^{-1} for NH$_3$(g). Values of the standard reaction enthalpy of (3.33) for 25°C, ΔPA, rather than the absolute values of PA, which are compiled in [53] are shown in Table 3.10 to facilitate the ranking of the solvent molecule basicities.

The acid dissociation of protic or protogenic solvent molecules in dilute solution in water according to

$$SH(aq) + H_2O \rightleftarrows S^-(aq) + H_3O^+(aq) \qquad (3.34)$$

TABLE 3.10 The Standard Molar Gibbs Energy for Protonation of Gaseous Solvent
Molecules Relative to HCl(g), $\Delta\Delta G_A$, their Proton Affinities (enthalpies) Relative to
Ammonia, ΔPA, and Differences of the Logarithmic Equilibrium Constants of Solvents
in Dilute Solutions in Water for Protonation, ΔpK_a, and Deprotonation ΔpK_b, Relative
to the Autoprotolysis Constant of Water, and the Logarithmic Constants pK_S for
Autoprotolysis of the Neat Solvents

Solvent	$\Delta\Delta G_A/kJ\cdot mol^{-1}$	$\Delta PA/kJ\cdot mol^{-1}$	ΔpK_a	ΔpK_b	pK_S
Water	72	−157	0.00	0.0	17.51
Methanol	44	−93	−2.42	−1.5	16.91
Ethanol	16	−66	−1.61	−1.6	19.10
n-Propanol	11	−56	−1.41		19.40
i-Propanol	8	−54		−0.3	21.08
t-Butanol	5	−44	1.49	0.1	26.80
Trifluoroethanol	−53	−147	−5.14		
Ethylene glycol		−25	−2.44		15.84
Tetrahydrofuran		−19		−2.7	35.50
1,4-Dioxane		−44		−1.9	
Acetone	−22	−31	6.7	−0.7	32.50
Propylene carbonate					
γ-Butyrolactone					
Formamide		−24		−2.3	16.80
N-Methylformamide		7		−1.9	10.74
N,N-Dimethylformamide	105	30		−1.9	23.10
N,N-Dimethylacetamide	0	51			23.95
Tetramethylurea		79			
N-Methylpyrrolidinone		53		−2.8	25.60
Ammonia	122	0		−8.25	32.50
Pyridine	67	70		−12.3	
Acetonitrile	−10	−67	6.6		32.20
Benzonitrile		−34			
Nitromethane	−62	−104			
Nitrobenzene		−45			
Chloroform	106				
1.1-Dichloroethane					
1,2-Dichloroethane					
Dimethylsulfoxide	−2	−20		−2.0	31.80
Tetramethylenesulfone				−2.2	25.45
Tetrahydrothiophene		2		1.0	
Trimethylphosphate		33			
Hexamethyl phosphoric triamide		94			20.56

is described by the equilibrium constant $K_a = [S^-][H_3O^+]/[SH]$ on the M scale, the
constant concentration of the water being included in the value of K_a. Values of its
negative logarithm, pK_a for 25°C from [1], from which the autoprotolysis constant of
water (see below) $pK_{SW} = 17.51$ is subtracted, ΔpK_a, are listed in Table 3.10. Negative
values of $pK_a - pK_{SW}$ denote solvents to be more acidic in water than water itself.

The protonation of solvents in dilute solution in water according to:

$$S(aq) + H_2O \rightleftharpoons SH^+(aq) + OH^-(aq) \qquad (3.35)$$

is described by the equilibrium constant $K_b = [SH^+][OH^-]/[S]$ on the M scale, the constant concentration of the water being included in the value of K_b. Values of ΔpK_b, i.e., its negative logarithm, pK_b for 25°C from [1], from which $pK_{SW} = 17.51$ is subtracted, are listed in Table 3.10. Negative values of $pK_b - pK_{SW}$ denote solvents to be more basic in water than water itself. Some solvents of very small basicity require a stronger acid than water to protonate them, and dilute sulfuric acid in water is used for this purpose.

Another quantity that characterizes the acid–base properties of protic and protogenic solvents, SH, is their autoprotolysis constant. This is the equilibrium constant K_S for the reaction:

$$2SH \rightleftharpoons SH_2^+ + S^- \qquad (3.36)$$

where all the species are dissolved in the solvent SH itself. The negative logarithm, pK_S, of this constant $K_S = [SH_2^+][S^-]/[SH]^2$ on the molar scale at 25°C is also shown in Table 3.10, taken from [3].

3.3.4 Miscibility with and Solubility in Water

Aqueous solutions of electrolytes that dissociate into ions have been and are studied so extensively that it is natural to enquire about the solutions of electrolytes in binary mixtures of water and cosolvents. Mixtures of nonaqueous solvents among themselves, although important for other purposes, have not received much attention when solutions of electrolytes are concerned. Therefore, this section deals only with binary mixtures of water with cosolvents. Many of the solvents on the list are miscible with water at all proportions, whereas others exhibit limited, even rather low mutual solubility at ambient conditions. For these cases, solubility of water in the solvent, x_w, is larger than that of the solvent in water, x_S, on the mole fraction scale. Conversion to (approximate) values of the solubility on the molar scale is according to:

$$c_S/M \sim 1000 x_S / [V_w + x_S(V_S - V_w)] \qquad (3.37)$$

The approximation is due to neglecting the relatively small excess volume of mixing, V^E, of the two components. For water-miscible solvents, the maximal value of c_S is its value for the neat solvent, the reciprocal of its molar volume: $c_{Sneat}/M = 1000/(V_S/cm^3 \cdot mol^{-1})$.

Related to the mutual solubility of solvents with water is their lipophilicity or hydrophobicity that can be described by (the logarithm of) their partition coefficient between n-octanol and water, $\log P_w^\circ$. The more hydrophobic a substance is, the larger the value of $\log P_\mathrm{W}^\circ$ for its partition. A linear expression connects the solubility of the solvent in water with this partition coefficient:

$$\log c_\mathrm{S} = 0.850 - 1.214 \log P_\mathrm{W}^\circ \tag{3.38}$$

The mutual solubilities and the distribution coefficients of solvents at 25°C are shown in Table 3.11 adapted from [3], and it is noted that the majority of solvents

TABLE 3.11   Mole Fraction Solubilities of Solvents in Water, x_S, and of Water in Solvents, x_W, and the Logarithm of the Octanol/Water Partition Coefficients of the Solvents, $\log P_\mathrm{W}^\circ$, at 25°C

Solvent	$10^3 x_\mathrm{S}$	$10^3 x_\mathrm{W}$	$\log P_\mathrm{W}^\circ$
Methanol	Miscible		−0.70
Ethanol	Miscible		−0.25
n-Propanol	Miscible		0.28
i-Propanol	Miscible		0.13
t-Butanol	Miscible		0.36
Trifluoroethanol	Miscible		0.41
Ethylene glycol	Miscible		−2.27
Tetrahydrofuran	Miscible		0.46
1,4-Dioxane	Miscible		−0.42
Acetone	Miscible		−0.24
Propylene carbonate	36.1	339	
γ-Butyrolactone	Miscible		−0.64
Formamide	Miscible		−1.67
N-Methylformamide	Miscible		−0.97
N,N-Dimethylformamide	Miscible		−1.01
N,N-Dimethylacetamide	Miscible		0.34
Tetramethylurea	Miscible		
N-Methylpyrrolidinone	Miscible		−0.38 [54]
Ammonia	Miscible, but reacts		−1.49
Pyridine	Miscible		0.65
Acetonitrile	Miscible		−0.34
Benzonitrile	0.35	50	1.56
Nitromethane	35.5	67.4	−0.34
Nitrobenzene	0.278	16.2	1.85
Chloroform	1.24	6.1	1.94
1.1-Dichloroethane	0.955	5.25	1.79
1,2-Dichloroethane	1.48	10.2	1.63
Dimethylsulfoxide	Miscible		−1.35
Tetramethylenesulfone	Miscible		
Tetrahydrothiophene	42.3 [55]		1.79 [55]
Trimethylphosphate			−0.52
Hexamethyl phosphoric triamide	Miscible		0.28

generally used for the dissolution of electrolytes are completely miscible with water and are very hydrophilic, having negative values of $\log P_W^\circ$.

3.3.5   Spectroscopic and Electrochemical Windows

Solvent for electrolytes that are to be studied spectroscopically need themselves to be transparent for the light of the wavelengths (for UV and visible light) and wavenumbers (for infrared light and Raman spectroscopy) employed for this purpose. The solvents dealt with here are all colorless when pure but have a UV cut-off that ought to be noted. The functional groups of the solvent molecules have characteristic infrared absorption bands that ought to be avoided when solutes in the solvents are to be studied by infrared or Raman spectroscopy. Table 3.12, mostly adapted from [1], shows the appropriate windows at which the solvents are sufficiently transparent.

For electrochemical applications, the available window of voltages that can be applied without oxidation and reduction of the solvent itself depends on the working electrode (often the dropping mercury electrode or else a platinum electrode), reference electrode (often the saturated calomel electrode, SCE), and the background electrolyte used to make the solution conductive (often tetraethyl- or -butyl perchlorate), so that a table of values for the various variants used (apart from those often employed) cannot be shown. Water has a rather limited electrochemical window, spanning only ~3.5 V, compared with ~4.5 V available for nitromethane and dimethylsulfoxide, ~5 V available for acetonitrile, and as much as ~6 V available for propylene carbonate. Some information concerning such electrochemical windows is given in [3].

3.4   PROPERTIES OF BINARY AQUEOUS COSOLVENT MIXTURES

Mixtures of water with cosolvents are often used as solvents for electrolytes and ions so that those properties of the mixtures that are related to the (possibly preferential) solvation of the ions by the components of such mixtures need to be known. As for the neat solvents dealt with in the previous sections of this chapter, the discussion concerning those solvents marked as miscible with water in Table 3.10 involves physical and chemical properties, to be dealt with in turn. Much of the information below is adapted from the book by Marcus [56].

3.4.1   Physical Properties of Binary Aqueous Mixtures with Cosolvents

The composition of binary aqueous mixtures with cosolvents S is usually expressed in terms of the mole fractions x_S of the cosolvent and $x_W = 1 - x_S$ of the water. In some cases, because of the convenience of preparing mixtures by volume rather than by mass, where the mass fractions are w_S and w_W, the volume fractions φ_S and φ_W are employed instead. The relationships between these measures of the composition are:

$$x_S = w_S / \left[w_S + \left(1 - w_S\right) M_S / M_W \right] \tag{3.39}$$

TABLE 3.12   Spectroscopic Windows of Solvents

Solvent	UV cut-off/nm	IR window/cm^{-1}
Water	190	
Methanol	205	1520–2760, >3600
Ethanol	205	1500–2800, >3600
n-Propanol	210	1460–2800, >3400
i-Propanol	210	1540–2600, >3500
t-Butanol	215	1500–1800
Trifluoroethanol	190	
Ethylene glycol		1500–2660
Tetrahydrofuran	220	<850, 1200–2780, >3040
1,4-Dioxane	220	700–850, 920–1000, 1500–2700
Acetone	330	700–1050, 1800–3000
Propylene carbonate	280	
γ-Butyrolactone		1500–1700, 1880–2880
Formamide		790–1200, 1750–3040
N-Methylformamide		
N,N-Dimethylformamide	270	740–950, 1800–2700
N,N-Dimethylacetamide	270	610–980, 1760–2800
Tetramethylurea		
N-Methylpyrrolidinone	260	
Ammonia		
Pyridine	305	800–970
Acetonitrile	195	<10580, 1500–2220, >2240
Benzonitrile	300	760–1480, 1500–2200
Nitromethane	380	670–1350, >1620
Nitrobenzene		1630–3060, >3120
Chloroform	245	800–1200, >1300
1.1-Dichloroethane		720–960, >1450
1,2-Dichloroethane	230	780–1200, >1500
Dimethylsulfoxide	265	<940, 1090–1400, 1450–2900
Tetramethylenesulfone		1450–2850
Tetrahydrothiophene		1440–2820, >3000

No data were found for TMP and HMPT.

$$\varphi_S \approx w_S / \left[w_S + (1 - w_S) \rho_w / \rho_S \right] \qquad (3.40)$$

$$\varphi_S \approx x_S / \left[x_S + (1 - x_S) V_w / V_S \right] \qquad (3.41)$$

where M is the molar mass (Table 3.2), ρ is the density, and V is the molar volume (Table 3.3) of the neat cosolvent and of water. The approximations in Equations 3.40 and 3.41 are due to the use of the densities and the molar volumes of the neat solvents rather than the partial molar volumes. This represents the neglect of the excess molar volume, generally $\leq \pm 2 \, cm^3 \, mol^{-1}$, compared to the molar volume of the mixtures.

The physical properties of the binary mixtures are expressed by the general symbol Y that may represent extensive thermodynamic quantities such as molar

Gibbs energies, enthalpies, entropies, heat capacities, and volumes, as well as intensive properties, such as permittivities or viscosities. The excess functions of extensive properties over those for ideal mixtures of the components, symbolized by Y^E (or the respective increments for intensive quantities, symbolized by ΔY), are usually defined in terms of the mole fraction composition with respect to the pure components:

$$Y^E \left(\text{or } \Delta Y \right) = Y - \left(x_W Y_W + x_S Y_S \right) \tag{3.42}$$

where the expression in the parentheses represents the property Y of an ideal mixture of the two components (for which $Y^E = 0$). Rather rarely is the ideal mixture expressed in terms of the volume fraction of the components, $(\varphi_W Y_W + \varphi_S Y_S)$. The excess property is commonly expressed in terms of the Redlich–Kister equation, when its values over the entire span of the compositions are required:

$$Y^E \left(\text{or } \Delta Y \right) = x_S \left(1 - x_S \right) \left[y_0 + y_1 \left(1 - 2 x_S \right) + y_2 \left(1 - 2 x_S \right)^2 + y_3 \left(1 - 2 x_S \right)^3 + \cdots \right] \tag{3.43}$$

Usually three or four terms in $y_j \left(1 - 2 x_S \right)^j$ $(j > 0)$ in the square brackets suffice for agreement of the fitted values with the experimental ones within the experimental errors of the latter. If only a short part of the composition range is to be studied, for example, water-rich mixtures, then better accuracy of fitting the experimental data may be achieved in terms of a simple power series up to third or fourth order: $Y^E = \Sigma y_i x_S^i$.

3.4.1.1 Thermodynamic Properties of the Mixtures

In those cases where Y represents a molar thermodynamic property of the binary mixture of water and the cosolvent, the partial molar quantities of the components are of interest. Differentiation of Equation 3.43 with respect to the mole fractions yield the partial molar excess values. The excess partial molar value for water is:

$$y_W^E = y_0' x_S^2 + y_1' x_S^3 + y_2' x_S^4 + y_3' x_S^5 + \cdots \tag{3.44}$$

Note that in y_W^E there are no constant and first-order terms in x_S. The coefficients y' are related to the coefficients y of Equation 3.43 as follows (if the latter are limited to four):

$$\begin{aligned} y_0' &= y_0 + 3 y_1 + 5 y_2 + 7 y_3; \\ y_1' &= -4 \left[y_1 + 4 y_2 + 9 y_3 \right]; \\ y_2' &= 12 \left[y_2 + 5 y_3 \right]; \\ y_3' &= -32 y_3 \end{aligned} \tag{3.45}$$

Similarly, the excess partial molar value for the cosolvent starts with the second-order term in x_W and is:

$$y_S^E = y_0'' x_W^2 + y_1'' x_W^3 + y_2'' x_W^3 + y_3'' x_W^4 + \cdots \tag{3.46}$$

and the coefficient of this expression is related to the y coefficients as:

$$\begin{aligned}
y_0{}'' &= y_0 - 3y_1 + 5y_2 - 7y_3; \\
y_1{}'' &= -4[y_1 - 4y_2 + 9y_3]; \\
y_2{}'' &= 12[y_2 - 5y_3]; \\
y_3{}'' &= 32y_3
\end{aligned}$$

(3.47)

If Y denotes the molar Gibbs energy of mixing for the binary mixture, then the partial molar quantities are the chemical potentials of the components. These are related to their activity coefficients f_j and vapor pressures p_j, $y_j^E = \mu_j^E = RT \ln f_j \approx RT \ln(p_j / x_j p_j{}^\circ)$, where subscript j denotes water or the cosolvent and $p_j{}^\circ$ is the vapor pressure of the pure component at the temperature T (neglecting vapor phase association). If Y denotes the molar enthalpy of mixing, it equals the excess molar enthalpies of the mixtures, because there is no ideal enthalpy of mixing.

The coefficients y_i of the Redlich–Kister expression (3.43) for the molar excess Gibbs energies and enthalpies of the miscible aqueous cosolvents on the list are shown in Table 3.13 adapted from [56]. It should be noted that whereas the $G^E(x_S)$ curves for many aqueous cosolvent systems are fairly symmetrical, the $H^E(x_S)$ curves for some systems are quite skew, even changing sign from negative at water-rich compositions to positive beyond a certain x_S.

The excess molar entropies of mixing water with cosolvents are obtained from the corresponding enthalpies and Gibbs energies, $S^E(x) = [H^E(x_S) - G^E(x_S)]/T$, from the data in Table 3.13 (provided the same temperature is employed for both enthalpies and Gibbs energies).

The excess molar heat capacities of the binary mixtures of water with cosolvents, C_P^E, are again expressed in terms of Equation 3.43 with coefficients $c_i \equiv y_i$ that are listed in Table 3.14. Data could not be found for a few cosolvents on the list.

The excess molar volumes of the binary mixtures of water with co-solvents, V^E, are similarly expressed in terms of Equation 3.43 with the *negatives* of the coefficients $v_i \equiv y_i$ listed in Table 3.14. Note that the entries of $-v_0$ are all positive, that is, the excess molar volumes of the equimolar mixtures are negative for all the solvents on the list (which are relevant to the solvation of ions). This is valid for the majority of the solvents over nearly the entire composition range, but there are cases where the sign changes at extremes of this range.

The expansibilities, compressibilities, and surface tensions of the mixtures are intensive properties, so they should be expressed not in terms of excess quantities but just as deviations from the linear dependence on the (mole fraction) composition. Data for the isobaric expansibilities and the adiabatic compressibilities of binary mixtures of water with many cosolvents on the list are available in [56]. The isobaric compressibilities can then be calculated from such data by Equation 3.3, using also the molar volumes of the mixtures, $V = x_W V_W + x_S V_S + V^E$.

TABLE 3.13 The Coefficients of Equation 3.44 for the Excess Molar Gibbs Energy ($Y^E = G^E$ with the $y_i \equiv g_i$) and Enthalpies ($Y^E = H^E$ with the $y_i \equiv h_i$) in J·mol^{-1} of Aqueous Mixtures with Cosolvents at 25°C From [56] Unless Otherwise Noted

Solvent	g_0	g_1	g_2	h_0	h_1	h_2
Methanol	1200	−87	−330	−3120	2040	−2213
Ethanol	2907	−777	494	−1300	−3567	−4971
n-Propanol[a]	3733	−1095		638	−2192	2598[b]
i-Propanol	3843	−984	−98	854	5167	−7243
t-Butanol	4150	−1308	879	170	−4129	1136[c]
Trifluoroethanol	3449	−1725	287[d]	2749	−2341	663[e]
Ethylene glycol	−558	164	−189	−2776	1933	−1172
Tetrahydrofuran	5484	47	1371	−136	7443	−4229
1,4-Dioxane	4560	−973	−421	611	6006	−1712
Acetone	4560	−163	1140	569	−5408	−1838
γ-Butyrolactone				2742	−5963	790[f]
Formamide	−5099	4367	−2681[g]	1074	152	381[h]
N-Methylformamide	−970	−834	−425[i]	−3635	−2438	−529[j]
N,N-Dimethylformamide	−978	−653	222	−7616	7751	−1904
N,N-Dimethylacetamide	1585	−701	−55[k]	−10447	7077	3580[l]
Tetramethylurea						
N-Methylpyrrolidinone	−487	−40	206[m]	−9983	8917	−2496
Pyridine	2404	−1212	1873	−5600	−1020	1756
Acetonitrile	5253	−639	1316	4640	2922	−1028[n]
Dimethylsulfoxide	−4909	2168	−5	−10372	6922	−2466
Tetramethylenesulfone[n]	4165	822	1822	4372	−1447	1254
Hexamethyl phosphoric triamide	−4673	4185	−4270	−11367	−10085	−11288

[a] G^E data from Ref. [63].
[b] Data from Ref. [62] for 30°C, and two further terms with $h_3 = 4267$ and $h_4 = -15109$ J·mol^{-1} are needed.
[c] Two further terms with $h_3 = -7698$ and $h_4 = -13530$ J·mol^{-1} are needed.
[d] Data from [61] and two further terms with $g_3 = -395$ and $g_4 = 361$ J·mol^{-1} are needed.
[e] Data from [58] and three further terms with $h_3 = 5705$, $h_4 = -4100$, and $h_5 = -9288$ J·mol^{-1} are needed.
[f] Data from [59] and two further terms with $h_3 = 7960$ and $h_4 = -6675$ J·mol^{-1} are needed.
[g] A further term with $g_3 = 1063$ J·mol^{-1} is needed
[h] A further term with $h_3 = 273$ J·mol^{-1} is needed.
[i] Two further terms with $g_3 = 948$ and $g_4 = 1439$ J·mol^{-1} are needed.
[j] A further term with $h_3 = -108$ J·mol^{-1} is needed.
[k] Values at 40°C from [60].
[l] Data from [60] and two further terms with $h_3 = -583$ and $h_4 = 799$ J·mol^{-1} are needed.
[m] The values are for 30°C and a further term with $g_3 = 444$ J·mol^{-1} is needed.
[n] Data from [57], coefficients of H^E are for 30°C.

The surface tensions of the mixtures are best described in terms of the following expression:

$$\sigma = \frac{x_W \sigma_W + x_S \sigma_S + x_W x_S \sigma_0}{(1 - \sigma_1 x_W)} \tag{3.48}$$

TABLE 3.14 The Coefficients of Equation 3.45 for the Excess Molar Heat Capacity ($Y^E = C_P^E$ with the $y_i \equiv c_i$) in J·K^{-1}·mol^{-1} and Volumes ($Y^E = V^E$ with the $y_i \equiv v_i$) in cm^3·mol^{-1} of Aqueous Mixtures with Cosolvents at 25°C from [56] Unless Otherwise Noted

Solvent	c_0	c_1	c_2	$-v_0$	$-v_1$	$-v_2$	$-v_3$
Methanol	15.8	37.7	−15.5	4.08	0.43	−0.69	−1.20
Ethanol	41.9	40.2	58.7	4.37	1.70	0.74	−1.64
n-Propanol[a,b]	52.6	4.9	−9.3[a]	2.63	0.30	0.41	1.85
i-Propanol	168	−587	742	3.59	2.28	2.49	
t-Butanol	296	−1166	1256	2.18	2.08	4.78	
Trifluoroethanol[c]				2.42	0.85	0.47	4.35
Ethylene glycol	1.36	4.85	3.24[d]	1.46	0.56	−0.22	
Tetrahydrofuran	−22.4	−200.4	−160.7	3.09	2.35	1.25	
1,4-Dioxane	25.9	10.8	−11.1[e]	2.45	1.90	0.88	0.54
Acetone	37.4	−40.4	117.4	5.79	2.24	0.34	
γ-Butyrolactone[f]				0.66	0.45	0.71	1.41
Formamide	−4.6	−8.0	−1.1	0.51	0.51	0.31	
N-Methylformamide	11.3	17.1	15.9	2.13	1.05	−0.14	
N,N-Dimethylformamide	38.1	34.0	12.2	4.70	1.22	−1.71	
N,N-Dimethylacetamide	41.0	31.2	−25.8[g]	6.14	3.02	0.49	−2.69[h]
Tetramethylurea[i]				6.18	−0.40	11.44	−9.31
N-Methylpyrrolidinone	51.3	37.2	79.9	4.59	1.93	−0.52	
Pyridine[j]				2.93	−0.97	−0.85	
Acetonitrile	35.0	−8.5	8.9[k]	1.99	2.04	1.32	
Dimethylsulfoxide	−39.8	−52.9	−61.0	3.90	2.09	−1.20	−2.04
Tetramethylenesulfone[l]	25.0	3.4	−4.6	0.35	0.84	−0.49	0.50[l]
Hexamethyl phosphoric triamide	19.7	−34.2	118.3[m]	5.43	4.65	4.43	

[a] Heat capacity data from [67] at 35°C and two further terms are needed with $c_3 = -101.1$ and $c_4 = 116.3$.
[b] Volume data from [72] with a further term with $v_4 = -2.58$, valid at $x_s \geq 0.055$.
[c] Data from [71].
[d] Two further terms are needed with $c_3 = 15.49$ and $c_4 = 18.15$.
[e] A further term is needed with $c_3 = 66.9$.
[f] Data from [69].
[g] Data from [66] and two further terms are needed with $c_3 = 31.5$ and $c_4 = 49.5$.
[h] Data from [65] and a further term is needed with $v_4 = -2.43$.
[i] From [68], but the data extend only to $x_s = 0.48$.
[j] Data from [70].
[k] A further term is needed with $c_3 = 35.1$.
[l] Heat capacity data from [64] for which a further term is needed with $c_3 = -24.8$ and volume data from [57] for which a further term is needed with $v_4 = -1.43$.
[m] A further term is needed with $c_3 = 218.7$.

The coefficients σ_0 and σ_1 of the last term are shown in Table 3.15 for most of the solvents on the list, adapted from [56] and valid for 25°C. However, for a few of the cosolvents (methanol, ethanol ethylene glycol, dimethylformamide, and dimethyl-sulfoxide) Equation 3.43 with four terms in the square brackets, as noted in the table,

TABLE 3.15 The Coefficients of Equation 3.49 for the Surface Tension and of Equation 3.44 for the Permittivity and the Viscosity at 25°C Adapted from [56]

Solvent	σ_0	σ_1	ε_0	ε_1	ε_2	η_0	η_1	η_2	η_3
Methanol[a]	−68	−38	−30.7	−1.8		2.50	2.72	1.14	−0.18
Ethanol[a]	−97	−196	−58.7	−24.5		3.92	5.03	4.60	0.97
n-Propanol[b,c]	−48.4	0.991	−82.9	−56.4	−40.4	3.96	5.97	5.40	1.84
i-Propanol	−49.1	0.983	−89.8	−46.1		4.02	5.01	4.61	3.04
t-Butanol	−50.6	0.992	−111.6	−64.6		9.21	10.13	5.06	
Trifluoroethanol[d,e]	−43.8	0.990	−68.9	−19.9		1.46	2.95	6.20	−1.79
Ethylene glycol[a]	−28.5	−12.1	−28.0	−11.2		−6.40	−0.20	0.89	−1.25
Tetrahydrofuran	−43.1	0.983	−97.6	−50.9		0.82	2.61	5.29	4.92
1,4-Dioxane	−35.1	0.964	−121.5	−99.4		2.44	4.28	3.01	0.62
Acetone	−40.4	0.981	−72.0	−32.3		0.61	2.89	3.80	0.31
γ-Butyrolactone[f]	−26.3	0.955				1.92	1.82	0.65	1.05
Formamide	−11.3	0.754	53.7	29.2		−1.97	0.66	−0.20	−0.18
N-Methylformamide[g]			−51.7	16.1	−4.1	2.87	2.51	0.08	−0.85
N,N-Dimethylformamide[a]	−49.2	−17.0	−30.8	−0.7		4.13	8.48	4.12	−7.24
N,N-Dimethylacetamide[h,i]	−28.4	0.884	−43.7	−13.8	−39.3	−1.74	55.29	−190.0	219.1
Tetramethylurea[j,k]	−32.9	0.963	−55.8	−5.3	−5.9[l]	12.18	20.35	21.25	9.34

N-Methylpyrrolidinone[i]	−22.5	0.945	−36.68	−9.9	12.47	13.72	1.39	−1.87
Pyridine[b]	−23.7	0.993	−98.28	−20.7	4.90	3.66	1.01	
Acetonitrile	−41.6	0.953	−40.2	−13.2	−0.27	0.46	0.90	0.88
Dimethylsulfoxide[a]	−37	4	−1.0	5.9	7.96	−10.38	−1.80	9.26
Tetramethylenesulfone[m,n]	−19.1	0.978	−44.9	−26.4	−4.74	1.22	0.16	−0.22
Hexamethyl phosphoric triamide			−102.7	−74.1	13.57	23.76	28.99	8.44

[a] For these cosolvents, Equation 3.44 applies for the surface tensions with the additional coefficients $\sigma_2 = -57$ and $\sigma_3 = -106$ for methanol, $\sigma_2 = 746$ and $\sigma_3 = -1155$ for ethanol, $\sigma_2 = -27.8$ and $\sigma_3 = -54.2$ for ethylene glycol, $\sigma_2 = -60.32$ and $\sigma_3 = -130.9$ for dimethylformamide, and $\sigma_2 = 35$ and $\sigma_3 = -135$ for dimethylsulfoxide.

[b] Surface tension data from [73] and viscosity data from [15, 74].

[c] Permittivity from [35].

[d] Surface tension and viscosity data from [76].

[e] Permittivity from [35] and surface tension from [77].

[f] Surface tension and viscosity data from [75], both at 30°C.

[g] Permittivity from [35] and viscosity from [78, 79].

[h] Permittivity and viscosity from [40].

[i] Surface tension data from [80].

[j] Surface tension data from [10], read from a figure.

[k] Viscosity from [10].

[l] Permittivity from [35] with a term in $\epsilon_3 = -29.5$ needed.

[m] Surface tension and viscosity data from [57].

[n] Permittivity from [35].

describes better the deviations of $\sigma(x_S)$ from linearity. In all the cases, the surface tension falls drastically when the cosolvent is added to water at low concentrations and then more moderately as the fraction of the cosolvent increases.

3.4.1.2 *Some Electrical, Optical, and Transport Properties of the Mixtures*   The relative permittivity and the dynamic viscosity of binary mixtures of water with cosolvents are also relevant to the solvation and behavior of electrolytes and ions in these mixtures. These, again, are intensive properties, so that rather than dealing with "excess quantities" deviations from ideal behavior according to eq. (3.43), with ΔY replacing Y^E, should be used.

The relative permittivities obey Equation 3.43 with two or three terms, the values of which for 25°C, adapted from [56], except where noted, are shown in Table 3.15. Except for aqueous formamide and *N*-methylformamide, in which the pure cosolvents have larger relative permittivities than that of water, the permittivities drop with increasing cosolvent contents, but not linearly.

The variation of the refractive index, n_D, between that of water and that of the neat water-miscible cosolvents does not exceed 13% (Table 3.5). Hence for some purposes, linear interpolation according to the mole fractions provides a sufficiently good approximation for n_D of the mixture. A better one is obtained when n_D is back-calculated from the molar refraction, assumed to be independent of the arrangements and bonding of the atoms, according to:

$$R_D = x_S R_{DS} + x_W R_{DW} = x_S V_S \left(n_{DS}^2 - 1\right) / \left(n_{DS}^2 + 2\right) + x_W V_W \left(n_{DW}^2 - 1\right) / \left(n_{DW}^2 + 2\right) \quad (3.49)$$

However, if the refractive index of the mixtures is to be used for calculation of the composition, as is sometimes done, then direct calibration is preferable to the use of Equation 3.49.

The viscosities of the binary aqueous mixtures with cosolvents generally increase with the content of the latter, but may diminish at larger contents, exhibiting a maximum. They obey Equation 3.44 with three or four terms, the values of which for 25°C, adapted from [56] except where noted, are shown in Table 3.15.

3.4.2   Chemical Properties of Binary Aqueous Mixtures with Cosolvents

3.4.2.1 *Structuredness*   The "openness" of a solvent is defined as the difference between its molar volume and its intrinsic volume. The relative free volume, $(V-V_X)/V$, (Section 3.3.1) of the binary aqueous mixtures with cosolvents may be practically prorated according to the mole fractions of the components, because the excess molar volumes are only small fractions of the molar volumes of the mixtures.

The "stiffness" of a solvent is defined as $U/V-P_{int}$, where U/V is the cohesive energy density and P_{int} is the internal pressure (Section 3.3.1). Of the water-miscible solvents, only the two ethers, tetrahydrofuran and 1,4-dioxane, and the amide HMPT are "non-stiff" (entries in *italics* in the second column in Table 3.8). This property bears on the energetics of introducing a solute (an ion) into a cavity that has to be formed in the solvent to accommodate the solute—the larger the stiffness, the more energy has to be invested in the formation of the cavity.

TABLE 3.16 The internal Pressure Increment, ΔP_{int}/MPa, in A 1 mol·dm^{-3} Solution of the Cosolvent in Water, the Maximal Partial Molar Excess Volume of Water, $V_{W\,max}^{E}$/ cm^3·mol^{-1}, in Water-rich Mixtures, the Incremental Maximal Partial Molar Excess Heat Capacity of Water, $\Delta C_{pW\,max}^{E}$/J·K^{-1}·mol^{-1}, in Water-rich Mixtures, and the Volume Corrected Preferential Solvation Parameters of Water Near Water, $\delta' x_{WW}$, and of Cosolvent Near Water, $\delta' x_{WS}$ (for the symbols see the text)

Solvent	ΔP_{int}	$V_{W\,max}^{E}$	$\Delta C_{pW\,max}^{E}$	$\delta' x_{WW}$	$\delta' x_{WS}$
Methanol	4	0.0206	1.7	+	–
Ethanol	2	0.015	11.3	++	–
n-Propanol	23	0.0064	4.1	++	– –
i-Propanol		0.027	14.9	++	–
t-Butanol	51	0.050	15.8	++	– –
Trifluoroethanol				++	– –
Ethylene glycol	25	<0	–5.1	–	0
Tetrahydrofuran	40			++	– –
1,4-Dioxane	62	~0	–5.8	++	–
Acetone	48		2.5	++	–
γ-Butyrolactone				+	–
Formamide	40	<0	<0	+	–
N-Methylformamide	38	<0	–4.1	–	+
N,N-Dimethylformamide	61	0.0333	7.8	– +	0
N,N-Dimethylacetamide		0.0208	0.8		
Tetramethylurea				+	–
N-Methylpyrrolidinone		0.0480		0	0
Pyridine				+	–
Acetonitrile	53	<0	–4.0	++	– –
Dimethylsulfoxide	41	0.022	–5.1	–	0
Tetramethylenesulfone				++	–
Hexamethyl phosphoric triamide				++	–

The stiffness of aqueous mixtures with cosolvents need not be a linear function of those of the components, but data on the internal pressures of the mixtures over the entire composition range are not available [30]. This is not the case concerning dilute solutions of cosolvents in water: these have been studied by Dack, Conti, and Zaichikov and their respective coworkers [81–84]. The values of $\Delta P_{int} = P_{int}(1M) - P_i(W)$, the internal pressure increment at 1 M of the cosolvent over that of pure water at 25°C, are shown in Table 3.16.

Dack [81] showed a plot of ΔP_{int} against the molar volumes of the solutes and arbitrarily selected five solutes: urea, formamide, acetonitrile, 1,4-dioxane, and piperidine to represent "non-interacting" ones. Solutes (cosolvents in the present case) lying above a line defined by these five solutes are structure breakers, in the sense of the transfer of water molecules from bulky to compact domains, leaning on the two-structure model of water. Strong hydrogen bonding protic solutes, such as the amides and alkanols, and indeed all the cosolvents on the list for which there are data, have ΔP_{int} values below the line defined by Dack and should according to him be considered as water-structure makers when in water-rich mixtures.

Other measures of the effect of cosolvents on the structure of water do not confirm this general conclusion. The excess partial molar volume of water in water-rich mixtures with cosolvents is a clear indication, if positive, that the bulky, low-density structure of the water is enhanced as argued by Marcus [35]. When the excess molar volume of the binary aqueous mixtures with the cosolvents are expressed as a third-order polynomial in the water-rich region, $V^E (x_S \leq 0.3) = b_0 + b_1 x_S + b_2 x_S^2 + b_3 x_S^3$, the partial molar volume of the water component is given by:

$$V_W^E = V^E - x_S \left(\partial V^E / \partial x_S \right)_{T,P} = -b_2 x_S^2 - 2b_3 x_S^3 \qquad (3.50)$$

Only in cases where b_2 in expression (3.50) is negative may V_W^E be positive. Values of the maximal V_W^E, $V_{W\,max}^E / cm^3 \cdot mol^{-1}$, at 25°C are shown in Table 3.16. Note that the positive values of $V_{W\,max}^E / cm^3 \cdot mol^{-1}$, though definite and significant, are quite small compared to, say, the absolute excess molar volumes of the equimolar mixtures $V^E (x_S = 0.5) / cm^3 \cdot mol^{-1} = 0.25 v_0$ from Table 3.14. The water structure enhancement occurs for a limited group of cosolvents, which on the one hand have relatively small molecules that can be accommodated in the voids in the structure of pure water and on the other form hydrogen bonds with surrounding water molecules. The presence of several methyl groups seems to induce such an enhancement, cf. t-butanol, dimethyl-formamide, and dimethylsulfoxide. Excluded from this group are cosolvents with longer hydrophobic groups than ethyl or those that interact too strongly with the water, like formamide and ethylene glycol, forming denser domains. These have $V_W^E < 0$ throughout the water-rich range and generally over the entire composition range.

The excess partial molar isobaric heat capacity of water, C_{PW}^E, in water-rich mixtures with cosolvents is another measure of the possible enhancement of the water structure [35]. The heat capacity of the fully hydrogen bonded water molecules is larger than that of non-hydrogen-bonded water molecules. Excitation of internal modes of the water molecules has to be deducted in part: $\Delta C_{PW\,max}^E = C_{PW\,max}^E - 0.25 C_{PW}$ (i.g.). The remainder indicates, if positive, that the relative extent of fully hydrogen bonded domains of the water is enhanced. The results, shown in Table 3.16 for 25°C, mostly agree with the conclusions from the partial molar excess volumes of the water in the solutions, but there are cases of disagreement. According to the excess partial molar heat capacity of the water, methanol hardly enhances the water structure and dimethyl sulfoxide does not do it at all, contrary to the conclusions from the corresponding excess volume.

The structuredness of the binary mixtures treated here can be further discussed in terms of their preferential mutual or self solvation and possible microheterogeneity. The surroundings of a molecule of the water, W (or of the cosolvent, S), generally differ in terms of the relative amounts of W and S molecules from the bulk composition due to preferential solvation. Fluctuation theory, in terms of the Kirkwood-Buff integrals derived from thermodynamic data, is well suited for studying the preferential solvation, provided the data are sufficiently accurate [85]. When the relative sizes of the molecules are taken into account, this approach yields the interactions among the components. A large number of binary aqueous-organic solvent mixtures has been studied by this method by Marcus [56, 85, 86]. The preferential solvation occurs in some cases beyond the first solvation shell, for example, in aqueous mixtures of 1-propanol, t-butanol, tetrahydrofuran, dioxane, acetonitrile, and sulfolane, where self interactions of the water molecules far outweigh those between water and organic cosolvent molecules.

The Kirkwood-Buff integrals diverge if the quantity

$$D = 1 + x_S x_W \left[\frac{\partial^2 \left(G^E RT \right)}{\partial x_S^2} \right]_{T,P} \tag{3.51}$$

is smaller than unity, and if D is negative phase separation takes place. For water-miscible cosolvents, $0 < D < 1$ denotes that microheterogeneity occurs. This is the case, for instance, for aqueous acetonitrile—on increasing x_S from dilute solutions, the onset of $D < 1$ occurs near $x_B = 0.25$ at all the temperatures studied, but the range where $D < 1$ becomes narrower as the temperature increases. The upper limit of the microheterogeneity is $x_S \sim 0.65$ at 5°C, 0.60 at 15°C, 0.55 at 25°C, 0.50 at 40°C, and 0.45 at 50°C. This composition range agrees substantially with that corresponding to the occurrence of microheterogeneity derived from various other measurements.

The volume-corrected preferential solvation parameters denote the difference between the local composition of the solvation shell around a given molecule and the composition of the bulk binary mixture. The quantity $\delta x_{WW}'$ pertaining to W molecules around a central W molecule and $\delta x_{WS}'$ pertaining to S molecules around this central W one describe the preferred self- or mutual-interactions of the molecules of the two components, noting that $\delta x_{SW}' = -\delta x_{WW}'$ and $\delta x_{SS}' = -\delta x_{WS}'$. Table 3.16 shows in symbolic form the magnitudes of these preferential solvation parameters: the extrema in the $\delta x_{ij}(x_S)$ curves are marked as (++) or (− −) if larger than 0.1, as + or − if larger than 0.01, all in the absolute sense, as zero otherwise, or as −+ if they change sign as the concentration varies. Those mixtures for which $\delta x_{WW}'$ is marked as ++ exhibit at least some microheterogeneity over some of the composition range. Domains in which water is self-associated and the cosolvent is practically excluded exist in mixtures having such bulk compositions.

3.4.2.2 Properties Related to the Ion Solvating Ability

The probes that are useful for measuring properties such as polarity and hydrogen bond and electron pair donicities of neat solvents may also be used for obtaining quantitative values for the aqueous binary mixtures with the cosolvents treated here. However, the *caveat* that was expressed regarding the ability of a single probe to represent the property concerning any arbitrary solute in the neat solvent is even more strongly needed in the case of the mixtures because of the possibility of preferential solvation of the given probe and the solute by one of the components.

The curves of the polarity and donicity indices as functions of the cosolvent mole fractions are generally nonlinear. The deviations ΔY in terms of Equation 3.42, where Y represents the Dimroth-Reichardt $E_T(30)$ and the Kamlet-Taft π^*, α, and β (see Section 3.3.2), then express the properties of the solvent mixture that depend on the self- and mutual-interactions of its molecules. Expressions similar to (3.43) may then be used to fit the experimental ΔY values. The coefficients of this expression for $E_T(30)$ and π^* are shown in Table 3.17 and those for β and α are shown in Table 3.18, the values being adapted from data in compilations by Marcus [56, 87].

TABLE 3.17   Coefficients of Equation 3.43 for the Polarity increments $\Delta Y = \Delta E_T(30)$ with $y_i \equiv e_i$ in Kcal·mol^{-1} (1 cal = 4.148 J) and the Polarity/polarizability Increments $\Delta Y = \Delta \pi^*$ with $y_i \equiv p_i$

Solvent	e_0	e_1	e_2	p_0	p_1	p_2
Methanol	−8.8	−9.4	3.5	0.26	0.21	0.39
Ethanol	−13.6	−12.9	−2.9	−0.11	0.66	1.44
n-Propanol	−13.6	−23.6	−25.2			
i-Propanol	−18.2	−33.6	−13.5	−0.33	−0.37	0.68
t-Butanol	−11.7	−33.7	−28.0	−0.28	−0.96	−0.88
Ethylene glycol	−6.8	−3.8	−3.0			
Tetrahydrofuran	−1.7	−38.5	−50.1	−0.47	−0.45	0.57
1,4-Dioxane	−3.6	−39.4	−15.8	−0.18	−0.01	0.34
Acetone	−1.4	−31.4	−14.8	−0.24	0.12	0.35
Formamide				0.16	0.13	0.18
N,N-Dimethylformamide	−14.9	−20.4	−0.2	0.03	0.38	0.41
N-Methylpyrrolidinone	−19.1	−16.7	−13.9			
Pyridine	−13.2	−27.1	−23.4	0.30	0.35	0.28
Acetonitrile	4.1	−31.8	6.8	−0.29	−0.07	0.52
Dimethylsulfoxide	−16.4	−10.1	−8.1	0.06	0.28	0.26

No data were found for trifluoroethanol, γ-butyrolactone, N-methylformamide, N,N-dimethylacetamide, tetramethylurea, sulfolane, and hexamethyl phosphoric triamide.

TABLE 3.18   Coefficients of Equation 3.43 for the Electron Pair Donicity Increments $\Delta Y = \Delta \beta$ with the $y_i \equiv b_i$ and the Hydrogen Bond Donicity Increments $\Delta Y = \Delta \alpha$ with $y_i \equiv a_i$

Solvent	b_0	b_1	b_2	a_0	a_1	a_2
Methanol	0.42	0.58	0.10	−0.41	−0.61	−0.29
Ethanol	0.24	0.65	0.52[a]	−0.32	−0.40	−0.25
i-Propanol	0.29	0.25	0.80	−0.30	−0.21	−0.59
t-Butanol	0.37	0.85	0.40			
Ethylene glycol	0.23	−0.23	−0.16			
Tetrahydrofuran	0.50	0.60	1.12	−0.40	−1.78	−2.62
1,4-Dioxane	0.70	−0.52	0.09	0.54	−1.18	0.37[b]
Acetone	0.63	0.52	0.18	0.40	−1.22	0.18
Formamide	0.12	0.12	0.02	−0.52	−0.61	−0.58
N,N-Dimethylformamide	0.39	0.17	0.08	−0.89	−1.68	−1.04
Pyridine	0.39	−0.08	0.04	−1.38	−1.41	−1.62
Acetonitrile	0.48	0.33	0.88	0.71	−1.77	0.80
Dimethylsulfoxide	0.24	0.34	0.13	−1.04	−1.51	−1.41
Hexamethyl phosphoric triamide				−1.10	0.10	−0.11

[a] Data from [87] and a term in $b_3 = -0.58$ is needed.
[b] Data from [87] and a further term in $a_3 = -16.4$ is needed.
No data were found for n-propanol, trifluoroethanol, γ-butyrolactone, N-methylformamide, N,N-dimethylacetamide, tetramethylurea, N-methylpyrrolidinone, and sulfolane.

The solvent index $E_T(30)$ depends on both the polarity and the hydrogen bond donicity of the solvent or binary aqueous mixture with a cosolvent. In all the mixtures, except aqueous acetonitrile, this index is smaller than expected for ideal mixtures, as the negative e_0 values indicate for the equimolar mixtures. The polarity/polarizability index π^* may be smaller (for most mixtures) or larger (for aqueous pyridine, formamide, DMF, and DMSO) than expected for ideal mixtures, as the sign of p_0 shows for the equimolar mixtures. The latter option $(p_0 > 0)$ occurs where the cosolvent is aromatic (pyridine) or has the $=O$ group (except acetone) that contributes to the polarizability.

The electron pair donicity index β of the mixtures is generally larger, that is, they are more basic, than expected for ideal ones $(b_0 > 0)$ due to the relatively small value of β for pure water. On the contrary, the hydrogen bond donicity of the mixtures is generally smaller, that is, they are less acidic, than expected for ideal ones $(a_0 < 0)$, except for the two ethers, due the large value of α for water itself. Mixtures that are microheterogeneous over some composition range exhibit a reduced slope of the $Y(x_S)$ curve in this range than in the water- and cosolvent-rich regions.

The acid–base interactions in the binary aqueous mixtures with cosolvents are described by the following equilibria with the equilibrium quotient in parentheses:

$$H_2O + H_2O \rightleftarrows H_3O^+ + OH^- \left(K_w\right)$$

$$HS + HS \rightleftarrows H_2S^+ + S^- \left(K_S\right)$$

$$H_2O + HS \rightleftarrows H_3O^+ + S^- \left(K_{bWS}\right)$$

$$H_2O + HS \rightleftarrows OH^- + H_2S^+ \left(K_{aWS}\right)$$

The relationship between these constants requires that $K_{aS} \cdot K_{bS} = K_w \cdot K_S$, so that only three independent constants are needed. The overall autoprotolysis constant of the mixtures, K_{ap}, according to Fonrodona et al. and Kiliç and Aslan [88, 89] is:

$$K_{ap} = x_w^2 K_w + x_S^2 K_S + x_w x_S \left(K_{aWS} + K_{bWS}\right) \tag{3.52}$$

But the equilibrium quotients themselves depend on the solvent composition: the products are ions, hence the permittivity of the mixtures plays an important role. The negative logarithms of the equilibrium quotients, their pKs, are linear with the mixture composition:

$$pK_w = x_w pK_{w(w)} + x_S pK_{w(S)} \tag{3.53}$$

$$pK_{abWS} = x_w pK_{abWS(w)} + x_S pK_{abWS(S)} \tag{3.54}$$

$$pK_S = x_w pK_{S(w)} + x_S pK_{S(S)} \tag{3.55}$$

Here the solvent subscript in parentheses denotes the neat solvent in which the pK is measured and $K_{abWS} = (K_{aWS} + K_{bWS})$. In addition to $pK_{w(w)}$ and $pK_{S(S)}$ from Table 3.10, the four equilibrium quotients shown in Table 3.19 taken from Roses et al. [90, 91] permit the calculation of pK_{ab} of the aqueous mixtures according to Equations

TABLE 3.19　The Equilibrium Quotients for the Acid–base Interactions According to Equations 3.52–3.55, Adapted from [56]

Solvent	$pK_{W(S)}$	$pK_{abWS(S)}$	$pK_{abWS(W)}$	$pK_{S(W)}$
Methanol	20.82	12.98	14.56	15.09
Ethanol	22.56	13.16	16.92	15.90
n-Propanol	24.52	13.10	17.77	16.10
i-Propanol	25.47	13.27	18.53	
t-Butanol	32.88	12.96	21.25	19.00
Ethylene glycol	19.43	12.52	13.92	15.07
Tetrahydrofuran	28.48	13.09	21.19	14.8
1,4-Dioxane	38.80	12.71	25.96	15.6
Acetone	23.52	13.57	20.28	24.20
Formamide		10.45	13.11	15.2
N-Methylformamide	25.86	12.64	18.57	15.6
N,N-Dimethylformamide	27.44	12.14	22.18	15.6
N,N-Dimethylacetamide	22.52	12.09	21.18	
Dimethylsulfoxide	25.52	13.57	23.3	15.5

No data for trifluoroethanol, γ-butyrolactone, tetramethylurea, N-methylpyrrolidinone, pyridine, acetonitrile. Sulfolane and hexamethyl phosphoric triamide were found.

3.52–3.55. The value for a given aqueous cosolvent mixture may then be compared with $pK_{W(W)}$ and $pK_{S(S)}$ in order to decide whether the mixture is more acidic or basic than either of the components.

REFERENCES

[1] J. A. Riddick, W. B. Bunger, T. K. Sakano, Organic Solvents, Wiley-Interscience, New York, 4th ed. (1986).

[2] DIPPR, The DIPPR Pure Component Data Compilation, Technical Database Services Inc., New York, version 12.4 (1997). Replaced since 2009 by the DIPPR 801 database of the American Institute of Chemical Engineers (AIChE).

[3] Y. Marcus, The Properties of Solvents, Wiley, New York (1998).

[4] Y. Marcus, G. Hefter, J. Solution Chem. **28**, 575 (1999).

[5] M. H. Abraham, J. C. McGowan, Chromatographia **23**, 243 (1987).

[6] Y. Marcus, J. Phys. Chem. **95**, 8886 (1991).

[7] J.-I. Kim, Z. Phys. Chem. (Frankfurt) **113**, 129 (1978).

[8] K.-S. Kim, H. Lee, J. Chem. Eng. Data **47**, 216 (2008).

[9] R. Naejus, D. Lemordant, R. Coudert, P. Willmann, J. Chem. Thermodyn. **29**, 1503 (1997).

[10] K. R. Lindfors, S. H. Opperman, M. E. Glover, J. D. Seese, J. Phys. Chem. **75**, 3313 (1971).

[11] T. Avraam, G. Moumouzias, G. Ritzoulis, J. Chem. Eng. Data **43**, 51 (1998).

[12] J. R. Langan, G. A. Salmon, J. Chem. Eng. Data **32**, 420 (1987).

[13] Y. Uosaki, S. Kitaura, T. Moriyoshi, J. Chem. Eng. Data **49**, 1410 (2004).

[14] Y. Uosaki, S. Kitaura, T. Moriyoshi, Rev. High Press. Sci. Technol. **7**, 1216 (1998).

[15] S. Trasatti, Electrochim. Acta **32**, 843 (1987).

[16] L. I. Krishtalik, N. M. Alpatova, E. V. Ovsyannikova, J. Electroanal. Chem. **329**, 1 (1992).

[17] V. I. Parfenyuk, Coll. J. Russ. Acad. Sci. **66**, 466 (2004).

[18] V. I. Parfenyuk, Russ. J. Electrochem. **44**, 50 (2008).

[19] V. I. Parfenyuk, Coll. J. **64**, 588 (2002).

[20] P. W. Selwood, Magnetochemistry, Interscience, New York, 2nd ed. (1956).

[21] Y. Marcus, J. Phys. Chem. B **109**, 18541 (2005).

[22] Z. Koczorowski, I. Zagorska, A. Kalinska, Electrochim. Acta **34**, 1857 (1989).

[23] S. Görlish, Proceedings of the IVth International Conference Surface Active Substances, Brussels, 52 (1964).

[24] Y. Marcus, J. Chem. Phys. **137**, 154501 (2012).

[25] D. C. Grahame, J. Chem. Phys. **21**, 1054 (1953).

[26] O. Redlich, J. Phys. Chem. **44**, 619 (1940).

[27] Y. Marcus, J. Solution Chem. **25**, 455 (1996).

[28] Y. Marcus, Fluid Phase Equil. **154**, 311 (1999).

[29] Y. Marcus, J. Solution Chem. **21**, 1217 (1992).

[30] Y. Marcus, Chem. Rev. **113**, 6536 (2013).

[31] W. Beckermann, F. Kohler, Int. J. Thermophys. **16**, 455 (1995).

[32] O. V. Dorofeeva, L. V. Gurvich, J. Phys. Chem. Ref. Data **24**, 13, 52 (1995).

[33] J. H. Hildebrand, R. H. Lamoreaux, Proc. Natl. Acad. Sci. USA **69**, 3428 (1972).

[34] Y. Marcus, Ion Solvation, Wiley, Chichester (1985).

[35] Y. Marcus, J. Mol. Liq. **158**, 23 (2011); **166**, 62 (2012).

[36] Y. Marcus, Chem. Soc. Rev. **22**, 409 (1993).

[37] J.-L. M. Abboud, R. Notario, Pure Appl. Chem. **71**, 645 (1999).

[38] C. Reichardt, Solvents and Solvent Effects in Organic Chemistry, VCH, Weinheim, 2nd ed. (1988).

[39] M. J. Kamlet, R. W. Taft, Acta Chem. Scand. B **39**, 611 (1985).

[40] K. Dimroth, C. Reichardt, T. Siepmann, F. Bohlmann, Liebigs Ann. Chem. **661**, 511 (1966).

[41] J. Hormodaly, Y. Marcus, J. Phys. Chem. **83**, 2843 (1979).

[42] J. P. Ceron-Carrasco, D. Jacquemin, C. Laurence, A. Planchart, C. Reichardt, K. Sraidi, J. Phys. Org. Chem. **27**, 512 (2014).

[43] M. J. Kamlet, J.-L. M. Abboud, R. W. Taft, J. Am. Chem. Soc. **99**, 6027 (1977).

[44] C. Laurence, P. Nicolet, M. T. Dalati, J.-L. M. Abboud, J. Phys. Chem. **98**, 5807 (1994).

[45] P. Nicolet, C. Laurence, J. Chem. Soc. Perkin Trans. **2**, 1071 (1986).

[46] V. Gutmann, E. Vychera, Inorg. Nucl. Chem. Lett. **2**, 257 (1966).

[47] Y. Marcus, J. Solution Chem. **13**, 599 (1984).

[48] M. J. Kamlet, R. W. Taft, J. Am. Chem. Soc. **98**, 377, 2886 (1976).

[49] U. Mayer, V. Gutmann, W. Gerger, Monatsh. Chem. **106**, 1235 (1975).

[50] H. Schneider, Y. Badrieh, Y. Migron, Y. Marcus, Z. Phys. Chem. **177**, 143 (1993).

[51] Y. Marcus, J. Phys. Chem. **91**, 4422 (1987).

[52] Y. Marcus, G. Hefter, T. Chen, J. Solution Chem. **29**, 201 (2000).

[53] S. G. Lias, J. E. Bartmess, J. F. Leibman, J. L. Holmes, R. D. Levin, and G. W. Mallard, J. Phys. Chem. Ref. Data **17**, Suppl. 1 (1988).

[54] J. Jiskra, H. A. Claessens, C. A. Cramrs, R. Kaliszan, J. Chromatogr. A **977**, 193 (2002).

[55] Z. Wu, B. Ondruschka, J. Phys. Chem. A **109**, 6521 (2005).

[56] Y. Marcus, Solvent Mixtures, Dekker, New York (2002).

[57] E. Tommila, E. Lindell, M.-L. Virtalaine, R. Laakso, Suom. Kemist. **42**, 95 (1969).

[58] M. Denda, H. Touhara, K. J. Nakanishi, J. Chem. Thermodyn. **19**, 539 (1987).

[59] D. H. S. Ramkumar, A. P. Kudchadker, D. D. Deshpande, J. Chem. Eng. Data **30**, 491 (1985).

[60] A. M. Zaichikov, Y. G. Bushuev, Zh. Fiz. Khim. **69**, 1942 (1995).

[61] A. Cooney, K. W. Morcom, J. Chem. Thermodyn. **20**, 735 (1988).

[62] S. R. Goodwin, D. T. M. Newsham, J. Chem. Thermodyn. **3**, 325 (1971).

[63] A. Apelblat, Ber. Bunsenges. Phys. Chem. **94**, 1128 (1990).

[64] M. Mundhwa, S. Elmahmudi, Y. Maham, A. Henni, J. Chem. Eng. Data **54**, 2895 (2009).

[65] P. Scharlin, K. Steinby, J. Chem. Thermodyn. **35**, 279 (2003).

[66] R. F. Checoni, P. L. O. Volpe, J. Solution Chem. **39**, 259 (2010).

[67] G. C. Benson, P. J. D'Arcy, J. Chem. Eng. Data **27**, 439 (1982).

[68] G. Jakli, W. A. Van Hook, J. Chem. Eng. Data **41**, 249 (1996).

[69] D. H. S. Ramkumar, A. P. Kudchadker, J. Chem. Eng. Data **34**, 459 (1989).

[70] J.-I. Abe, K. Nakanishi, H. Touhara, J. Chem. Thermodyn. **10**, 483 (1978).

[71] C. H. Rochester, J. R. Symonds, J. Fluor. Chem. **4**, 141 (1974).

[72] G. C. Benson, O. Kiyohara, J. Solution Chem. **9**, 791 (1980).

[73] G. Vazquez, E. Alvarez, J. M. Navaza, J. Chem. Eng. Data **40**, 611 (1995).

[74] S. Z. Mikhail, W. R. Kimel, J. Chem. Eng. Data **8**, 323 (1963).

[75] D. H. S. Ramkumar, A. P. Kudchadker, J. Chem. Eng. Data **34**, 463 (1989).

[76] G. Gente, C. La Mesa, J. Solution Chem. **29**, 1159 (2000).

[77] A. Kundu, N. Kishore, Biophys. Chem. **109**, 427 (2004).

[78] B. Garcia, R. Alcalde, J. M. Leal, J. S. Matos, J. Phys. Chem. B. **101**, 7991 (1997).

[79] M. N. Islam, M. A. Ali, M. M. Islam, M. K. Nahar, Phys. Chem. Liq. **41**, 271 (2003).

[80] L. A. Tsvetkova, E. A. Indeikin, N. G. Savinskii, O. M. Koroleva, Zh. Nauch. Priklad. Fotogr. **44**, 28 (1999).

[81] M. R. J. Dack, Austr. J. Chem. **29**, 771 (1976).

[82] G. Conti, E. Matteoli, Z. Phys. Chem. (Leipzig) **262**, 433 (1981).

[83] A. M. Zaichikov, J. Struct. Chem. **48**, 94 (2007).

[84] A. M. Zaichikov, M. A. Krestwyanikov, J. Struct. Chem. **49**, 285 (2008).

[85] Y. Marcus, in P. E. Smith, J. P. O'Connell, and E. Matteoli, eds., Fluctuation Theory of Solutions: Applications in Chemistry, Chemical Engineering, and Biophysics, CRC Press, Boca Raton, FL, 65 (2013).

[86] Y. Marcus, Monatsh. Chem. **132**, 1387 (2001).

[87] Y. Marcus, J. Chem. Soc. Perkin Trans. **2**, 1751 (1994).

[88] G. Fonrodona, C. Rafols, E. Bosch, M. Roses, Anal. Chim. Acta **335**, 291 (1996).

[89] E. Kiliç, N. Aslan, Microchim. Acta **151**, 89 (2005).

[90] M. Roses, C. Rafols, E. Bosch, Anal. Chem. **65**, 2294 (1993).

[91] C. Rafols, E. Bosch, M. Roses, A. G. Asuero, Anal. Chim. Acta **302**, 355 (1995).

4

ION SOLVATION IN NEAT SOLVENTS

4.1 THE SOLVATION PROCESS

It is impossible to follow by measurements an actual experimental process of the transfer of a single ion from its isolated state in the gas phase to its fully solvated state in a solution. Such a process, however, may be dealt with as a thought process and theoretical considerations may be applied to it.

Ben-Naim and Marcus [1] discussed a process termed "solvation" that applies to a particle of a (non-ionic[1]) substance transferring from its isolated state in the gas phase into a liquid irrespective of the concentration. The particle would then be surrounded by solvent molecules only in an ideally dilute solution (infinite dilution), or by solvent molecules as well as by molecules of its own kind at any arbitrary mole fraction with regard to the solvent, and by molecules identical with itself only on condensation into its own liquid. The interactions involved and their thermodynamics are all covered by the same concept of solvation. The solvation process of a solute S is defined [1] as the transfer of a particle of S from a fixed position in the (ideal) gas phase (superscript G) to a fixed position in a liquid (superscript L) at a given temperature T and pressure P. Statistical mechanics specifies the chemical potential of S in the ideal gas phase as:

$$\mu_S^G = k_B T \ln q_S^{-1} + k_B T \ln \rho_S^G \Lambda_S^3 = \mu_S^{*G} + k_B T \ln \rho_S^G \Lambda_S^3 \tag{4.1}$$

[1]The complications involved with charged particles are dealt with in Section 4.2.1.2.

Ions in Solution and Their Solvation, First Edition. Yizhak Marcus.
© 2015 John Wiley & Sons, Inc. Published 2015 by John Wiley & Sons, Inc.

and similarly in the liquid phase as:

$$\mu_S^L = k_B T \ln q_S^{-1} + k_B T \ln \rho_S^L \Lambda_S^3 = \mu_S^{*L} + k_B T \ln \rho_S^G \Lambda_S^3 \tag{4.2}$$

Here q_S is the internal partition function of S ($\equiv 1$ for a structureless particle, such as a noble gas atom), ρ_S is its number density, and Λ_S^3 is its momentum partition function. The quantity μ_S^* in either phase pertains to the particle at a fixed position where it is devoid of the translational degrees of freedom, and $k_B T \ln \rho_S \Lambda_S^3$ expresses its liberation from this constraint. The Gibbs energy of solvation, is expressed as the change in the chemical potential of S at equilibrium, where $\mu_S^L - \mu_S^G$, is then:

$$\Delta \mu_S^* = \mu_S^{*L} - \mu_S^{*G} = k_B T \ln \left(\frac{\rho_S^G}{\rho_S^L} \right)_{eq} \tag{4.3}$$

The quantity $\Delta \mu_S^*$ is called also the coupling work of the solute S with its surroundings. If the internal partition function q_S is modified by the presence of the solvent in the liquid, then these changes are absorbed into $\Delta \mu_S^*$ and are part of the solvation interactions measured by it. The molar Gibbs energy of solvation is $\Delta G_S^* = N_A \Delta \mu_S^*$ and the corresponding molar entropy and enthalpy of solvation are:

$$\Delta S_S^* = -N_A \left(\frac{\partial \Delta \mu_S^*}{\partial T} \right)_P \tag{4.4}$$

$$\Delta H_S^* = \Delta G_S^* + T \Delta S_S^* \tag{4.5}$$

It is important now to point out that these thermodynamic quantities of solvation are *not* the same as the *standard* molar quantities. These pertain to the transfer of S (possibly an ion but irrespective of it being charged) from the ideal gas phase at $P° = 0.1 \, MPa$ to the aqueous solution at a standard state of $1 \, mol \cdot dm^{-3}$ but with hypothetical properties of a solution at infinite dilution. Such quantities, expressed by the general symbol $\Delta_h Y^\infty$ when the liquid is water and the process is hydration, include contributions from the change in the space available to the solute that are extraneous to the solvation (interactions with surrounding) process proper. The differences between $\Delta_h Y^*$ and $\Delta_h Y^\infty$ are fairly small compared with the absolute (ionic) values and in the cases of the Gibbs energy and enthalpy near the uncertainties of the values, but still ought to be noted, as follows [2]:

$$\Delta_h G_S^* = \Delta_h G^\infty + RT \ln \left(\frac{RT}{0.001 P°} \right) = \Delta_h G^\infty + 7.93 \ \text{kJ} \cdot \text{mol}^{-1} \tag{4.6}$$

$$\Delta_h H_S^* = \Delta_h H^\infty + RT \left(1 - \alpha_{pw} T \right) = \Delta_h H^\infty + 2.29 \ \text{kJ} \cdot \text{mol}^{-1} \tag{4.7}$$

$$\Delta_h S_S^* = \Delta_h S^\infty + R \left[1 - \alpha_{pw} T - \ln \left(\frac{RT}{0.001 P°} \right) \right] = \Delta_h S^\infty - 18.9 \ \text{J} \cdot \text{K}^{-1} \cdot \text{mol}^{-1} \tag{4.8}$$

The numerical values pertain to 25°C and the adjustments are independent of the solute, whether an ion or an uncharged species. These differences are not generally acknowledged and the standard quantities $\Delta_h Y^\infty$ are generally employed for discussion of the solvation of the solutes instead of the proper ones, $\Delta_h Y^*$.

4.2 THERMODYNAMICS OF ION HYDRATION

Conventional individual ionic standard molar thermodynamic quantities of hydration, $\Delta_h Y_I^{\infty\text{conv}}$, are obtained from measured values for complete electrolytes $\Delta_h Y^\infty \left(C_p^{z+} A_q^{z-} \right)$, and are based on assigning the value zero to that of the hydrogen ion, $\Delta_h Y^{\infty\text{conv}} (H^+) = 0$, at all temperatures. Conventional values for anions are obtained from the measured values for acids: $\Delta_h Y^{\infty\text{conv}} (A^{z-}) = \Delta_h Y^\infty (H_{z-}A^{z-})$. With these, the conventional values for cations are obtained as $\Delta_h Y^{\infty\text{conv}} \left(C^{z+} \right) = \left[\Delta_h Y^\infty \left(C_p^{z+} A_q^{z-} \right) - q\Delta_h Y^{\infty\text{conv}} \left(A^{z-} \right) \right] / p$, from the measured values for salts.

Absolute values for ions are obtained from the conventional ones when an absolute value is assigned to the hydrogen ion: $\Delta_h Y^\infty (C^{z+}) = \Delta_h Y^{\infty\text{conv}} (C^{z+}) - z_+ \Delta_h Y^\infty (H^+)$ for cations and $\Delta_h Y^\infty (A^{z-}) = \Delta_h Y^{\infty\text{conv}} (A^{z-}) + z_- \Delta_h Y^\infty (H^+)$ for anions.

4.2.1 Gibbs Energies of Ion Hydration

A method to obtain the absolute values of $\Delta_h G^\infty (H^+)$ has recently been proposed by Kelly et al. [3, 4], based on the cluster pair approximation of Tuttle and Malaxos [5]. The operative expression for water as the solvent is:

$$\frac{1}{2}\left[\Delta_h G^{*\text{con}} \left(C^+ \right) - \Delta_h G^{*\text{con}} \left(A^- \right) \right] + \frac{1}{2}\left[\Delta_{\text{clust}} G^\circ{}_n \left(C^+ \right) - \Delta_{\text{clust}} G^\circ{}_n \left(A^- \right) \right]$$
$$= \frac{1}{2}\left[\Delta_{\text{clust}} G^\circ{}_{n+1,\infty} \left(C^+ \right) - \Delta_{\text{clust}} G^\circ{}_{n+1,\infty} \left(A^- \right) \right] + \Delta_h G^* \left(H^+ \right) \quad (4.9)$$

The first couple of terms on the left-hand side are the conventional standard molar Gibbs energies of hydration of pairs of (univalent) cations C^+ and anions A^-. The superscript * for the standard state means that the same concentration, 1 mol·dm^{-3}, is used for the gas and solution phases, ensuring that the derived quantities pertain solely to the interaction of the ion with its surrounding solvent and does not involve the compression of the gaseous species (Section 4.1, where $k_B T \ln \left(\Lambda_S^{3\,G} / \Lambda_S^{3\,L} \right)$ replaces the right-hand side of Eq. 4.3). The second couple of terms on the left-hand side are the sums of the stepwise clustering Gibbs energies of the cation and anion with $n \leq 6$ water molecules in the gas phase, also experimentally or computationally available quantities (Section 2.1.2). The essence of the cluster pair approximation is to find pairs of a cation C^+ and an anion A^- for which the difference of the first couple of terms on the right-hand side vanishes: $\Delta_{\text{clust}} G^\circ{}_{n+1,\infty} (C^+) - \Delta_{\text{clust}} G^\circ{}_{n+1,\infty} (A^-) = 0$. This procedure, then, leads to the value $\Delta_h G^* (H^+) = -1113 \pm 8$ kJ·mol^{-1} at 25°C or $\Delta_h G^\infty (H^+) = -1121 \pm 8$ kJ·mol^{-1} according to Equation 4.6.

However, this excessively negative value, $\Delta_h G^\infty (H^+) = -1121 \pm 8 kJ \cdot mol^{-1}$, is not consistent with the preferred quantity derived from the enthalpy and entropy of hydration at 25°C (Sections 4.2.2 and 4.2.3): $\Delta_h G^\infty (H^+) = \Delta_h H^\infty (H^+) - 298.15\Delta_h S^\infty (H^+) = -1103 - 0.29815(-131) = -1064 \pm 6 kJ \cdot mol^{-1}$ This value is compatible with the value $-1059 \pm 5 kJ \cdot mol^{-1}$ from the "real" electrode potentials obtained by Gomer and Tyson [6] (Section 4.2.1.2) and $-1066 \pm 17 kJ \cdot mol^{-1}$ resulting from a critical examination of methods for obtaining individual ionic thermodynamic quantities by Conway [7]. These values, or a slightly less negative value $-1056 \pm 6 kJ \cdot mol^{-1}$ but in agreement with them within their uncertainties, have been employed by Marcus [7, 8] in the calculation of the standard molar Gibbs energies of hydration of ions according to the following considerations.

4.2.1.1 Accommodation of the Ion in a Cavity   A particle transferred from its isolated state in the gas phase to a solution requires a free space in the solution to accommodate it. This pertains to a neutral species as well as to an ion. A cavity in the water having an effective radius r_{eff} is formed in which the particle can be accommodated and allowed to interact with its surroundings. There is then a positive contribution to the Gibbs energy of hydration of the particle, resulting from the work done against the cohesive forces of the solvent for the creation of the cavity. This quantity, $\Delta_h G_{neut}$, is obtained experimentally according to Abraham and Liszi [9] for a noble gas atom or some inert molecule such as methane as:

$$\Delta_h G_{neut} / kJ \cdot mol^{-1} = 41 - 87 r_{eff} \tag{4.10}$$

In the case that the particle is an ion, the effective radius pertains to it with its first hydration shell of immobilized water molecules $r_{eff} = r_I + \Delta r$, where r_I is the crystal ionic radius and Δr is the width of the hydration shell. In this shell, $n = 0.36 \, |z|/r_I$ water molecules of diameter $d_w = 0.276$ nm are immobilized as suggested by Marcus [8]. Hence the addend Δr is obtained from the volume of this hydration shell and the number of water molecules immobilized:

$$\Delta r = \left[r_I^3 + 0.36 \, |z| \, d_w^3 / 8 r_I \right]^{1/3} - r_I \tag{4.11}$$

4.2.1.2 Electrostatic Interactions   The thought process described above for the solvation of solutes is as follows: their transfer from a fixed position in the ideal gas phase to a fixed position in the liquid solvent works very well for neutral solutes S but cannot be applied simply to ions $I^{\pm z}$. For a single ion $I^{\pm z}$, such a process involves the passage of the ion through the gas-solvent interface and is connected with not well-defined electrostatic consequences. Once the individual ion is in solution, its properties depend on its location with respect to the surface and to the walls of the vessel, due to its long-range electric field.

To circumvent this difficulty, the Born process has traditionally been used for the estimation of individual ionic standard molar Gibbs energy of ion hydration, $\Delta_h G_I^\infty$, dealing, however, only with the direct electrostatic effects involved. It results from the following idealized process. An isolated ion in the gaseous phase, $I^{z\pm}$ (g), is discharged to produce a neutral particle, $I(g)$. Its electric self energy, E_{self} $(I^{z\pm}, g)$, Equation 2.2, must be provided for this stage of the process. The neutral particle $I(g)$ is then transferred into the bulk of liquid water, this stage occurring without any electrostatic energetic component for crossing the gas/liquid boundary. Then the neutral particle is charged up to the original value, producing the infinitely dilute aqueous ion, $I^{z\pm}$ $(aq)^{\infty}$. This charging depends on the permittivity of the water $\varepsilon = 4\pi\varepsilon_0\varepsilon$, where ε is the temperature- and pressure-dependent relative permittivity ($\varepsilon = 78.4$ at 298.15 K and ambient pressure). The electrostatic energy of interaction with the surrounding water is thereby released in this stage. The net effect of this idealized process representing the hydration of the ion is:

$$\Delta_h G_I^{\infty}\left(\text{Born}\right) = -\left(\frac{N_A e^2}{4\pi\varepsilon_0}\right)z_I^2 r_I^{-1}\left(1-\frac{1}{\varepsilon}\right) \tag{4.12}$$

This quantity depends on the square of the charge number of the ion and is inversely proportional to its radius r_I. In the case of ion hydration, the large relative permittivity of water causes the term $(1-1/\varepsilon) = 0.987$ at ambient conditions to be inconsequential.

There are problems with this mode of calculation of $\Delta_h G_I^{\infty}$ for individual ions. The calculated values, summed for a cation and an anion producing an electrolyte, do not agree with the experimental value for the electrolyte. One problem is the use of the same value of the radius for the isolated ion and for that in the aqueous solution, the crystal ionic radius r_I obtained experimentally from diffraction data (Section 2.2), criticized by Stokes [10]. For the isolated ion, the radius is an ill-defined quantity (see the discussion leading to Eq. 2.2). Another problem, pointed out by Stokes and by Noyes [10, 11], is the use of the relative permittivity of pure water for the description of the electrostatic interaction of the ion with its immediate surroundings, where dielectric saturation, due to the high electric field of the ion, occurs (see the discussion following Eq. 2.14).

Various schemes have been proposed to correct these problems. The scheme according to Marcus [8] involves the splitting of the ion hydration process into two spatial regions: one adjacent to the ion, where dielectric saturation occurs and $\varepsilon \approx n_D^2$ (the square of the refractive index), and the other beyond this, where the bulk value ε prevails (see Fig. 4.1). Furthermore, the addition of the empirical quantity Δr (Section 4.2.1.1) to the ionic radius meets the first objection mentioned above.

The electrostatic effect of the ion hydration in the hydration shell requires replacement of $r_I^{-1}(1-1/\varepsilon_r)$ in the Born expression, Equation 4.12, by $[\Delta r/r_I(r_I+\Delta r)](1-n_D^2)$:

$$\Delta_h G_{I\,el\text{-shell}}^{\infty} = -\left(\frac{N_A e^2}{4\pi\varepsilon_0}\right)z_I^2\left[\frac{\Delta r}{r_I\left(r_I+\Delta r\right)}\right]\left(1-n_D^2\right) \tag{4.13}$$

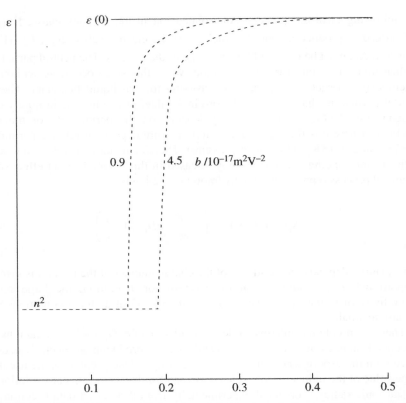

FIGURE 4.1 The relative permittivity of a solvent having a bulk permittivity $\varepsilon(0)$, a high-field value of the permittivity of n^2 (the square of its refractive index), and b coefficient of the field dependence of the permittivity (Eq. 3.16) as indicated, as a function of the distance r/nm from a point charge producing the electric field (From Ref. 12, with permission from the publisher, Wiley).

The modification of the ionic radius applies to the electrostatic effect of the ion hydration in the bulk water region beyond the hydration shell and for this region $(r_I + \Delta r)^{-1}$ should replace r_I^{-1} in Equation 4.12:

$$\Delta_h G_{I\,el\text{-}bulk}^\infty = -\left(\frac{N_A e^2}{4\pi\varepsilon_0}\right) z_I^2 \left(r_I + \Delta r\right)^{-1}\left(1 - \frac{1}{\varepsilon_r}\right) \qquad (4.14)$$

Estimation of reasonably correct $\Delta_h G_I^\infty$ values [8] results from the sum of the electrostatic terms, Equations 4.13 and 4.14, and the cavity formation term, Equation 4.10. Such values are shown in Table 4.1. It should be noted that the ionic standard molar Gibbs energies of hydration need to be compatible with the corresponding enthalpies and entropies of hydration (see the following text) according to $\Delta_h G_I^\infty = \Delta_h H_I^\infty - T\Delta_h S_I^\infty$. The latter quantities are more directly available from experimental data, so that values adopted from Ref. 8 are presented in Table 4.1

TABLE 4.1 Standard Molar Enthalpies of Hydration [13] of Ions, and the Corresponding Entropies [13] and Gibbs Energies of Hydration at 25°C

Ion	$-\Delta_h H_I^\infty/\text{kJ}\cdot\text{mol}^{-1}$	$-\Delta_h S_I^\infty/\text{J}\cdot\text{K}^{-1}\cdot\text{mol}^{-1}$	$-\Delta_h G_I^\infty/\text{kJ}\cdot\text{mol}^{-1}$
H^+	1103	131.1	1064
Li^+	531	141.8	489
Na^+	416	111.2	383
K^+	334	74.3	312
Rb^+	308	65.1	289
Cs^+	283	58.6	266
Cu^+	585	142.7	542
Ag^+	483	116.9	448
Au^+	587		
Tl^+	339	72.0	318
H_3O^+	231	215.0[a]	(167)
NH_4^+	329	111.6	301
$C(NH_2)_3^+$	170[b]	102	(140)
Me_4N^+	209	144.1	(166)
Et_4N^+	193	222	(127)
Pr_4N^+	210	327	(113)
Bu_4N^+	214	418[c]	(89)
Ph_4P^+	47		
Ph_4As^+	47	302	−43
Be^{2+}	2510	310.4	2417
Mg^{2+}	1949	331.2	1837
Ca^{2+}	1602	252.4	1527
Sr^{2+}	1470	241.7	1398
Ba^{2+}	1332	208.2	1270
Ra^{2+}	1321	167.0	(1271)
V^{2+}	1690	296.2	(1602)
Cr^{2+}	1933	308.1	1841
Mn^{2+}	1874	291.8	1787
Fe^{2+}	1972	362.4	(1864)
Co^{2+}	2036	336.9	(1936)
Ni^{2+}	2119	351.4	(2014)
Cu^{2+}	2123	320.0	2028
Zn^{2+}	2070	317.6	1975
Pd^{2+}	2054	324	(1957)
Cd^{2+}	1833	285.4	1748
Sn^{2+}	1577	245.1	(1504)
Sm^{2+}	1464	253.5	(1388)
Eu^{2+}	1481	241.3	1409
Yb^{2+}	1602	264.6	(1523)
Pt^{2+}	2064[d]	318.0	(1969)
Hg^{2+}	1853	251.7	1778
Hg_2^{2+}		251.9	
Pb^{2+}	1572	209.4	(1510)
UO_2^{2+}	1363	400.5	(1244)

(*continued*)

TABLE 4.1 (Continued)

Ion	$-\Delta_h H_1^\infty$/kJ·mol^{-1}	$-\Delta_h S_1^\infty$/J·K^{-1}·mol^{-1}	$-\Delta_h G_1^\infty$/kJ·mol^{-1}
Al^{3+}	4715	538.5	4554
Sc^{3+}	3967	478.0	3824
V^{3+}	4450	551.8	(4285)
Cr^{3+}	4670	514.6	(4517)
Fe^{3+}	4462	556.1	(4296)
Co^{3+}	4691	551	(4527)
Ga^{3+}	4709	559.5	(4542)
Y^{3+}	3594	482.5	(3450)
In^{3+}	4127	385.7	4012
Sb^{3+}	3840	506	(3689)
La^{3+}	3312	454.7	3176
Ce^{3+}	3367	457.1	3231
Pr^{3+}	3411	464.5	3273
Nd^{3+}	3447	464.9	3308
Pm^{3+}	3467	467.6	3328
Sm^{3+}	3492	467.3	3353
Eu^{3+}	3535	469.6	3395
Gd^{3+}	3549	461.8	3411
Tb^{3+}	3580	486.1	3435
Dy^{3+}	3604	492.9	3457
Ho^{3+}	3639	489.6	(3493)
Er^{3+}	3674	506.5	3523
Tm^{3+}	3695	503.9	3545
Yb^{3+}	3742	495.1	3494
Lu^{3+}	3695	504.0	3545
Tl^{3+}	4125	434.0	(3996)
Bi^{3+}	3626	394.3	3508
Ac^{3+}	3307	428	(3179)
U^{3+}	3437	438.6	(3306)
Am^{3+}	3529	447.9	(3395)
Cm^{3+}	3571	449.5	(3437)
Zr^{4+}	6991	763.3	(6773)
Sn^{4+}	7656	374.3	(7544)
Ce^{4+}	6327	560.4	6160
Hf^{4+}	7159	728.1	(6942)
Th^{4+}	6057	688.3	5852
U^{4+}	6572	689.2	(6367)
Np^{4+}	6652	666.5	(6453)
Pu^{4+}	6769	668.5	(6570)
F$^-$	510	137.2	469
Cl$^-$	367	75.7	344
Br$^-$	336	58.8	318
I$^-$	291	35.9	280
OH$^-$	520	160.8	472
SH$^-$	340	98.0	(311)
ClO$^-$	432	152	387
BrO$^-$		163	
IO$^-$	493		
CN$^-$	346	80.4	(322)

TABLE 4.1   (Continued)

Ion	$-\Delta_h H_I^\infty/\text{kJ·mol}^{-1}$	$-\Delta_h S_I^\infty/\text{J·K}^{-1}\text{·mol}^{-1}$	$-\Delta_h G_I^\infty/\text{kJ·mol}^{-1}$
NCO^-	399	90.0	(372)
SCN^-	311	66.0	(291)
N_3^-	302	82.1	(278)
HF_2^-	642	96.6	(613)
HO_2^-	(499)	182.6	445
I_3^-	173	73.2	(151)
BO_2^-	538	230.8	469
ClO_2^-	475	133.5	435
NO_2^-	412	93.7	(384)
NO_3^-	312	86.4	(286)
ClO_3^-	299	79.8	(275)
BrO_3^-	376	94.8	(348)
IO_3^-	450	147.6	406
ClO_4^-	246	56.8	(229)
MnO_4^-	250	64.4	231
TcO_4^-		68.8	271[c]
ReO_4^-	244	60.6	226
BF_4^-	227	65.7	(207)
HCO_2^-	432	124.0	(395)
$CH_3CO_2^-$	425	169.4	(374)
BPh_4^-	47	308.2	−(45)
HCO_3^-	384	137.3	343
HSO_4^-	368	129.0	330
$H_2PO_4^-$	522	166.0	473
S^{2-}	1348	122.3	(1312)
CO_3^{2-}	1397	245.2	(1324)
$C_2O_4^{2-}$	1261	205.1	(1200)
SO_3^{2-}	1376	249.3	(1302)
SO_4^{2-}	1035	200.4	(975)
SeO_4^{2-}	964	182.8	(909)
CrO_4^{2-}	1012	186.8	(956)
MoO_4^{2-}		219.5	1404[e]
WO_4^{2-}		211.6	1406[e]
$S_2O_3^{2-}$		179.7	
SiF_6^{2-}	981	143.3	(938)
$Cr_2O_7^{2-}$		73.4	
HPO_4^{2-}		272.1	1366[e]
PO_4^{3-}	2879	421.8	(2753)
AsO_4^{3-}		385.2	3033[e]
$Fe(CN)_6^{3-}$		154.7	2367[e]
$Co(CN)_6^{3-}$		165.6	
$Fe(CN)_6^{4-}$		286.0	4944[e]

[a] From Ref. 14.

[b] From Ref. 15.

[c] From Ref. 16.

[d] The $\Delta_h H^\infty$ was calculated in Ref. 13 from a semi-empirical equation found in Ref. 17, which with $\Delta_f H(Pt^{2+}, g)$ yielded $\Delta_f(Pt^{2+}, aq)$.

[e] Estimated from anion extraction data [18], which differ for univalent anions by −30 to 30 kJ·mol from the values shown here, an uncertainty to be multiplied by the charge number of the anion for multivalent anions.

as they are (with corrections to rounding errors up to $\pm 2\,kJ\cdot mol^{-1}$) if compatible as stipulates, but otherwise reported within parentheses as calculated from the enthalpies and entropies.

As an alternative, "real" standard molar Gibbs energies of hydration of individual ions, $\Delta_h G_I^{\infty R}$, are obtained from the experimentally measurable electromotive force of a suitable cell. This consists of a downward flowing jet of aqueous solution in the center of a vertical tube and of another solution along the inner surface of the tube concentric with the jet, with a small air (vapor) gap between them. The "real" $\Delta_h G_I^{\infty R}$ values differ from the thermodynamic ones $\Delta_h G_I^{\infty}$ by a quantity depending on the surface potentials of the aqueous solutions, $\Delta\chi$ (Section 3.2.3):

$$\Delta_h G_I^{\infty R} = \Delta_h G_I^{\infty} + z_I F \Delta\chi \qquad (4.15)$$

The algebraic value of the ionic charge z_I is to be used and $F = 96485.3\,C\cdot mol^{-1}$ is Faraday's constant. The surface potential pertains to the crossing of the ion through the vapor/liquid boundary mentioned in the thought process described at the beginning of this section. The value of $\Delta\chi$ is afflicted by considerable uncertainties pointed out by Parfenyuk [19] and has a probable error of about $\pm 0.05\,V$ (causing a probable error of about $\pm 5\,kJ\cdot mol^{-1}$ in the resulting $\Delta_h G_I^{\infty}$ for a univalent ion). This makes the use of the measurable "real" standard molar Gibbs energies of hydration unattractive for obtaining individual ionic values of $\Delta_h G_I^{\infty}$ [20, 21].

4.2.2 Entropies of Ion Hydration

The standard molar entropy of hydration of an ion is $\Delta_h S_I^{\infty} = S_I^{\infty} - S_I^{\circ}$, the difference between its standard molar entropy in the aqueous solution (Table 2.8) and the standard molar entropy of the isolated ion in the ideal gas phase (Table 2.3). The latter, S_I°, are calculated from the third law of thermodynamics and spectroscopic data without invoking any extra-thermodynamic assumptions. The former, S_I^{∞}, do involve the assumptions leading to $\Delta S^{\infty}(H^+, aq) = -22.2 \pm 2\,J\cdot K^{-1}\cdot mol^{-1}$ for the hydrogen ion (Section 2.3.1.2). With $S^{\circ}(H^+, g) = 108.9\,J\cdot K^{-1}\cdot mol^{-1}$, the standard molar entropy of hydration of the hydrogen ion is then $\Delta_h S^{\infty}(H^+) = -22.2 \pm 2 - 108.9 = -131.1 \pm 2\,J\cdot K^{-1}\cdot mol^{-1}$. The standard molar entropies of hydration of ions are shown in Table 4.1, derived from $\Delta_h S_I^{\infty} = S_I^{\infty} - S_I^{\circ}$ but also obtainable from the conventional values by use of the absolute value of the hydrogen ion. They are related to the effect that ions have on the structure of water according to various approaches. This aspect is fully dealt with in Section 5.1.1.7.

4.2.3 Enthalpies of Ion Hydration

A widely accepted estimate of the standard molar enthalpy of hydration of the hydrogen $\Delta_h H^{\infty}(H^+)$ ion is based on the differences $\Delta\Delta_h H^{\infty conv}$ between the conventional values of cations and anions having the same radii $r_I = r_+ = r_-$. The values of $\Delta_h H^{\infty conv}(C^+, r_+)$ and $\Delta_h H^{\infty conv}(A^-, r_-)$ are plotted against $(r_I + d_W)^{-3}$ and the differences $\Delta\Delta_h H^{\infty conv}$ are taken for given values of r_I. When these differences are extrapolation to a zero value of $(r_I + d_W)^{-3}$ (infinite ion sizes), $2\Delta_h H^\infty (H^+)$ is obtained as the intercept on the ordinate according to Halliwell and Nyburg [22]. The resulting value $\Delta_h H^\infty (H^+) = -1091 \pm 10 \ kJ \cdot mol^{-1}$ from this plot is the absolute enthalpy of hydration of the hydrogen ion.

As an alternative, consider the standard molar enthalpy of hydration of tetraphenylphosphonium tetraphenylborate, $\Delta_h H^\infty$ (TPTB). This is obtainable from the molar heat of solution of this salt extrapolated to infinite dilution and its standard molar enthalpy of formation. The values of $\Delta_h H_I^\infty$ depend strongly on the electrostatic interactions of the ions with the solvent and water, and these diminish considerably as the sizes of the ions increase. The ionic enthalpy of hydration for the bulky tetraphenyl ions is therefore small compared with those of ordinary, small ions. Therefore, the uncertainty involved in equating the values for these very similar reference ions, $\Delta_h H^\infty (Ph_4 P^+) = \Delta_h H^\infty (BPh_4^-) = 47 \pm 5 \ kJ \cdot mol^{-1}$, is also small, as shown by Marcus [13]. This equality leads from the conventional values to the absolute value for the hydrogen ion, based on the additivity of the standard values: $\Delta_h H^\infty (H^+) = -1103 \pm 7 \ kJ \cdot mol^{-1}$. This value is compatible within the uncertainties with $-1091 \pm 10 \ kJ \cdot mol^{-1}$ mentioned above and $-1104 \pm 17 \ kJ \cdot mol^{-1}$ resulting from a critical examination by Conway of methods for obtaining individual ionic thermodynamic quantities [7]. The resulting values for ions are shown in Table 4.1.

4.3 TRANSFER THERMODYNAMICS INTO NONAQUEOUS SOLVENTS

Having established in the previous sections the thermodynamics of ion hydration, attention may now be turned to the thermodynamics of the solvation of the ions in nonaqueous solvents. As mentioned before, it is expedient to deal not with the solvation process of ions in the ideal gas phase going into the nonaqueous solvents, but to deal with their transfer from water (W) as the source solvent to the nonaqueous solvent as the target solvent (S):

$$\Delta_t Y^\infty (I^{z\pm}, W \to S) = \Delta_{solv} Y (I^{z\pm}, S) - \Delta_h Y^\infty (I^{z\pm}, W) \qquad (4.16)$$

The transfer thermodynamics can be measured directly in many cases and more accurately than the solvation thermodynamics proper, and $\Delta_t Y^\infty (I^{z\pm})$ is then a fairly small difference between two large numbers. If required, $\Delta_{solv} Y (I^{z\pm}, S) = \Delta_h Y^\infty (I^{z\pm}, W) + \Delta_t Y^\infty (I^{z\pm}, W \to S)$ can be readily reconstituted.

4.3.1 Selection of an Extra-Thermodynamic Assumption

The individual ionic thermodynamic quantities for hydration, listed in Table 4.1, are based on two extra-thermodynamic assumptions: for the enthalpies, they are based on the tetraphenylphosphonium tetraphenylborate (TPTB), $\Delta_h H^\infty (Ph_4 P^+) = \Delta_h H^\infty (BPh_4^-)$, and for the entropies on the temperature derivative of the electromotive force of thermocells or of the potential of a mercury electrode at the point of zero charge (Section 2.3.1.2). The latter determined the value for the hydrogen ion, $S^{\infty}(H^+, aq)$, hence from $\Delta_h S_I^{\infty} = S_I^{\infty} - S_I^{\circ}$ (Section 4.2.2) for the other ions. The standard molar Gibbs energy of hydration of individual ions is then obtained from $\Delta_h G_I^{\infty} = \Delta_h H_I^{\infty} - T \Delta_h S_I^{\infty}$ but could be calculated also from the cavity formation (Section 4.2.1.1) and electrostatic (Section 4.2.1.2) terms in generally good agreement (values not in parentheses in Table 4.1).

4.3.2 Thermodynamics of Transfer of Ions into Nonaqueous Solvents

4.3.2.1 Gibbs Energies of Transfer Contrary to the indirect estimation of the standard molar Gibbs energy of hydration, the situation differs for the thermodynamics of transfer that are dominated by more direct estimates of $\Delta_t G_I^{\infty} = \Delta_t G^{\infty}$ ($I^{z\pm}$, W $\rightarrow$ S). These are obtained from the electromotive force of suitable double cells, polarography, or solubility measurements. The various extra-thermodynamic assumptions that have been applied for obtaining $\Delta_t G_I^{\infty}$ values were summarized by Marcus [23]. The individual ionic $\Delta_t G_I^{\infty}$ values obtained from such assumptions, of course, lead to the measurable values for the transfer of complete electrolytes if added, weighted by the stoichiometric coefficients.

The use of "real" Gibbs energies of hydration and solvation depends on the difference between the surface potentials of W and of S: $\Delta\Delta\chi = \Delta\chi(S) - \Delta\chi(W)$ (see Eq. 4.15), but each of these surface potentials is known to no better than $\pm 0.1 V$ (Section 3.2.3 and Table 3.6), and hence the applicability of this method is to a few solvents of interest only and results in uncertainties of ca. $\pm 14 kJ \cdot mol^{-1}$ in $\Delta_t G_I^{\infty}$.

A negligible liquid junction potential was assumed when a salt bridge of 0.01 M tetraethylammonium picrate (Et_4NPic) in any of a set of bridge solvents S_{br} is used in a cell such as:

$$Ag | 0.01M \, AgClO_4 \, in \, S_1 \| 0.01M \, Et_4NPic \, in \, S_{br} \| 0.01M \, AgClO_4 \, in \, S_2 | Ag \qquad (4.17)$$

where $\|$ denotes the liquid junction at the salt bridge and $|$ denotes a phase boundary. The same electromotive forces of the cells were obtained according to Alexander et al. [24] within $\pm 0.01 V$ when the bridge solvent S_{br} was MeCN, DMSO, $MeNO_2$, MeOH, acetone, DMF, and formamide, lending credence to the assumption, in view of the small and near equal conductivities of the selected bridge electrolyte. The $\Delta_t G^{\infty}(Ag^+, S_1 \rightarrow S_2)$ obtained from cell (4.17) is then beset with an uncertainty of $\leq 1 kJ \cdot mol^{-1}$. Only in the case where S_1 was water was the constancy of the electromotive force when different bridge solvents S_{br} were used within as much as $\pm 0.023 V$, corresponding in an uncertainty in $\Delta_t G^{\infty}(Ag^+, W \rightarrow S) \pm 2.3 \ kJ \cdot mol^{-1}$. A problem with this method is the need to establish the independence from the nature of S_{br} for any new solvent S_2 to be examined, and unexpected non-validity of the assumption has been documented in several cases, with no known causes.

The use of a model, akin to the treatment of ion hydration in Sections 4.2.1.1 and 4.2.1.2, leads to the expression:

$$\Delta_t G^\infty \left(I^{z\pm}, W \rightarrow S \right) = RT \ln D_{\text{neut}} \left(r_{\text{eff}} \right)_W^S + \left(\frac{N_A e^2}{4\pi\varepsilon_0} \right) z_I^2 \left[\left(r_{IS}\varepsilon_S \right)^{-1} - \left(r_{IW}\varepsilon_W \right)^{-1} \right] \quad (4.18)$$

Here $D_{\text{neut}} \left(r_{\text{eff}} \right)_W^S$ is the distribution ratio between S and W of a neutral analog to the ion $I^{z\pm}$ with the same effective radius, r_{eff}, as that of the ions, but this may differ in the two solvents: $r_{IS} \neq r_{IW}$. This difficulty of selecting unknown radius addends Δr causes considerable uncertainties in the application of this method. A modification of this approach plots the $\Delta_t G^\infty$ (E, W $\rightarrow$ S), where E is a series of electrolytes with ions (say cations) of increasing radii r_I and a given counter-ion (say an anion), against the reciprocal of the radius to r_I^{-1} and extrapolates this to infinitely large radii, that is, $r_I^{-1} = 0$. This extrapolated value of $\Delta_t G^\infty$ (E, W $\rightarrow$ S) should yield $\Delta_t G^\infty$ ($I^\pm$, W $\rightarrow$ S), where $I^\pm$ is the counter-ion of the series (an anion in the example cited).

If a reference ion I_{ref} could be found that has (near) zero standard molar Gibbs energies of transfer into any target solvent S, then its "conventional" $\Delta_t G^{\infty\text{conv}}$ (I_{ref}, W $\rightarrow$ S) = 0 values could be used to establish a conventional set of $\Delta_t G^{\infty\text{conv}}$ ($I^{z\pm}$, W $\rightarrow$ S) values that would not depart appreciably from the "absolute" values. Such a reference ion appears to be Me_4N^+, where the deviations of $\Delta_t G^\infty$ (Me_4N^+, W $\rightarrow$ S) from zero appear to be no larger than 5 kJ·mol^{-1} according to other, reliable methods (see the following text). If a ligand, such as cryptand(222) (tricyclo-N,N'-tris(3,6-dioxo-1,8-octadiyl), did envelop a cation completely and shielded it from interaction with the solvent, then the distribution of a salt of such a complexed cation between W and S, corrected for the distribution of the ligand itself, should correspond to the use of the reference ion, as suggested by Villermaux and Delpuech [25]. This idea appears to work for transfer between aprotic solvent but not when water is used as the source solvent and it has not been tested sufficiently widely against other reliable methods.

A method that has been extensively used is the reference ion/molecule redox couple using (mainly) polarographic measurements. A reference ion I_r^{n+} that can be reduced to $I_r^{(n-1)+}$ and is complexed by a sufficiently large ligand L that shields it from interaction with the solvent is used in both the oxidized and reduced forms as a reference electrode in a cell such as:

$$Ag \left| AgClO_4, L_p I_r^{n+} + L_p I_r^{(n-1)+}, I^\pm \left(\text{in } S_1 \text{ or } S_2 \right) \right| Hg \quad (4.19)$$

In this cell Ag/AgClO$_4$, $L_p I_r^{n+} + L_p I_r^{(n-1)+}$ constitutes the reference electrode and $I^\pm$/Hg is the working (dropping mercury) electrode reversible with respect to the ion $I^\pm$ to be studied, and measurements are carried out in solvents S_1 and S_2. The half-wave potential in the polarographic measurement, E_{hwp}, represents sufficiently accurately the standard electromotive force, E^∞, if, for instance, $I^\pm$/Pt were the working electrode. For a proper selection of the reference ion I_r^{n+} and its ligand L, the potential of the reference electrode is independent of the solvent, whether S_1 or S_2, so that the standard molar Gibbs energy of transfer of $I^\pm$ from S_1 to S_2 is:

$$\Delta_t G^\infty \left(I^\pm, S_1 \rightarrow S_2 \right) = F \left[E^\infty \left(S_2 \right) - E^\infty \left(S_1 \right) \right] \approx F \left[E_{\text{hw}} \left(S_2 \right) - E_{\text{hwp}} \left(S_1 \right) \right] \quad (4.20)$$

In practice, due to the ligands that could be used, water is not a good choice for the source solvent S_1 and acetonitrile is usually employed instead. The two best studied sets $L_p I_r^{n+} + L_p I_r^{(n-1)+}$ are the fic/foc (ferricinium/ferrocene), that is, dicyclopentadienyl-Fe(III) + dicyclopentadienyl-Fe(II)) proposed by Strehlow and coworkers [26] and bis(biphenyl)Cr(I) + bis(biphenyl)-Cr(0) preferred by Gritzner [27]. The constancy of the reference electrode potential with these two redox couples has been widely tested. The cobalticinium/cobalticene and bis(o-dicarbolide)Ni^{2-})/$-1/0$ reference electrodes, studied by Krishtalik et al. [28], have received less attention.

There are still problems with this approach. One is that cells such as (4.19) can be applied only to ions reducible at the dropping mercury electrode in solvents that are inert to being affected electrochemically. More serious is the apparent insufficient shielding of the reference ion I_r by the ligands and the insufficient size of the complex formed, so that electrostatic effects are present for $L_p I_r^{n+}$ that are not compensated for in $L_p I_r^{(n-1)+}$. Thus, if acetonitrile is used as the source solvent S_1, the target solvent S_2 should have relative permittivities in the range $30 \leq \varepsilon_r \leq 45$ in order to minimize the electrostatic effects. Most serious is the fact that this method has not been tested with regard to the additivity of cation and anion $\Delta_t G^\infty (I^\pm, S_1 \to S_2)$ values to yield the measurable electrolyte value. The independence of the reference electrode potential from the nature of the solvent should still take place if silver perchlorate were exchanged with another soluble silver salt, such as the nitrate or fluoride, but this has not been tested either.

An extensively employed and tested method involves a reference electrolyte, the cation and anion of which are large and of the same size, have globular form with peripheries that are inert to interactions with solvents, and which differ only in the sign of the charge. Approximations to such an electrolyte are TPTB and tetraphenylarsonium tetraphenylborate (TATB), of which mainly the latter has been widely used. Solubility (s) measurements are generally employed, ignoring the activity coefficients of the only very slightly soluble salts in most solvents. The extra-thermodynamic assumption is that:

$$\Delta_t G^\infty \left(Ph_4 As^+, W \to S \right) = \Delta_t G^\infty \left(BPh_4^-, W \to S \right)$$
$$= 0.5RT \ln \left[\frac{s(TATB, W)}{s(TATB, S)} \right] \quad (4.21)$$

Replacement of TATB by TPTB yields results within the experimental error of $\pm 0.4 \, kJ \cdot mol^{-1}$. There are two minor problems with this assumption pointed out by Kim [29], one depending on the not exactly equal sizes of the cation and the anion and the other on the (unknown) sign-dependent interactions of the ions with the quadrupole of the solvent. The ratio of the van der Waals radii of the TATB ions is $r_{vdW} (Ph_4 As^+)/r_{vdW} (BPh_4^-) = 1.0122$ (it is somewhat nearer unity for the ions of TPTB). This discrepancy causes difference of 1.2 and 2.2% in the electrostatic and neutral Gibbs energies (see the radius dependencies in Sections 4.2.1.2 and 4.2.1.1). Once the $\Delta_t G^\infty (Ph_4 As^+, W \to S)$ is established by Equation 4.21 or measurements other than solubility, the anion transfer $\Delta_t G^\infty (A^-, W \to S)$ is obtained from measurements on $\Delta_t G^{\infty}$ ($Ph_4As^+A^-$, W → S) and similarly for the cation transfer $\Delta_t G^{\infty}$ (C^+, W → S) from measurements on $\Delta_t G^{\infty}$ ($C^+BPh_4^-$, W → S).

On the whole, conceptually the reference electrolyte extra-thermodynamic assumption for $\Delta_t G^{\infty}$ is sound, and its implementation by means of the TATB assumption has been deemed to be reliable by a number of authors, [23, 24, 29, 30] among others, as the least objectionable one. The validity of other assumptions might be and has been tested against the TATB assumption, although there is no certainty in its own validity. An estimate of its possible reliability is $\pm 2\,kJ \cdot mol^{-1}$ from independent determinations by several authors for a given ion/target solvent system as shown by Marcus [31].

The resulting $\Delta_t G^{\infty}$ ($I^{\pm}$, W → S) values on the M-scale for the transfer of numerous ions from water to non-aqueous solvents at 25°C are shown in Table 4.2, adapted from Refs. 31–35. Values shown to one decimal have an estimated uncertainty of $\pm 0.5\,kJ \cdot mol^{-1}$, those shown as integral values have an estimated uncertainty of $\pm 2.0\,kJ \cdot mol^{-1}$, and those shown in bold font have been more recently recommended [34, 35] as the most probably correct values. The uncertainties of the latter values are derived from independent determinations of the $\Delta_t G^{\infty}$ ($I^{z\pm}$, W → S) values by several authors and are understood to be beyond the uncertainty due to the preferred extra-thermodynamic assumption dealt with in the previous paragraph.

The standard molar Gibbs energies of transfer of ions from water to nonaqueous solvents were submitted by Marcus et al. [32] to a statistical treatment in terms of the linear solvation energy relationship:

$$\Delta_t G^{\infty}\left(I^{z\pm}, W \rightarrow S\right) = \sum P_{Ij}\left(P_{Sj} - P_{Wj}\right) \tag{4.22a}$$

where P_I is a property of the ion relevant (j) to a corresponding property of the solvent or water, P_S or P_W. It was necessary to consider separately three classes of ions: small cation, large ions (tetraalkyl- or tetraphenylonium ones), and anions.

Small cations required three solvent properties, namely the polarity/polarizability ($\Delta\pi^*$), hydrogen bond donation ($\Delta\alpha$) and electron pair donation ($\Delta\beta$), and the Δ signifying $P_S - P_W$ (Section 3.3.2). Other properties examined: $\Delta\varepsilon^{-1}$, $\Delta\delta_H^2$, and ΔV (all normalized to have values in the range from 0 to 1, to be commensurate with $\Delta\pi^*$, $\Delta\alpha$, and $\Delta\beta$) were found not to be required by the statistical criteria of goodness of fit of the data matrix. An exception is the hydrogen ion, for which ΔV may replace $\Delta\pi^*$ profitably. The relevant cation properties (Section 2.1.1) are $-z_I^2/r_I$ and its softness σ_I for $\Delta\pi^*$, the ionic volume $v_I = (4\pi/3)r_I^3$ for $\Delta\alpha$, and z_I^2/r_I and the (negative of the) molar refractivity $-R_D$ for $\Delta\beta$. The operative expression for the standard ionic molar Gibbs energy of transfer on the M scale at 25°C according to Marcus [36] is then:

$$\Delta_t G^{\infty}\left(small\,I^{z+}, W \rightarrow S\right) / kJ\,mol^{-1}$$
$$= \left[-3.26z_I^2/r_I + 30.3\sigma_I\right]\Delta\pi^* + 225r_I^3\Delta\alpha + \left[-3.72z_I^2/r_I - 3.78R_D\right]\Delta\beta \tag{4.22b}$$

TABLE 4.2 Ionic Standard Molar Gibbs Energies of Transfer from Water into Nonaqueous Solvents, $\Delta G^\infty(\mathrm{I}^\pm, \mathrm{W} \rightarrow \mathrm{S})/\mathrm{kJ \cdot mol^{-1}}$, at 25°C[a]

Ion\solvent	MeOH	EtOH	1-PrOH	1-BuOH	TFE	EG	Me₂CO	PC	NH₃[b]	FA	NMF	DMF	DMA
H⁺	**8.7**	**11.1**	9	3		**6.3**	**10**	50	−97			**−14.4**	
Li⁺	**3.0**	**10.3**	11			**3.3**	**10**	23.8	−35	−10	−20	**−22.4**	−22
Na⁺	**7.0**	**14.9**	17	19		**1.3**	**4**	14.6	−17	−8	−7	**−10.6**	−12.1
K⁺	**8.4**	**16.4**	17	20	39	**1.6**	**4**	5.3	−12	−4.3	−6	**−9.1**	−11.7
Rb⁺	9.6	**16.6**	19	23			**4**	−1.0	−13	−5	−8	**−10.2**	−8
Cs⁺	**8.7**	**16.1**	17	19			**4**	−7.0	−15	−6.0	−7	**−11.9**	−7
Ag⁺	**8.5**	**4.5**	1		50	1	**9**	18.8	−100	−15.4	−15	**−17.2**	−29.0
Tl⁺	4.1	7	7	12			**3**	11.0		−1	−10	**−11.5**	−13
NH₄⁺	**11.3**	7	7	12					−28				
Me₄N⁺	**7.0**	**15.6**	11	12			**3**	−11		−1.3		−5.3	
Et₄N⁺	1	**3.5**	5	7				−13		−7.1		−8.0	
Pr₄N⁺	−8.8	**−2.0**	−6	−7				−22				−17.0	
Bu₄N⁺	**−23.1**	**−12.0**	−17	−12				−31		−23.9	−33	−29	
Ph₄P⁺	**−23.4**												
Ph₄As⁺	**−23.4**	−22.4	−25	−20		−21.3	−32	−36.0				**−39.1**	−38.7
Mg²⁺	2.0												
Ca²⁺	11.2												
Sr²⁺	7.1												
Ba²⁺	16.9								−66			−21.4	
Cu²⁺	25.2	**19.1**	43			6.2	43.5	73	−181	−4	−4	**−14.4**	−27
Zn²⁺	27.4	**36.0**					83	81	−149			−29.6	−7
Cd²⁺	32.9	22					61	70	−152	−28	−32	−33.5	−22
Hg²⁺	48	64					2	47	−228			−44	
Pb²⁺	**10.3**	**21.8**					78	47		−12		−34	−36

Ion\solvent	DEA	NMPy	MeCN	MeNO$_2$	PhNO$_2$	Py	1-CNPy	DMSO	TMS	HMPT	1,1DCIE	1,2DCIE
H$^+$		−25	**44.8**	95	33	−25		**−19.4**		40		
Li$^+$		−35	**28.5**	48	38			**−17.3**	6			
Na$^+$	−2	−15	**11.9**	31.6	36	16	24	**−12.7**	−3	−17	29	25
K$^+$	−10	−11	**6.5**	15.4	21	6	14	**−12.2**	−4	−16	30	26
Rb$^+$	−9	−8	**5.4**	11	19	13		**−10.2**	−9	−10	29	25
Cs$^+$	−11	−10	**4.5**	5.6	18	35		**−121.0**	−10	−7	28	24
Ag$^+$	−29	−26	**−24.1**	21		−57		**−32.0**	−4	−44		
Tl$^+$	−12	−15	**8.0**	16	15	−1		**−21.9**	−12	−26		
NH$_4^+$		−24			27							
Me$_4$N$^+$		−3	**3.3**	−4.6	4		−2	−2			18	16
Et$_4$N$^+$			**−7.0**	−10.3	−5			−9			11	5
Pr$_4$N$^+$			**−13.0**	−20.0	−16							
Bu$_4$N$^+$			**−32.0**		−31							
Ph$_4$As$^+$	−38.7	−40	**−34.2**	−32.6	−36	−38	−39	**−38.3**	−36	−39	−27	−23
Ba^{2+}			57.4					−26.6				−33
Cu^{2+}	−45		**66.8**	95		−50		**−42.2**	62	−52		
Zn^{2+}	−33	−8	**28.3**					**−40.3**	76	−86		
Cd^{2+}	−22	−22	42.2					**−60.4**	39	−36		
Hg^{2+}			42	31		−79		−48				
Pb^{2+}	−34	−35	**30.0**					**−65.7**	29	−38		

(continued)

TABLE 4.2 (Continued)

Ion\solvent	MeOH	EtOH	1-PrOH	1-BuOH	TFE	EG	Me$_2$CO	PC	NH$_3$[b]	FA	NMF	DMF	DMA
OH$^-$	12.4												
F$^-$	**20.1**	**26.4**				**15.8**		56		25		51	
Cl$^-$	**13.2**	**20.3**	26	29	−10	**8.4**	57	39.8	43	13.7		48.3	54.9
Br$^-$	11.1	**19.0**	22	24	−8	**6.6**	42	30.0	33	10.7		36.2	44.0
I$^-$	**7.3**	**14.0**	19	22	−8	**4.3**	25	13.7	25	7.3		20.4	21
I$_3^-$	−12.6							−17		−7		−17	−30
N$_3^-$	9.1	17.0				7	43	27		11		36	40
CN$^-$	**9.5**	7					48	36		13.3		40	
SCN$^-$	4.6	−3.6				5		7.0		7		18.4	21
NO$_3^-$	12.5	14											
ClO$_4^-$	**6.5**	10	17	22				−3		−12		4	
TcO$_4^-$	5.8	8.8	15	17		3.1	12.4			−1.6		3.3	5.0
CH$_3$CO$_2^-$	16.0	−6.6								20		66	70
CF$_3$SO$_3^-$	−16.0												
Pic$^-$	**−3.6**	**−0.2**	−7	−2		−6.8	−32	−6		−7		−7	
BPh$_4^-$	**−22.7**	**−21.3**	−25	−20		−21.7		−36.0		−23.9	−33	−38.5	−38.7
CO$_3^{2-}$						53							
SO$_4^{2-}$	34	46				37	103				54	105	
S$_2$O$_3^{2-}$		85											
Cr$_2$O$_7^{2-}$							49						
Fe(CN)$_6^{3-}$	44.2												

Ion\solvent	DEA	NMPy	MeCN	MeNO$_2$	PhNO$_2$	Py	1-CNPy	DMSO	TMS	HMPT	1,1DCIE	1,2DCIE
F^-			47		44							65
Cl^-		51	**42.1**	37.7	35	34	31	**38.9**	47	58	58	52
Br^-		37	31.3	29.0	29	21		**24.3**	35	46	43	38
I^-		19	16.8	18.9	18	19		**14.1**	21	30	31	25
I_3^-			-15		-23			-41	-14			
N_3^-		46	37	28				25.8	41	49		
CN^-			35	31				35				
SCN^-		18	14.4	15	6	20		9.7	22	20		
NO_3^-		-12	21	21	24	16	24					7
ClO_4^-			2	4.7	10		-5	-1		-7	22	16
TcO_4^-	10		6.9		5							
$CH_3CO_2^-$			61	56				50				
$CF_3SO_3^-$			-23			-14						
Pic^-			-4		-3	-5	-6					
BPh_4^-	-38.7	-40	**-33.6**	-32.6	-36	-38	-39	**-37.1**	-36	-39	-27	-33
$C_2O_4^{2-}$		56.4										
SO_4^{2-}		89.2										

Bold face entries are "recommended values."

[a] From Refs. 31–34.

[b] At 240 K.

Positive $\Delta_t G^\infty (I^{z\pm}, W \to S)$ values resulted for these cations for transfer into alkanols, acetone, PC, MeCN, Py, MeNO$_2$, PhNO$_2$, and the dichloroethanes—the cations prefer water over these solvents—and negative values resulted for transfer into the cation-preferred amides, liquid ammonia, DMSO, and TMS that have strong electron pair donor (β) properties. Exceptions are the soft cations Ag$^+$, Cd^{2+}, and Hg^{2+} that prefer the soft solvents MeCN and Py.

Large ions require only two solvent properties for a good statistical fit: $\Delta\pi^*$ and $\Delta\delta_H^2$. The relevant ion properties are $-z_I/r_I (= -z_I^2/r_I$ since these ions are all univalent) and $-R_D$ for $\Delta\pi^*$ and $-z_I/r_I$ and v_I for $\Delta\delta_H^2$. These hydrophobic ions from Pr$_4$N$^+$ and larger ones all prefer the less structured nonaqueous solvents (negative $\Delta_t G^\infty$) over the highly hydrogen bonded water, but Me$_4$N$^+$ and Et$_4$N$^+$ have small positive $\Delta_t G^\infty$ values for transfer into some of the solvents. The operative expression is [36]:

$$\Delta_t G^\infty \left(\text{large } I^\pm, W \to S \right)/\text{kJ} \cdot \text{mol}^{-1}$$
$$= \left[-5.29 z_I/r_I - 0.46 R_{DI} \right] \Delta\pi^* + \left[-2.1 z_I/r_I + 440 r_I^3 \right] \Delta\delta_H^2 \qquad (4.22c)$$

The transfer of anions requires three solvent properties, that is, $\Delta\pi^*$, $\Delta\alpha$, and ΔV, and the following relevant ion properties, that is, z_I/r_I, R_{DI}, σ_I, and r_I^3. The operative expression is [36]:

$$\Delta_t G^\infty \left(I^-, W \to S \right)/\text{kJ} \cdot \text{mol}^{-1}$$
$$= \left[3.0\Delta\pi^* + 7.5\Delta\alpha - 0.038\Delta V \right] z_I/r_I + 30\sigma_I\Delta\alpha - 11.7 r_I^3 \Delta V + R_{DI}\Delta\pi^* \qquad (4.22d)$$

The transfer of the smaller anions from water to nonaqueous solvents has positive $\Delta_t G^\infty$ values; they prefer aqueous environment, because of the dominant $\Delta\alpha < 0$ term in Equation 4.22d (an exception is TFE, that has $\Delta\alpha > 0$). As anions become larger, they may prefer the nonaqueous environment, because then the term $-v\Delta V$ may become dominant, the effect of the structured water "squeezing" out the large anion into the less structured solvent being then relatively more important. This is the case for I_3^-, CF$_3$SO$_3^-$, and picrate (Pic$^-$) ions and for a few solvents also ClO$_4^-$.

4.3.2.2 Enthalpies of Transfer

Calorimetric measurements of the heats of solution of electrolytes in different solvents yield the most accurate data for their standard molar enthalpies of transfer between the solvents. Less accurate are the temperature derivatives of the solubilities and of the electromotive force (half-wave potential) of electrochemical cells.

The splitting of the electrolyte values into the individual ionic contributions can be made with the extrapolation method applied for ion hydration (Section 4.2.2), as suggested by Somsen and Weeda [37]. This is implemented by the selection of a cation C$^+$ and anion A$^-$ of equal size (radius r), and then:

$$\left[\Delta_t H^\infty \left(A^- \right) + \Delta_t H^\infty \left(Na^+ \right) \right] - \left[\Delta_t H^\infty \left(C^+ \right) - \Delta_t H^\infty \left(Na^+ \right) \right] = ar^{-3} + br^{-4} \qquad (4.23)$$

The terms on the left-hand side correspond to neutral species, and hence are available from measurements, and those on the right-hand side pertain to the ion-dipole and ion-quadrupole interactions. Plots of the left-hand side against r^{-3} extrapolated to infinitely large ion size, although not linear (due to the br^{-4} term), then yield $2\Delta_t H^\infty$ (Na$^+$, W → S) as the ordinate intercept.

A much more commonly used method for splitting electrolyte enthalpies of transfer into the individual ionic contributions is the TATB method, according to which, in analogy with its use for the Gibbs energies, $\Delta_t H^\infty$ (Ph$_4$As$^+$, W → S) = $\Delta_t H^\infty$ (BPh$_4^-$, W → S), practiced by Arnett and McKelvey and by Friedman [38, 39]. The uncertainty involved in the application of the TATB method was estimated as ± 1 kJ·mol^{-1} by Cox and Parker [40]. Indeed, if the TATB assumption is accepted for $\Delta_t G^\infty$ at 25°C, at which temperature it is generally applied, there is no good reason for not accepting it at any other temperature, and hence the TATB assumption should be valid also for the $\Delta_t H^\infty$ of ions. The TPTB assumption has in more recent years replaced to some extent the TATB one, but with hardly any effect on the results.

The standard molar enthalpies of transfer of ions from water into nonaqueous solvents, $\Delta_t H^\infty$ (I$^\pm$, W → S), at 25°C are shown in Table 4.3, adapted from the earlier report by Marcus [41] and from the more recent compilation by Hefter et al. [42]. In addition to the solvents shown, there are data for only a few ions in other solvents: liquid ammonia (at 240 K), NMPy, and pyridine, which were given full statistical weight in the critical compilation [41]. The transfer enthalpies of divalent cations into MeOH, EtOH, and DMSO have also been examined critically [41, 42]. The enthalpies of transfer of small cations from water are negative but turn positive for the larger, hydrophobic cations. The transfer enthalpies from water of small anions are positive and turn negative as their size increases. These trends reflect the interactions more fully discussed in Section 4.3.2.1 [32].

The dependence of the ionic standard molar enthalpies of transfer on the properties of the solvents and the ions are expressed in a manner similar to Equation 4.22a. For small cations, the operative expression for values at 25°C is [36]:

$$\Delta_t H^\infty \left(\text{small I}^{z+}, \text{W} \to \text{S}\right)/\text{kJ mol}^{-1} =$$
$$\left[1.00 z_I^2 / r_I - 0.61 R_{DI}\right]\Delta\pi * + \left[0.96 z_I / r_I + 3.49 R_{DI}\right]\Delta\alpha$$
$$+ \left[-4.96 z_I / r_I - 8.0 \sigma_I + 3.49 R_D\right]\Delta\beta \qquad (4.24)$$

For large, hydrophobic ions, the operative expression is [36]:

$$\Delta_t H^\infty \left(\text{large I}^\pm, \text{W} \to \text{S}\right)/\text{kJ} \cdot \text{mol}^{-1} =$$
$$\left[-0.60 z_I / r_I - 0.38 R_{DI}\right]\Delta\pi * + \left[-8.15 z_I / r_I + 360 r_I^3\right]\Delta\delta_H^2 \qquad (4.25)$$

For anions, the operative expression is:

$$\Delta_t H^\infty \left(\text{I}^-, \text{W} \to \text{S}\right)/\text{kJ} \cdot \text{mol}^{-1} =$$
$$\left[-0.65\Delta\pi * + 2.5\Delta\alpha - 0.055\Delta V\right]z_I / r_I + 15.5\sigma_I \Delta\pi *$$
$$+ \left[0.18\Delta\alpha - 0.196\Delta V\right]r_I^3 \qquad (4.26)$$

TABLE 4.3 Ionic Standard Partial Molar Enthalpies of Transfer from Water into Nonaqueous Solvents, $\Delta_t H^\infty(I^\pm, W \rightarrow S)/kJ\cdot mol^{-1}$, at 25°C[a]

Ion\Solvent	MeOH	EtOH	1-PrOH	EG	PC	FA	DMF	DMA	Py	MeCN	DMSO	TMS[b]	HMPT	1,2-DCIE
H+	-15.5		-17.4	-22.0	44				-43.9	3.0	-25.5	73		
Li+	-18.9	-20.2	-18.4		2.8	-5.7	-25.4			-13.0	-27.8	12	-57.4	
Na+	-23.1	-19.4	-18.4	-21.5	-10.5	-14.4	-34.1	-39.0	-30.3	-11.4	-29.8	-16	-50.6	-25.1
K+	-21.2	-19.6	-18.3		-22.5	-16.6	-42.2	-44.2	-24.1	-22.9	-36.5	-26	-46.6	-28.0
Rb+	-19.3				-24.9	-17.4	-41.3		-27.8	-24.6	-38.2	-28	-43.8	-27.6
Cs+	-17.6	-11.8	-12.7		-27.5	-17.4	-39.0	-46.2	-28.1	-26.1	-34.1	-26	-45.0	-27.2
Ag+	-20.9				-11.1		-35		-106	-52.7	-51.5	-14		
NH4+	-23.7	-26.5	-28.0		-19.8	-14.0	-38.5		-36.4	-12.0	-37.9	-16	-63.8	-15.9
Me4N+	0.3	0.2	0.9		-18	0.1	-18.8			-2.1	-16.6		-34.1	-10.0
Et4N+	7.1		11.6				-2.2		-3.7	10.5	3.9	17	-6.1	
Pr4N+	15		11.8								16.0		1.6	
Bu4N+	20.4	21.8	19.8		21		15.4	10.5	7.9	18.2	23.3	26	5.3	9.0
Pe4N+							14.6			21.1				
Ph4P+	-1.1	0.0	4.4	-6.1		-1.2	-18.0	-16.5			-11.2		-25.8	
Ph4As+	-1	0.0			-13.1	-0.5	-17.2	-13.7	-22.8	-12.1	-11.2	-11		-22.6
Ba2+	-60.6		-21.8			-40.2	-85.5			-8.5	-78.5		-122.2	
Zn2+	-45.6					-23.9	-62.7	-53.2		20.1	-62.2		-74.8	
Cd2+	-40.4					-27.5	-63.3			8.2	-70.8		-93.5	
Hg2+									-160		-76			
F-	15.7					23.4	5.7							
Cl-	7.9	10.2	7.3	17.8	26.2	-3.4	21.4	35.6	28.2	18.4	20.0	27	38.2	16.3

Br	5.4	5.9	*15.2*	−1.2	5.5	17.2	10.9	7.6	4.7	*13*	17.7	*0.8*
I⁻	−0.1	−3.6	*−1.6*	−8.6	−11.8	−0.7	−7.3	−9.3	−10.9	*−8*	−5.8	*−15.5*
SCN⁻	−2.5		−10.0	−10.0	−10.0		−4.7	−6.6	−4.9			−4.6
N₃⁻	0.5		*16.7*	−1.0	−1.0		8.8	−2.5	−0.5	*15*		
NO₃⁻	2.8		*−4.4*				2.8		−0.5			7.7
ClO₄⁻	−3.1	0.0	*−16.3*	−23.4	−15.3	−18.8	−16.6	−19.2			−20.7	*−24.3*
BF₄⁻	3.6		−20.8				−12.9	−14.5		*−18*		
CF₃SO₃⁻	4.6		−3.9				0.8	2.1			−0.5	
Pic⁻	−0.9		−12.0					−12.4	−12.0			
BPh₄⁻	−1.1	4.4	*−13.2*	−1.2	−18.0	−16.5	−22.8	−12.1	−11.2	*−11*	−25.8	−22.6

ᵃFrom Refs. 32, 41 (in italics) and from Ref. 42.
ᵇAt 30°C.

4.3.2.3 Entropies of Transfer Thermodynamic consistency requires the entropies of transfer to be obtained from $\Delta_t S^{\infty}(I^{\pm}, S_1 \rightarrow S_2) = [\Delta_t H^{\infty}(I^{\pm}, S_1 \rightarrow S_2)$ $-\Delta_t G^{\infty}(I^{\pm}, S_1 \rightarrow S_2)]/T$, provided the enthalpy and the Gibbs energy are obtained by the same method.

Independent estimates of $\Delta_t S^{\infty}(I^{\pm}, S_1 \rightarrow S_2)$ have been made for the thermocell method with 0.1 M Bu_4NClO_4 supporting electrolyte being used in two half-cells at different temperatures, separated by a liquid junction involving this electrolyte. The temperature derivative of its junction potential was assumed to be negligible. The uncertainty of the method claimed by Hörzenberger and Gritzner [43] was $\pm 4 \, J \cdot K^{-1} \cdot mol^{-1}$.

Another, quite empirical, method assumes that $\Delta_t S^{\infty}(I^{\pm}, W \rightarrow S) = a(S)$ $+ b(S) S^{\infty}(I^{\pm}, W)$, that is, a linear function of the standard molar entropy of the aqueous ion with different coefficients a and b for each solvent, proposed by Criss et al. [44]. This requires estimates of $\Delta_t S^{\infty}(I^{\pm}, W \rightarrow S)$ from other sources to establish a and b for a given solvent and may be used, for this solvent, for ions for which only $S^{\infty}(I^{\pm}, W)$, but not the entropy of transfer, is known. There being no theoretical basis for the linear relationship, this linearity has to be reestablished for each new solvent.

The entropies of transfer, $\Delta_t S^{\infty}(I^{\pm}, W \rightarrow S)$, are shown in Table 4.4 and were adapted from the aforementioned compilations [41] and [42]. It should be noted that the former were obtained from several extra-thermodynamic assumptions that were acceptable but the latter invariably from the TPTB one. Therefore, there are some inconsistencies (within $\pm 10 \, J \cdot K^{-1} \cdot mol^{-1}$) and the later values should be preferred. The entropies of transfer from water to nonaqueous solvents are negative for small ions and positive only for the largest, hydrophobic, ions: Bu_4N^+, Ph_4P^+, Ph_4As^+, and BPh_4^-. It should be noted that for the alkali metal cations, the magnitudes of $\Delta_t S^{\infty}(I^{\pm}, W \rightarrow S)$ show a most negative value somewhere in the middle of the series, but for the tetraalkylammonium and halide ions, the values are monotonous with the ionic sizes. On the whole, the standard molar entropies of small ions, whether cations or anions and irrespective of the charge and sign, show a pronounced uniformity.

4.3.2.4 Ionic Heat Capacities in Nonaqueous Solvents The standard molar ionic heat capacities in water of the alkali metal, halide, perchlorate, tetraalkylammonium, and tetraphenylonium salts are reported in Table 2.8 and also by Abraham and Marcus [45], based on the TPTB assumption. Based on the same assumption, their $C_P^{\infty}(I^{\pm}, S)$ in S=MeOH, EtOH, 1-PrOH, PC, NMF, DMF, MeCN, MeNO$_2$, and DMSO are reported by Marcus and Hefter [46]. The reservation expressed in Section 2.3.1.1, that slight differences in the sizes and induced partial charges in the phenyl rings cause a large uncertainty in equating the TPTB ions, applies also to the nonaqueous solutions. The values of $C_P^{\infty}(I^{\pm}, S)$ are listed as far as available in Table 4.5.

The $\Delta C_P^{\infty}(I^{\pm}, W \rightarrow S)$ can be readily calculated from these data, with the uncertainties borne in mind. Correlation expressions for the large tetraalkylammonium ions, relating the $C_P^{\infty}(I^{\pm}, W \text{ or } S)$ linearly to number of carbon atoms of the ions were presented by Abraham and Marcus [45]. The molar heat capacity change of solvation

TABLE 4.4 Ionic Standard Partial Molar Entropies of Transfer from Water into Nonaqueous Solvents, $\Delta_t S^\infty$(I$^\pm$, W→S)/J·K^{-1}·mol^{-1}, at 25°C[a]

Ion\Solvent	MeOH	EtOH	EG	PC	FA	DMF	MeCN	DMSO	1,2DClE
H$^+$	-81		-87				-288		
Li$^+$	-73	-101		-65	13		-99	-35	-155
Na$^+$	-101	-108	-75	-75	-21	-79	-78	-57	-167
K$^+$	-96	-113		-89	-28	-111	-89	-82	-180
Rb$^+$	-103	-103		-77	-32	-101	-100	-94	-176
Cs$^+$	-89	-92		-66	-38	-91	-96	-71	-172
Ag$^+$	-95			-92		-8	-98	-65	
NH$_4^+$	-100								
Me$_4$N$^+$	-18	-36			5	-19		-24	-105
Et$_4$N$^+$	31					45	39		-50
Pr$_4$N$^+$									
Bu$_4$N$^+$	146					165		88	
Ph$_4$P$^+$	75	75	50			71	74	91	
Ph$_4$As$^+$				71	80	71	75		34
F$^-$	-20								-100
Cl$^-$	-27	-33		-43	-32	-99	-87	-34	-126
Br$^-$	-28	-43		-51	-39	-106	-89	-70	-130
I$^-$	-27	-46		-64	-47	-119	-88	-79	-138
N$_3^-$	-34			-43		-125	-72		
ClO$_4^-$	-26	-46			-40	-100	-72	-69	-138
BPh$_4^-$	75	75	50	71	80	71	75	91	34

[a]From Ref. 41 (in *italics*) and from Ref. 42.

TABLE 4.5 Ionic Standard Partial Molar Heat Capacities in Non-aqueous Solvents, $C_P^\infty(I^\pm, S)/J\cdot K^{-1}\cdot mol^{-1}$, at 25°C[a]

Ion\Solvent	MeOH	EtOH	1-PrOH	PC	NMF	DMF	MeCN	MeNO$_2$	DMSO
Li$^+$	-20	3	-89	27	-6	-14	52	200	-46
Na$^+$	42	91	13	57	47	32	81	134	-26
K$^+$	56	129		38	51	35	31	30	-27
Rb$^+$				36		34	35		
Cs$^+$	43	122		36	49	29	33	1	-35
Me$_4$N$^+$	138			159					
Et$_4$N$^+$	300	291	280	267		257	266		
Pr$_4$N$^+$	412	418	396	383		385			
Bu$_4$N$^+$	557	589	577	496		517	513		
Pe$_4$N$^+$	670	821	753	610			626		
Ph$_4$P$^+$	555	594	539	498	572	480	490	475	515
Ph$_4$As$^+$	596			512			501		
F$^-$	-131	-194						71	60
Cl$^-$	-102	-185	-247	44	10	108	55	61	58
Br$^-$	-76	-170	-211	44	27	84	38	107	95
I$^-$	-56	-93	-123	41	37	58	29	64	114
SWCN$^-$	-6[b]						-170[b]		
ClO$_4^-$	53			103		68	76		107
BPh$_4^-$	555	594	539	498	572	480	490	475	515

[a]From Ref. 46.
[b]From Ref. 35.

per $-CH_2-$ group (in $J \cdot K^{-1} \cdot mol^{-1}$) is 69.5 in water, 14.6 for the alkanols, and 5.3 for the aprotic solvents. The large value for water signifies the building-up of a quasi-clathrate structure of the water around the ion [45]. Explicit expressions relating C_P^∞ ($I^\pm$, W or S) to properties of the ions combined with those of the solvents for the small ions are also given in Ref. 45, but these cannot be readily deconvoluted to represent separately the ion and solvent property effects.

The dependence of the heat capacities of transfer of small ions on the properties of the ions and the solvents at 25°C are expressed by the following operative equations [36] analogous to (4.22a). For transfer into aprotic solvents:

$$\Delta_t C_P^\infty \left(\text{small } I^{z\pm}, W \to S\right)/J \cdot K^{-1} \cdot mol^{-1} =$$
$$[-0.45\Delta\delta_H - 19\Delta\beta]z_I + [3.5\Delta\delta_H - 170\Delta\beta]r_I^{-1}$$
$$+ [-0.08\Delta\delta_H - 7.1\Delta\beta]R_{DI} \tag{4.27}$$

whereas for transfer into protic solvents, the coefficients of the ion properties differ for cations:

$$\Delta_t C_P^\infty \left(\text{small } I^{z\pm}, W \to S\right)/J \cdot K^{-1} \cdot mol^{-1} =$$
$$[-1.8\Delta\delta_H + 1.5\Delta\kappa_T]z_I + [0.11\Delta\delta_H - 6.1\Delta\kappa_T]r_I^{-1}$$
$$+ [-0.48\Delta\delta_H - 15.5\Delta\beta]R_{DI} \tag{4.28}$$

where for anions the $15.5\Delta\beta$ in the last term is replaced by $11.9\Delta\alpha$, expressing the abilities of the protic solvents to form hydrogen bonds with the anions. A solvent property not required for the thermodynamic quantities described earlier is the difference between the isothermal compressibility of the target solvent and water, $\Delta\kappa_T$.

4.3.2.5 Ionic Volumes in Nonaqueous Solvents

The standard partial molar volumes of ions in water are discussed in Section 2.3.1.5, based on assigning to the hydrogen ion certain temperature-dependent values. The assignment of individual ionic standard partial molar volumes in nonaqueous solvents is dealt with extensively by Hefter and Marcus in Refs. 47 and 48. Several methods have been examined there, the most reliable being the "direct" method based on ultrasound vibration potentials (UVP), the adjustment of conventional (based on $V^{\infty \text{ conv}}(H^+, S) = 0$) values of cations and anions to lie on a single curve when plotted against r_I^3, and the splitting of the TATB and TPTB standard partial molar volumes *unequally* between the ions, taking differences in their ionic sizes into account. Each of these has its problems and drawbacks and the least objectionable one being the use of $V^\infty (Ph_4P^+, S) - V^\infty (BPh_4^-, S) = 2 \pm 2$ cm$^3 \cdot mol^{-1}$ in all solvents S (at 25°C). The use of this assumption leads to $V^\infty(H^+, W) = -5.8 \pm 2$ cm$^3 \cdot mol^{-1}$ that is well compatible with the value given in Section 2.3.1.5, -5.5 ± 0.2 cm$^3 \cdot mol^{-1}$. It is therefore possible to obtain the standard molar ionic volumes of transfer, ΔV^∞ ($I^\pm$, W $\to$ S) from the values in Table 2.8 and the ionic volumes in nonaqueous solvents, V^∞ ($I^\pm$, S), listed in Table 4.6, being based on essentially the same extra-thermodynamic assumption.

TABLE 4.6 Ionic Standard Partial Molar Volumes in Nonaqueous Solvents, $V^\infty(I^\pm, S)$/cm^3·mol^{-1}, at 25°C[a]

Ion/solvent	MeOH	EtOH	EG	Me$_2$CO	PC	FA	NMF	DMF	DMA	MeCN	MeNO$_2$	DMSO	HMPT
H$^+$	-17	-9	-11	-62	-9	3	-10	-8		-20	-18	-4	1
Li$^+$	-19	-18	-8			-7	-3	-2		-17	-13	3	6
Na$^+$	-19	-6	-3	-18	-2	-3		6		-9	-8	11	22
K$^+$	-8	3	6		7	7	5			-6		17	
Rb$^+$	-4	9	12		11	11		10				23	
Cs$^+$	3	17	21		17	17	14	17		1	3		
Ag$^+$	-8				-8			-11		-20		-6	
NH$_4^+$	2	12				12	10	11	8	0		14	
Me$_4$N$^+$	65	76	79	125	84	83			77			85	
Et$_4$N$^+$	125	137	138	137	143	142	140	142	141	132		143	
Pr$_4$N$^+$	199	207	206	198	213	211	211	213	210	206		212	212
Bu$_4$N$^+$	263	276	275	271	283	279	280	281	278	276		282	296
Pe$_4$N$^+$	336	341			344	349		352		341		342	343
Ph$_4$P$^+$	262	263	292	255	286	294	286	284		271	285	289	285
Ph$_4$As$^+$	268		5		292	301		291		281		295	
F$^-$	-1	-17	24			13					13	-2	
Cl$^-$	15	12	31	-8	16	25	26	3	24	3	14	9	
Br$^-$	23	15	41	25	26	32	30	9	33	10	21	17	0
I$^-$	30	23		31	33	44	41	24	39	25	34	31	20
SCN$^-$	36			27		48	46			29			
NO$_3^-$	30	22				37	37	19		15		25	
ClO$_4^-$	41			20	44	54	50	35	39	34		44	
BPh$_4^-$	260	261	284	253	284	292	284	282	286	273	283	287	277

[a]From Ref. 48.

Values of V^{∞} ($I^{z\pm}$, S) for some other ions, for example, some divalent cations (in PC, FA, DMF, and DMSO) and in some other solvents, are available in the review by Marcus and Hefter [48].

Statistical correlations have been established by Marcus et al. [49] between V^{∞} ($I^{\pm}$, S) and the properties of the ions and those of the solvents, but these cannot be deconvoluted to separate dependences on such properties. For the tetraalkylammonium ions (from Et_4N^+ onwards), there is a linear dependence of V^{∞} ($I^{\pm}$, S) on the intrinsic volumes of the ions, meaning that there is a constant volume increment for each $-CH_2-$ group, $16\,cm^3 \cdot mol^{-1}$, irrespective of the solvent. For small ions, the dependence of V^{∞} ($I^{\pm}$, S) is about equally shared by the intrinsic ionic volumes and their basicities (for anions) and acidities (for cations), the dominant solvent property being their tightness, measured by their solubility parameter.

Theoretical considerations of the values of V^{∞} ($I^{\pm}$, S), in terms of the intrinsic volumes according to Marcus et al. [50], the electrostriction, and solvent structural effects are presented by Marcus [51]. The electrostriction can be calculated according to the principles described in Section 2.3.1.5, Equation 2.24, involving the compressibility and the electric field-dependent permittivity of the solvent in addition to the size (radius) of the ion. The solvent structural effects are $V_{str}/cm^3 \cdot mol^{-1} = 6.9 \pm 0.6$ per $-CH_2-$ group in tetraalkylammonium cations in the solvents tested (MeOH, EtOH, EG, Me_2CO. PC, FA, NMF, DMF, MeCN, and DMSO), but only $5.4\,cm^3 \cdot mol^{-1}$ in water, 23.4 ± 1.0 per $-C_6H_5$ group for the tetraphenylonium ions (for these solvents, including water, but only 18.0 for MeOH and EtOH) and is more solvent-dependent for each $-CH_3$ group in the ions (ranging from 0.0 for methanol to 7.5 for water).

The standard molar volumes of transfer at 25°C of small ions are represented by the operative expression reported by Marcus [36]:

$$\Delta_t V^{\infty} \left(\text{small } I^{z\pm}, W \rightarrow S\right)/cm^3 \cdot mol^{-1} =$$
$$\left[0.016\Delta V + 0.08\Delta\delta_H\right]r_I^3 - \left[0.88\Delta\kappa_T - 0.79\Delta g\right]AB_I \qquad (4.29)$$

The solvent property included in Equation 4.29 that is not defined in the previous sections is g, the Kirkwood dipole orientation parameter (Section 3.3.1). The corresponding ion property is AB_I, the ability of the ion to partake in Lewis acid-base interactions (Table 2.2 for anions). For large, hydrophobic ions the corresponding expression is:

$$\Delta_t V^{\infty} \left(\text{large } I^{\pm}, W \rightarrow S\right)/cm^3 \cdot mol^{-1} =$$
$$\left[0.146\Delta V + 0.064\Delta\delta_H - 1.037\Delta\alpha_P\right]r_I^3 - \left[0.55\Delta V - 4.73\Delta\alpha_P\right]R_{DI} \qquad (4.30)$$

where α_P is the polarizability of the solvent.

4.4 THE STRUCTURE OF SOLVATED IONS

The solvation (including hydration) of ions in the gas phase is described in Section 2.1.2 in terms of the stepwise Gibbs energies and enthalpies of the formation of ion/solvent clusters. The absolute values of these quantities diminish as the number of solvent molecules in the clusters, n, increases, and beyond the third step, the values of $\Delta_{n-1,n}G°$ are approximately proportional to $(n-1)^{-1}$. This holds for the univalent ions for which data are available from $n = 1$, but solvated clusters of multivalent ions are not stable with $n < 5$ or 6. No preference for a certain number n is apparent for the gas-phase clusters.

In solution, on the contrary, the numbers of solvent molecules around ions may show preferences for certain values, the coordination number. A cation has the solvent molecules oriented toward it with their electron-pair donor atoms, carrying a fractional negative charge, directed at the cation, possibly resulting in a coordinate bond. Small multivalent cations tend to form such bonds with water, with definite coordination numbers (6) and geometries (regular octahedron) in the first hydration shell (e.g., for Mg^{2+}). These water molecules are polarized by the charge of the cation and are hydrogen-bond donors to water molecules in a second hydration shell that may remain with the cation as it moves in the solution. Such definite solvent coordination is not confined to water but occurs with many other solvents with large donicities (Section 3.3.2.2), such as DMF and DMSO.

An anion in aqueous solutions has the water molecules pointing one of their hydrogen atoms toward it, resulting in hydrogen bonds. Anions tend to be large and have a relatively small electric field and have no definite coordination number of water molecules hydrating them, generally ≤2 for univalent anions. Multivalent anions, such as CO_3^{2-} or SO_4^{2-}, bind more water molecules by accepting hydrogen bonds from them, and anions such as HSO_4^- or $H_2PO_4^-$ also donate hydrogen bonds to adjacent water molecules.

Ions with hydrophobic groups in their periphery around a buried charge, such as $(C_6H_5)_4B^-$ or $(C_4H_9)_4N^+$, are generally only poorly solvated. However, $(C_2H_5)_4N^+$ and larger tetraalkylammonium ions may have the water in clathrate-like or enhanced tetrahedral ice-like structures around them as for nonionic hydrophobic solutes.

Solvation numbers, h_{IS}, of ions are the time-average numbers of solvent molecules residing in their first solvation shells (and in the second, if formed) and may be fractional rather than integral. If only non-directional electrostatic association takes place, then geometric constraints may occur, smaller ions having smaller solvation numbers than larger ions, although the solvent molecules are bonded more energetically to the former.

Consider an ion $I^{z\pm}$ placed at the origin of coordinates with solvent species S (or water, W) surrounding it. The number of particles of these species in a spherical shell of thickness dr at a distance r from the center of the ion is:

$$dn_{IS}(r, dr) = 4\pi r^2 g_{IS}(r)\, \rho_S° dr \tag{4.31}$$

where $\rho_S°$ is the number density (n_S/V) of the solvent in the bulk and $g_{IS}(r)$ is the conditional probability of finding a molecule of S at the distance r from the ion, the pair correlation function. There is no correlation between the particles at large distances, and hence $g_{IS}(r \to \infty) = 1$, but at very small distances $g_{IS}(r < d_{IS}) = 0$. Here d_{IS} is the distance between the center of the ion and the center of the nearest atom of the solvent molecule and the large repulsion of the electronic shells of the atoms prevent their overlapping. Generally $4\pi r^2 g_{IS}(r)\rho_S^\circ$ has a maximum at a distance somewhat beyond d_{IS}, and undulates further out, possibly having a second maximum and eventually reaching $4\pi r^2 \rho_S^\circ$ asymptotically. The solvation number h_{IS} is obtained from the integral:

$$h_{IS} = 4\pi\rho_S^\circ \int_0^{r'} g_{IS}(r) r^2 dr \tag{4.32}$$

The choice of the upper limit of the integration, r', is somewhat arbitrary: commonly it represents the distance at which $4\pi r^2 g_{IS}(r)\rho_S^\circ$ reaches its minimum after the first peak, but an alternative is to use the peak distance as r' and take the coordination number as twice the integral (4.32) up to this point, assuming symmetry of the peak. The distance of the ion from its next-nearest neighbors is given by the second peak and a corresponding second solvation number is obtained from integration to the second trough in the curve; see Figure 4.2.

The study of the constitution of the solvation shells of ions can be made along three lines. (1) Pair correlation functions and coordination numbers are obtained from diffraction of x-rays or neutrons. These refer to the geometry of solvent

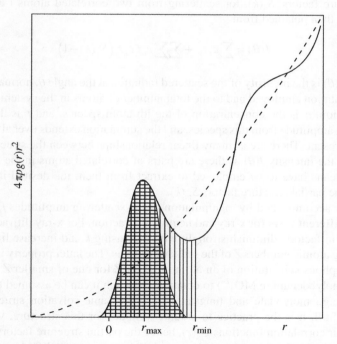

FIGURE 4.2 Coordination numbers obtained from the pair correlation function $g(r)$: the dashed curve represents $4\pi\rho r^2$, the continuous thick curve is $4\pi\rho g(r) r^2$. The cross-hatched area, symmetric with respect to r_{max} of the first peak, and the area with vertical lines up to r_{min} of the first trough are two versions of the coordination number of a solvent around an ion (Taken from Ref. 52, with permission from the publisher, Wiley).

molecules that are accommodated around an ion, whether strongly bound to it or not. (2) Such quantities can also be obtained from computer simulations, most notably from combinations of quantum mechanical and molecular dynamics calculations. (3) Certain bulk properties of solutions of electrolytes, such as the compressibility, the electrostriction, and the entropy of solvation. Such numbers tend to be smaller than the geometrical values and pertain to solvent molecules sufficiently strongly bound to the ion so that they move along with it in the solution.

4.4.1   Hydration Numbers from Diffraction Studies

The interference of radiation, x-rays or neutrons, scattered from correlated atoms yields eventually experimental values of the pair correlation function $g_{IS}(r)$. The diffraction methods yield the intensity I of the beam of the radiation at a fixed wavelength λ diffracted at various angles θ for the defined variable k:

$$k = \lambda^{-1} 4\pi \sin\left(\frac{\theta}{2}\right) \tag{4.33}$$

Structure factors, $S_{ij}(k)$, for scattering from two correlated atoms i and j in the liquid are then obtained from

$$I(\theta) = \sum c_j f_j^2 + \sum \sum c_i f_i c_j f_j \left(S_{ij}(k) - 1\right) \tag{4.34}$$

Here $I(\theta)$ is the intensity of the scattered radiation at the angle θ, normalized to the instrumentation employed and to the total number of atoms in the system exposed to the radiation, c_j is the concentration of the jth atom species, and f_j is the coherent scattering amplitude from this species, and the summation extends over all the atomic species present. There are as many linear relationships between the structure factors $S_{ij}(k)$ and the intensity $I(\theta)$ as there are pairs of correlated atoms in the system and special means have to be employed to extract from them the desired information, namely the partial structure factors $S_{IS}(k)$.

This is accomplished by manipulation of the scattering amplitudes f_j and this is done in different ways for x-ray and neutron diffraction. For x-ray diffraction, the f_j, called form factors, diminish strongly with increasing k and increase linearly with increasing atomic numbers Z of the scattering atoms. The latter property is employed in isomorphous substitution of an atom of larger Z for one of smaller Z (e.g., tungsten for molybdenum in MO_4^{2-}) to obtain $S_{IS}(k)$ when it can be assumed that the two species (e.g., molybdate and tungstate) have the same solvation structure in the solution. Still, it is the practice to assume a model for this structure, yielding the partial pair correlation functions $g_{IS}(r)$, hence the partial structure factors $S_{IS}(k)$ (see the following text), and compare them with the experimental $I(\theta)$ data. In the case of neutron diffraction, isotopic substitution is resorted to (NDIS), because the scattering takes place from the nuclei of the atoms, and the f_j, called coherent scattering lengths, are independent of k but depend on the isotopic composition of the scattering atoms. In some favorable cases, the f_j are negative (for ^7Li, ^{62}Ni, and ^1H)

whereas they are positive but specific for the majority of nuclei. Differences in the scattering intensities from two isotopically different atoms are obtained, but again the structures are refined by empirical potential structure refinement computer calculations of model structures of the solvated ions. More detailed descriptions of the methods involved in these diffraction methods are in the review by Ohtaki and Radnai [53].

Once the partial structure factors are available from the diffraction experiments, they are submitted to Fourier transformation to yield the partial pair correlation functions:

$$g_{IS}(r) = 1 + \left(2\pi^2 \rho_S^\circ r\right)^{-1} \int_0^\infty \left(S_{IS}(k) - 1\right) k \sin(kr)\, \mathrm{d}k \qquad (4.35)$$

Then the solvent-coordination number of the ion in the solution, h_{IS}, is obtained by the integration according to Equation 4.32. Second hydration shells have been definitely ascribed to divalent and trivalent cations from x-ray diffraction measurements. The coordination number h_{IW} for water molecules in this second shell is generally *assumed* to be 12, the number then being corroborated by the diffraction data.

Salt molalities $\geq 1.0\,\mathrm{mol\cdot kg^{-1}}$, and hence ion molalities $\geq 2.0\,\mathrm{mol\cdot kg^{-1}}$ for uni-univalent salts, and even larger total ion molalities for more highly charged salts, are used for diffraction measurements. The molar ratio of water-to-salt is generally in the range 4–40 and little "free" water exists in the solutions at these concentrations. Only when low concentrations ($\leq 0.5\,\mathrm{mol\cdot kg^{-1}}$) can be employed, corresponding to a mean distance apart of the ion centers of $d \geq 1.2\,\mathrm{nm}$, may some solvent molecules exist in between the solvation shells of the ions. At the high concentrations generally employed, ion pairing, in which the counter-ion penetrates into the first solvation shell of the ion to be investigated, may take place and should be taken into account.

The results of ion hydration studies by x-ray and neutron diffraction methods yielding h_{IW} values were summarized in Refs. 31 and 53. The values reported for a given ion by several authors using either method differ considerably, to some extent, due to the different concentrations that were used. Therefore, only the ranges of the reported h_{IW} values (mostly for univalent counter-ions) are shown in Table 4.7, augmented by a few more recent values from Ref. 54.

NDIS was applied to 1 m LiCl in D_2O (and also more concentrated solutions) [55] and it was concluded that the Li^+–water molecular interaction beyond the first hydration shell is not strong enough to form a well-defined second hydration shell, but other measurements do suggest its formation.

4.4.2 Hydration Numbers from Computer Simulations

Computer simulations are generally made with one ion and 256–1024 solvent molecules, that is, in fairly dilute solutions. They have the advantage that counter ions are absent, and the results pertain directly to the individual ions studied. Earlier studies of ion hydration by classical (Monte Carlo and molecular dynamics) computer simulations are summarized in Refs. 31 and 53, but recently results

TABLE 4.7   Ranges of Hydration Numbers of Ions Obtained x-ray and Neutron Diffraction [53]

Cation	$h_{IW(first)}$	$h^{\infty}_{IW(second)}$	Anion	h_{IW}
Li^+	4–6	Indicated	F^-	4–6
Na^+	4–8	None	Cl^-	5.3–8
K^+	6–8	None	Br^-	4.2–6.5
Rb^+	6–8^a		I^-	4.2–9.6
Cs^+	6–8	None	SCN^-	1.2–2.1^a
Ag^+	2–4	Present	NO_3^-	3–4.3
OH_3^+	3.1–4	None	ClO_4^-	4–5
NH_4^+	4–8	None	$H_2PO_4^-$	4.2–8.8
Be^{2+}	4	Indicated	$CH_3CO_2^-$	4^a
Mg^{2+}	6	Present	SO_3^{2-}	9
Ca^{2+}	6–7	Present	SO_4^{2-}	7.6–9.6
Sr^{2+}	8	Present	SeO_4^{2-}	8
Ba^{2+}	9.5	None	CrO_4^{2-}	12
V^{2+}	6	None	MoO_4^{2-}	12
Mn^{2+}	6	None	WO_4^{2-}	12
Fe^{2+}	6	Indicated	PO_4^{3-}	4 or 8
Co^{2+}	6	Indicated		
Ni^{2+}	5.5–7.1	Indicated		
Cu^{2+}	4+2	None		
Zn^{2+}	6	Indicated		
Cd^{2+}	6	Indicated		
Sn^{2+}	3–4	None		
Hg^{2+}	6–7	None		
Al^{3+}	6	Present		
Sc^{3+}	~7^a	None		
Cr^{3+}	6	Present		
Fe^{3+}	4.9–6.1	Indicated		
Y^{3+}	8.0	Present		
Rh^{3+}	6	Present		
In^{3+}	6	Indicated^a		
La^{3+}	8–9.1	None		
Pr^{3+}	9.2–10	None		
Nd^{3+}	8.0–10	None		
Sm^{3+}	8.0–9.3	None		
Eu^{3+}	8.3–8.9	None		
Gd^{3+}	8.0–9.9	None		
Tb^{3+}	8.2–9	None		
Dy^{3+}	6.8–9	Indicated		
Er^{3+}	6.3–8.2	Present		
Tm^{3+}	7.3–8.2	None		
Yb^{3+}	7.8–8	None		
Lu^{3+}	6.8–7.7	None		
Tl^{3+}	5–6	None		
Th^{4+}	5.5–8.1	Indicated		
U^{4+}	7.9–9.2	Present		

^{a}From Ref. 54.

were obtained from the more sophisticated application of quantum-mechanical combined with molecular-mechanical computer simulations [56, 57]. The quantum chemical treatment involves the first (more recently also the second) hydration shell of an ion, and the water beyond this is then simulated by means of molecular dynamics methods, the interface between these two regions being also carefully treated. Such calculations require large computation capacities and times and are confined to solvents that have few atoms: water and ammonia. Potentials of flexible water molecules are used in the quantum-mechanical as well as the molecular dynamics calculations and both pair potentials and three-body potentials are employed.

The primary results of these studies are appropriate $g_{IW}(r)$ pair correlation functions that, integrated according to Equation 4.25, yield the coordination (average hydration) numbers in the solutions h_{IW}. These are obtained with ≥ 499 water molecules per ion and should correspond to infinite dilutrion too, h_{IW}^{∞}. The resulting average coordination numbers of cation hydration [57] and subsequent studies are shown in Table 4.8. For most cations (except univalent ones), the second shell average hydration number could be determined. In recent years, this methodology has been applied also to anions with results also shown and referenced in Table 4.8. For some oxyanions, the average hydration number of the surface of the anion that is available for hydration is also given, it being somewhat smaller than the sum of the nearest neighbors of all the oxygen atoms of the anion. For formate and acetate, the water molecule neighbors of the carboxylate oxygen atoms were reported in Refs. 81 and 82, but are not shown in Table 4.8, because these do not represent the total hydration numbers of these anions.

4.4.3 Hydration Numbers from Bulk Properties

Solvation numbers obtained from bulk properties refer to those solvent molecules that are bound strongly to the ion and move along with it in the solution but may pertain to both the first and the second solvation shells (for multivalent ions). They are operationally defined according to the methods employed to obtain them, but when the numbers obtained from different methods agree with one another, they can be taken to have a more universal meaning. The bulk properties are measured at finite concentrations but are generally extrapolated to infinite dilution, and special considerations are required to obtain the values for individual ions. On the other hand, such measurements are also made in more concentrated solutions, and the concentration dependence of the solvation numbers can then be elucidated.

One method to employ bulk properties of electrolytes to obtain solvation numbers is to consider the electrostriction caused by the electric field of the ion, that is, the compression of the water in its hydration shell. The compression of electrostricted water *per mole of water* is $\Delta V_{Wel} = -3.5\,cm^3 \cdot mol^{-1}$ at 25°C, independently of the nature of the ion [83]. This value revised an earlier estimate of $-2.9\,cm^3 \cdot mol^{-1}$ by Marcus [84] based on less accurately established second pressure derivatives of the density

TABLE 4.8 Average Hydration Numbers of Ions at infinite Dilution Obtained by QM-MM Computer Simulations

Cation	$h^{\infty}_{IW(first)}$	$h^{\infty}_{IW(second)}$	Reference	Anion	h^{∞}_{IW}	Reference
Li^+	4.2		57	F^-	4.6	58
Na^+	5.4		57	Cl^-	5.6	58
K^+	8.3		57	I^-	8.0	58
Rb^+	7.1		57	OH^-	3.7	59
Cs^+	7.8		57	HS^-	5.9	60
Ag^+	5.4		57	NO_3^-	12.8 $(7.9)^a$	61
Au^+	4.7		57	ClO_4^-	10	62
Tl^+	5.9	17.5	63	HCO_3	6.1 $(5.5)^a$	64
Be^{2+}	4	9.0	65	HSO_4^-	10.9 $(8.0)^a$	61
Mg^{2+}	6.0	18.3	57	CO_3^{2-}	12.8 $(8.9)^a$	66
Ca^{2+}	7.0	19.1	57	SO_3^{2-}	12.5	67
Sr^{2+}	9.0	20.4	57	SO_4^{2-}	11.3	68
Ba^{2+}	9.3	23.5	57	CrO_4^{2-}	13.8	62
V^{2+}	6.0	15.8	57	$HAsO_4^{2-}$	4.8	69
Mn^{2+}	6.0	15.9	57	PO_4^{3-}	13 $(29)^b$	70
Fe^{2+}	6.0	12.4	57			
Co^{2+}	6.0	15.9	57			
Ni^{2+}	6.0	13.7	57			
Cu^{2+}	6.0	12.7	57			
Zn^{2+}	6.0	14.7	57			
Ge^{2+}	3.3 + 2.5	5.8 + 9.4	71			
Pd^{2+}	4	12	72			
Cd^{2+}	6.0	11.7	57			
Sn^{2+}	8.0	23.7	57			
Pt^{2+}	5.4		73			
Hg^{2+}	6.2	21.7	57			
Hg_2^{2+}	3.3 + 3.5		74			
Pb^{2+}	9.0	24.3	57			
UO_2^{2+}	5		75			
Al^{3+}	6.0	11.8	56			
Sc^{3+}	7.0	15.0	76			
Ti^{3+}	6.0	11.0	57			
V^{3+}	6	13.5	56			
Cr^{3+}	6.0	15.4	57			
Fe^{3+}	6.0	13.6	56			
Co^{3+}	6.0	15.2	57			
Ga^{3+}	6.0	13.6	57			
Sb^{3+}	4 + 4	8 + 5	56			
La^{3+}	9.5	25.6	77			
Ce^{3+}	9	24.2	56			
Ir^{3+}	6	13.5	78			
Tl^{3+}	5.2	18.1	63			
Bi^{3+}	9	20.5	79			
Zr^{4+}	8	17.8	56			
Ce^{4+}	9	17.4	56			
Hf^{4+}	8	17.2	56, 80			
U^{4+}	9	17.4	56			

^{a} Coordination number of surface available for the solvent.
^{b} Second solvation shell.

and permittivity of water. A stepwise, shell-by-shell electrostatic calculation according to Marcus and Hefter [85] yields the molar (*per mole of ions*) electrostriction of the water by an ion at infinite dilution, V_{Iel}^{∞}. The ratio between these quantities can be construed to represent the time-average hydration number of the ion [84]:

$$h_{IWel}^{\infty} = \frac{V_{Iel}^{\infty}}{\Delta V_{Wel}} \tag{4.36}$$

Another way to use bulk properties of the solutions is to consider the ion and the water in its first hydration shell to be non-compressible by an external pressure, the huge electric field of the ion having already produced the maximal possible compression. The hydration number is then defined by using the standard molar ionic compression, $\left(\partial V_I^{\infty}/\partial P\right)_T$, that is a negative quantity, as:

$$h_{IWcomp}^{\infty} = 1 - \left(\partial V_I^{\infty}/\partial P\right)_T / \kappa_{TW} V_W \tag{4.37}$$

Experimental $\left(\partial V_E^{\infty}/\partial P\right)_T$ values for electrolytes are split into individual ionic values on the assumption by Matheison and Conway [86], that is, $\left(\partial V_I^{\infty}/\partial P\right)_T\left(Cl^-, aq\right) = -16.5 \pm 1.5 \text{ cm}^3 \cdot GPa^{-1} \cdot mol^{-1}$ at 25°C.

A further method to obtain solvation numbers depends on the immobilization of the solvent molecules near the ions as derived from the molar ionic entropies of solvation, $\Delta_{solv}S^{\infty}$ (Section 4.3.2.3). The method proposed by Marcus [23] specifies the molar entropy of solvent immobilization as:

$$\Delta_{immob}S_I^{\infty} = \Delta_{solv}S_I^{\infty} - \Delta_{comp}S - \Delta_{el}S_I^{\infty} \tag{4.38}$$

The term $\Delta_{comp}S = -26.7 \text{ J} \cdot K^{-1} \cdot mol^{-1}$ pertains to the compression of the space available to the ions in the gas phase, RT/P°, and in the solution, $1 \text{ dm}^3 \cdot mol^{-1}$, and is independent of the natures of the ion and the solvent. The term $\Delta_{el}S^{\infty}$ is the temperature derivative of the corresponding Born Gibbs energy for the solvent beyond the first solvation shell, Equation 4.14,

$$\Delta_{el}S_I^{\infty} = \left(\frac{N_A e^2}{4\pi\varepsilon_0}\right) z_I^2 \left(r_I + 2r_S\right)^{-1} \varepsilon_r^{-2}\left(\frac{\partial \varepsilon_r}{\partial T}\right)_P \tag{4.39}$$

at a distance $\geq r_I + 2r_S$ (ion radius + solvent diameter) from the center of the ion. Based on much earlier considerations by Ulich [87], the solvation numbers are then obtained from the ratio of the entropy of immobilization of the solvent to its molar entropy of freezing, $\Delta_{fr}S_S(T^{\circ})$. This should be adjusted for the temperature of interest:

$$\Delta_{fr}S_S\left(T^{\circ}\right) = \Delta_{fr}H_S/T_{Sfr} + \int_{T_{Sfr}}^{T^{\circ}}\left\{\left[C_P\left(S, l\right) - C_P\left(S, cr\right)\right]/T\right\}dT \tag{4.40}$$

from its normal freezing temperature, T_{Sfr}, to $T°$ (generally 25°C, at which the entropy data are obtained). Here the heat capacities refer to the liquid (l) and crystalline (cr) forms of the solvent S. Then the solvation number is the ratio:

$$h_{ISentr}^{\infty} = \frac{\Delta_{immob}S_I^{\infty}}{\Delta_{fr}S_S(T°)} \tag{4.41}$$

The hydration numbers at infinite dilution from these three methods are shown in Table 4.9 and are compared with the empirical approximation $h_{IWemp}^{\infty} = 0.360|z|/(r_I/nm)$ suggested by Marcus [13] that can be used for ions for which no other value of the hydration number is available.

Hydration numbers at temperatures other than 25°C, up to 200°C, have been reported by Marcus [54, 83, 88] for some twenty common ions, obtained according to the electrostriction method and Equation 4.36. The values of h_{IWel}^{∞} increase appreciably as the temperature is raised and the compressive effect of the ionic electric field increases with the diminishing permittivity and structure of the water.

Hydration numbers diminish as the concentration of the electrolyte increases, mildly at low concentrations but strongly at larger concentrations when the hydration shells of oppositely charged ions overlap [90]. An approximate relationship for the diminution of the hydration number of an electrolyte with the concentration according to Padova [91] is:

$$\frac{h_{EW}}{h_{EW}^{\infty}} = 1 - 1.85\left(h_{EW}^{\infty}\right)^{-1} c_E^{1/2} \tag{4.42}$$

This is derived from expression (4.37) for h_{Icomp}, the approximation being the use of the limiting Debye-Hückel slope instead of the actual value for the slope of the apparent molar compressibility with the square root of the concentration. Recently a detailed examination by Marcus [92] of the isothermal compressibility of aqueous electrolytes as a function of their concentration yielded hydration numbers $h_{EW}(c)$ that exhibit the expected decrease with raised concentrations. No manner of how to split h_{EW} into the ionic contributions at finite concentrations has been proposed.

The average distance between ions in a solution, d^{av}, is inversely proportional to the cube root of the concentration as shown by Marcus [90]:

$$d^{av} = N_A^{-1/3} \cdot \left(\sum v_I c_I\right)^{-1/3} = 1.1844 \cdot \left(\sum v_I c_I\right)^{-1/3} \tag{4.43}$$

The summation extends over all the ions present and the numerical value pertains to distances in nanometer and concentrations c_I in M multiplied by their stoichiometric coefficients v_I. The hydration shells start to overlap at a limiting electrolyte concentration c_{Elim} at which from Equation 4.43 $d^{av} \leq r_{+W} + r_{-W}$, where $r_{\pm W}$ is the radius of a hydrated ion, the sum of the ionic radius and the diameter of a water molecule. This concentration c_{Elim} is as low as 1.43 M for aqueous NaCl and is lower still for solutions of unsymmetrical multivalent electrolytes (e.g., 1:2). At finite concentrations below the overlap limit c_{Elim}, experimental values of $(\partial V_I/\partial P)_T$ may be used for the estimation of the hydration numbers from the expression (4.37) for h_{Icomp} given earlier [92].

TABLE 4.9 Hydration Numbers of Ions At infinite Dilution Obtained from Electrostriction, Equation 4.36 [84], Compressibility, Equation 4.37 [52], Entropy of Hydration, Equation 4.41 [52], and Empirically, $h_{IWemp}{}^{\infty} = 0.360 \, |z_I| \, /(r_I/nm)$

Cation	$h_{IWelec}{}^{\infty}$	$h_{IWcomp}{}^{\infty}$	$h_{ISentr}{}^{\infty}$	$h_{IWemp}{}^{\infty}$
H^+	2.6[a]		4.0	
Li^+	4.4[a]	2.9	4.3	5.2
Na^+	2.9[a]	3.9	3.0	3.5
K^+	2.0[a]	3.1	1.6	2.6
Rb^+	1.8[a]	2.8	1.1	2.4
Cs^+	1.5[a]	2.3	0.9	2.1
Cu^+			4.0	3.8
Ag^+	3.5	3.4	3.1	3.1
Tl^+		1.8[b]	1.2	2.7
NH_4^+	1.5[a]	0.4	3.4	2.4
Me_4N^+		0.4[b]	4.7	1.3
Et_4N^+		0.5[b]	4.8	1.1
Be^{2+}			10.2	
Mg^{2+}	11.4	8.0	12.2	10.0
Ca^{2+}	10.0	7.5	7.8	7.2
Sr^{2+}	11.8	9.9	7.3	6.4
Ba^{2+}	10.6	9.4	5.9	5.3
Ra^{2+}			4.3	5.0
Mn^{2+}	10.5	7.7	10.7	8.7
Fe^{2+}			11.3	9.2
Co^{2+}	13.0	8.8	10.7	9.6
Ni^{2+}	14.2	9.4	11.0	10.4
Cu^{2+}	13.6	9.2	9.8	9.9
Zn^{2+}	11.6	9.3	10.3	9.6
Cd^{2+}		6.0	9.0	7.6
Sn^{2+}			7.0	7.7
Hg^{2+}			7.7	7.1
Hg_2^{2+}			9.0	
Pb^{2+}	11.1		6.1	6.1
UO_2^{2+}			15.1	2.6
Al^{3+}			18.5	20.4
Sc^{3+}			17.3	14.4
Ti^{3+}			14.5	16.1
Fe^{3+}			18.8	16.6
Ga^{3+}			19.5	17.4
Y^{3+}			16.7	12.0
In^{3+}			12.9	13.7
La^{3+}		15.1[b]	21.5	10.3
Ce^{3+}		16.7[b]	21.2	10.7
Pr^{3+}		12.9	21.9	10.8
Nd^{3+}		12.5	21.8	11.0
Sm^{3+}		11.1	21.9	11.3
Eu^{3+}		10.5	21.6	11.4
Gd^{3+}		11.0	21.85	11.5
Tb^{3+}		11.7	22.7	11.7
Dy^{3+}		12.0	22.8	11.9

(*continued*)

TABLE 4.9   (Continued)

Cation	h_{IWelec}^{∞}	h_{IWcomp}^{∞}	h_{ISentr}^{∞}	h_{IWemp}^{∞}
Ho^{3+}		12.5	22.8	12.0
Er^{3+}		12.6	23.5	12.1
Tm^{3+}		12.8	23.3	12.3
Yb^{3+}		13.0	23.5	12.4
Lu^{3+}			23.3	12.6
Tl^{3+}			14.3	12.3
Bi^{3+}			12.9	10.6
Pu^{3+}			21.1	10.7
Ce^{4+}			19.0	18.0
Hf^{4+}				20.3
Th^{4+}			24.0	14.4
U^{4+}				15.0
Anion				
F^-	3.7	5.5	4.9	2.7
Cl^-	1.7	2.1	2.3	2.0
Br^-	1.2	0.8		1.8
I^-	0.5	0.6^b		1.6
OH^-	2.3^a	6.6	5.9	2.7
HS^-			2.8	1.7
CN^-			2.1	1.9
OCN^-			2.5	1.8
SCN^-	0.5	0.1^b	1.6	1.7
N_3^-			2.2	1.9
NO_2^-			2.6	1.9
NO_3^-	1.4	0.8	2.0	2.0
ClO_3^-		1.7^b	2.1	1.8
BrO_3^-		2.6^b	2.7	1.9
IO_3^-		6.7^b	4.8	2.0
ClO_4^-	0.8^a		1.2	1.4
MnO_4^-	0.7		1.5	1.5
BF_4^-	0.7		1.6	1.6
HCO_2^-			3.9	2.1
$CH_3CO_2^-$		4.0^b	5.7	2.2
HCO_3^-		1.7^b	4.4	2.3
HSO_4^-			4.1	1.9
$H_2PO_4^-$		6.2	5.6	1.8
S^{2-}			4.6	3.9
CO_3^{2-}	5.5	14.4^b	8.8	3.6
SO_3^{2-}	4.8		8.9	3.6
SO_4^{2-}	4.8^a	10.6	6.9	3.1
CrO_4^{2-}	3.5	12.4^b	6.4	3.0
HPO_4^{2-}		16.4	9.8	3.6
PO_4^{3-}	8.9	30.9	15.8	4.5
$Fe(CN)_6^{3-}$		2.7	4.7	2.5
$Fe(CN)_6^{4-}$		12.1	10.3	3.2

[a] From Refs. 83, 88.
[b] From Ref. 89.

4.4.4 Solvation Numbers in Nonaqueous Solvents

Ionic solvation numbers in nonaqueous solvents are available mainly for univalent ions at 25°C, obtained from bulk properties of the ions at infinite dilution and properties of the solvents by the application of two of the methods dealt with earlier: electrostriction volume and entropy of solvent immobilization.

The electrostriction volume, $V_{\mathrm{ISesI}}^{\infty}$, around ions in nonaqueous solvents and the electrostrictive compression per mole of solvent, ΔV_{Sels}, was dealt with by Marcus [84]. The solvation numbers could then be calculated from $h_{\mathrm{ISel}}^{\infty} = V_{\mathrm{ISels}}^{\infty}/\Delta V_{\mathrm{Sels}}$, as shown in Table 4.10.

The entropy of solvent immobilization, $\Delta_{\mathrm{immob}}S_{\mathrm{I}}^{\infty}$, Equations 4.38 and 4.39, was dealt with by Marcus [23], where the molar entropy of freezing, adjusted for $T^{\circ} = 25°C$, $\Delta_{\mathrm{fr}}S_{\mathrm{s}}(T^{\circ})$, was also obtained. The ratios $h_{\mathrm{ISentr}}^{\infty} = \Delta_{\mathrm{immob}}S_{\mathrm{I}}^{\infty}/\Delta_{\mathrm{fr}}S_{\mathrm{S}}(T^{\circ})$ then yield solvation numbers that are also recorded (in *italics*) in Table 4.10.

It is noted that in the cases where values of the solvation numbers are available from both methods, $h_{\mathrm{ISentr}}^{\infty} > h_{\mathrm{ISel}}^{\infty}$, and it is surmised that the latter, those obtained from the electrostriction volume, are more realistic.

4.5 THE DYNAMICS OF SOLVATED IONS

4.5.1 The Mobility of Ions in Solution

The conductivity of aqueous ions and their self-diffusion are dealt with in Sections 2.3.2.1 and 2.3.2.2. Here such information concerning ions in nonaqueous solvents is briefly presented.

Einstein [93] derived from the Navier–Stokes differential equations for a sphere moving in a viscous medium an expression for the diffusion coefficient D for such spheres that may be rewritten as:

$$D = \frac{k_{\mathrm{B}}T}{6\pi r\eta} \tag{4.44}$$

where r is the radius of the sphere and η is the viscosity of the medium. The coefficient 6 arises from the assumption of zero tangential motion of the fluid near the sphere. A point that has not found an ultimate solution is whether *sticking* or *slipping* conditions should be used, when the Einstein expression is applied to particles as small as ions moving in fluids, the molecules of which are commensurate with the ionic sizes. Sticking conditions mean that at least some of the solvent in the first solvation shell "sticks" to the ion and moves along with it, whereas slipping conditions refer to the movement of the bare ion, considered as a sphere, allowing a free tangential motion of the fluid relative to the sphere. Coefficients between 4 and 6 account for diffusion coefficients under these extreme conditions and conflicting theoretical results for this hydrodynamic boundary were suggested [94]. For microscopic translational motion, the "slip" condition might be more appropriate especially for

TABLE 4.10 Ionic Solvation Numbers At 25°C in Nonaqueous Solvents from Electrostriction (roman font) and Entropy of Immobilization (italic font)

Ion	MeOH	EtOH	PrOH	EG	PC	NH$_3$	FA	DMF	MeCN	HMPT	DMSO	TMS	1,2DCIE
ΔV_{Sel}	6.0		14.4	9.5	7.1				4.9		5.0		
$\Delta_{fr}S_S(T^\circ)$	−13.8	−37.3	−47.3		−30.1	−32.9	−34.8	−81.9	−35.9	−61.3	−49.9	−23.2	−35.3
Li$^+$	3.3 *12.4*	4.9	3.7	1.0	1.4 *5.8*	5.2	2.7	1.8	4.3 *5.3*		1.0 *3.0*		*5.8*
Na$^+$	3.7 *11.2*	4.4	3.2	0.6	0.7 *5.1*	4.9	3.0	1.7	4.1 *4.8*		0 *2.7*	*5.1*	*5.4*
K$^+$	2.5 *8.6*	3.6	2.8	0.1	0 *4.3*	4.3	2.3	1.5	3.3 *3.8*			*4.8*	*4.8*
Rb$^+$	2.5 *7.3*	3.1	2.5	0	0 *3.6*	3.8	2.1	1.3	3.5 *3.5*	3.0	*2.5*	*4.0*	*4.5*
Cs$^+$	3.2 *6.2*	2.6	2.1	0	3.1	3.6	1.8	1.2	3.1 *3.3*		2.0 *2.4*	*3.2*	*4.3*
Ag$^+$	*11.7*				1.7 *5.9*		3.1	1.7	4.9 *4.9*		2.0 *3.5*	*5.0*	
Me$_4$N$^+$	8.5	3.5	2.8		4.3	4.2	3.1	1.5	4.1		2.7		*5.0*
Et$_4$N$^+$	10.7	3.4			4.8		3.7	1.7	4.1		3.1		*5.7*
Ba^{2+}	*25.*							4.1	*9.6*		*6.7*		
Zn^{2+}	*31.2*							4.1	*10.9*	6.3	*6.2*		
Cd^{2+}	*29.1*							3.6	*8.7*	7.5	*5.9*		
F$^-$	1.5			0.3			*3.1*				2.0		*4.3*

The molar electrostrictive compression of the solvent, ΔV_{Sel}/cm³·mol⁻¹, and the entropy of freezing, $\Delta_{fr}S_S(T')$/J·K⁻¹·mol⁻¹ of the solvents are also shown.

Cl⁻	1.8	0.2	1.4		4.7		3.4	1.7
	3.1		2.8	1.5	3.0	1.6	5.7	3.4
Br⁻	1.5	0.1	0.8		4.5		3.0	1.5
	2.5		2.6	1.3	2.7	1.5	5.1	3.1
I⁻	2.2	0.2	1.4		3.7		2.4	1.3
	1.4		2.3	1.0	2.3	1.4	4.4	2.7
SCN⁻	1.2		3.1		2.9		1.8	1.1
N₃⁻	4.9							
							2.3	5.6
NO₃⁻	1.0				4.3		2.1	2.2
					3.1			
ClO₄⁻	1.5		2.2		2.5		2.2	
	2.6	1.3	2.8	1.4		1.4	1.9	3.4

aqueous ions than the "stick" one [95], although the latter is generally employed when no compulsive evidence for the contrary is known.

4.5.2   Rate of Solvent Exchange Near Ions

The solvation of ions is a dynamic process, and solvent molecules in the solvation shells of ions do move out from them in exchange with molecules that come in. The rate of exchange of water molecules between the hydration shells of ions and bulk water [96] indicates the strength of the hydration and indirectly the effects of the ions on the water structure. The temperature coefficients of the self diffusion coefficient of water in the electrolyte solution, $D_{W(E)}$, and of the ion mobility, u_I (Section 2.3.2.1), yield the activation Gibbs energy of the exchange, $\Delta G^{\neq}_{exch}$:

$$d\ln u_{ion}/dT + T^{-1} - d\ln D_{W(E)}/dT = \frac{\Delta G^{\neq}_{exch}/RT^2}{\left[1 + 0.0655\exp\left(\Delta G^{\neq}_{exch}/RT\right)\right]} \quad (4.45)$$

Positive values of $\Delta G^{\neq}_{exch} > 0$ characterized small and multivalent cations, such as Li^+, Na^+, Mg^{2+}, and Ca^{2+}, and were called "positively hydrated." Large univalent cations and most anions, such as K^+, Cs^+, Cl^-, Br^-, and I^-, have negative $\Delta G^{\ddagger}_{exch} < 0$ and were designated as "negatively hydrated" [96]. These terms are no longer in common use.

The residence time of water molecules near each other in bulk water, that is, the average time it takes for a water molecule to diffuse away from its neighbors, is obtained from the diameter of a water molecule, d_W, and the diffusion coefficient of neat water, D_W^* : $\tau_W^* = 0.5 d_W^2 / D_W^* = 17$ ps at 25°C. The ratio of the average residence time of a water molecule in the hydration shell of the ion, τ_{WI}, to that in the bulk was obtained from the activation Gibbs energy of the exchange

$$\frac{\tau_{WI}}{\tau_W^*} = \exp\left(\frac{\Delta G^{\neq}_{exch}}{RT}\right) \quad (4.46)$$

The unimolecular rate constants, k_r, for water release from the hydration shells of cations [12] are expected to correspond with the values of τ_{WI} deduced from Equation 4.46. These rate constants, obtained from ultrasound absorption [97, 98], and NMR line widths [99–102] depend on the competition between water molecules and anions for sites in the coordination shell of the cations. These rate constants at 25°C span nearly 17 orders of magnitude (from K^+, the fastest to Rh^{3+}, the slowest), and hence the logarithms $\log(k_r/s^{-1})$ are shown in Table 5.4. Considerably less experimental information is known for the rate of desolvation pertaining to the first solvation shell of cations in nonaqueous [12] solvents. Hardly any experimental rate constants regarding the rates of dehydration of anions are available. Computer simulations fill this gap concerning ions of both signs in aqueous solutions (Section 5.2.2).

4.6 ACID/BASE PROPERTIES OF IONS IN SOLUTION

Ions in solution act as Lewis acids if positively charged (cations) and as Lewis bases if negatively charged (anions). The former accepts electron pairs from donor atoms in the surrounding solvent and the latter donates non-bonding electron pairs to the hydrogen atoms of protic solvents to form hydrogen bonds. The strengths of these interactions can be given quantitative measures as follows.

There exists an analogy between the transfer of ions from water to nonaqueous solvents (Section 4.3) and the corresponding transfer of non-electrolytes. The standard molar Gibbs energies of transfer of organic non-electrolytes N have been related to their properties and the properties of the target solvents by Marcus [103] by the expression:

$$\Delta_{tr}G^\infty\left(N,W \to S\right)/kJ\cdot mol^{-1} = \text{cavity term} - 43.8\alpha_N\Delta\beta_S - 26.4\beta_N\Delta\alpha_S \quad (4.47)$$

The cavity term is irrelevant to the present discussion of acid/base properties of the ions, the α_N and β_N are the Lewis acidity and basicity of the non-electrolyte, and the $\Delta\alpha_S$ and $\Delta\beta_S$ are the differences between the Lewis acidity and basicity of the non-aqueous target solvents and those of the water source solvent (Section 3.3.2). For cations, the analogous expression is according to Marcus et al. [36]:

$$\Delta_{tr}G^\infty\left(I^{z+},W \to S\right) = \text{other terms} - A_+\Delta\beta_S \quad (4.48)$$

where the "other terms" deal with quantities that do not involve the electron pair donicity (Lewis basicity) of the target solvent and $A_+ = \left[3.72z_I^2/r_I + 3.78R_D\right]$ according to Equation 4.22b. Therefore, the Lewis acidity index of the cation is obtained from Equations 4.47 and 4.48 as:

$$\alpha_+ = \alpha\left(I^{z+}\right) = \frac{A_+}{43.8} \quad (4.49)$$

In the same vein, for anions:

$$\Delta_{tr}G^\infty\left(I^{z-},W \to S\right) = \text{other terms} - B_-\Delta\alpha_S \quad (4.50)$$

where $B_- = 7.5z_I/r_I + 30\sigma_I$ according to Equation 4.22d and therefore the Lewis basicity index of the anion is:

$$\beta_- = \frac{B_-}{26.4} \quad (4.51)$$

according to Equation 4.47. These α_+ values for cations and β_- values for anions are shown in Table 4.11.

Alternatively, acceptor numbers of cations, AN_+, were obtained directly by the use of solvatochromic indicators in inert solvents. The indicator used by Linert et al.

TABLE 4.11 Lewis Acidity and Basicity indexes of Ions from the Standard Molar Gibbs Energies of Transfer and from indicator Measurements in Poorly Solvating Solvents

Cation	α_+	α_+ (from AN_+)	Anion	β_-	β_- (from DN_-)
Li^+	1.24	1.97	F^-	2.88	
Na^+	0.89	0.88	Cl^-	1.67	1.01
K^+	0.85		Br^-	1.25	0.71
Rb^+	0.92		I^-	0.71	0.14
Cs^+	1.11		OH^-	2.13	0.85
Cu^+	1.15		SH^-	0.62	
Ag^+	1.18		CN^-	1.01	−0.07
Tl^+	1.56		CNO^-	0.58	1.49
NH_4^+	0.98		SCN^-	0.35	0.49
Mg^{2+}	4.66	3.09	N_3^-	0.69	0.78
Ca^{2+}	3.54	2.67	I_3^-	−0.41	
Sr^{2+}	3.24		NO_2^-	1.31	
Ba^{2+}	2.93		NO_3^-	1.55	
Mn^{2+}	4.28	4.02	ClO_3^-	1.38	
Fe^{2+}	4.54	4.73	BrO_3^-	2.01	
Co^{2+}	4.71	4.49	IO_3^-	3.21	
Ni^{2+}	5.06	3.41	ClO_4^-	1.52	
Cu^{2+}	4.77	5.03	ReO_4^-	1.55	
Zn^{2+}	4.65	6.04	BF_4^-	1.56	
Cd^{2+}	3.85		$CH_3CO_2^-$	2.01	
Hg^{2+}	3.86		S^{2-}	1.82	
Pb^{2+}	3.91		CO_3^{2-}	3.75	
Al^{3+}	14.32		SO_4^{2-}	2.91	
Cr^{3+}	12.6		PO_4^{3-}	4.46	
Fe^{3+}	12.04				
Ga^{3+}	12.37				
Y^{3+}	8.71				
In^{3+}	9.83				
La^{3+}	7.52				
Gd^{3+}	8.55				
Lu^{3+}	9.23				
Tl^{3+}	8.88				
Bi^{3+}	8.19				
Th^{4+}	14.07				

[104] was bis-*cis*-1,10-phenantrolino-dicyanoiron(II) in the poorly solvating nitromethane solvent. Similarly, the donor number of anions, DN_-, were obtained by Linert et al. [105] by the use of the indicator N,N,N',N'-tetramethylethylenediaminocopper(II) tetraphenyl-borate in the poorly solvating 1,2-dichloroethane solvent. These acceptor and donor numbers of ions are related linearly to the Lewis acidity and basicity indexes dealt with above, that is, $\alpha_+ = (AN_+ - 27.7)/9.0$ and $\beta_- = (DN_- - 27.7)/8.5$ as in Ref. 89 and are also shown in Table 4.11.

REFERENCES

[1] A. Ben-Naim, Y. Marcus, J. Chem. Phys. **81**, 2016 (1984).

[2] Y. Marcus, Biophys. Chem. **51**, 111 (1994).

[3] C. P. Kelly, C. J. Cramer, D. G. Truhlar, J. Phys. Chem. B **110**, 16066 (2006).

[4] C. P. Kelly, C. J. Cramer, D. G. Truhlar, J. Phys. Chem. B **111**, 408 (2007).

[5] T. R. Tuttle, Jr., S. Malaxos, J. V. Coe, J. Phys. Chem. B **106**, 925 (2002).

[6] R. Gomer, G. Tyson, J. Chem. Phys. **66**, 4413 (1977).

[7] B. E. Conway, J. Solution Chem. **7**, 721 (1978).

[8] Y. Marcus, J. Chem. Soc. Faraday Trans. **87**, 2995 (1991).

[9] M. H. Abraham, J. Liszi, J. Chem. Soc. Faraday Trans. 1 **74**, 1604, 2858 (1978).

[10] R. H. Stokes, J. Am. Chem. Soc. **86**, 979 (1964).

[11] R. M. Noyes, J. Am. Chem. Soc. **84**, 513 (1962).

[12] Y. Marcus, Ion Solvation, Wiley, Chichester (1985).

[13] Y. Marcus, J. Chem. Soc. Faraday Trans. 1 **83**, 339 (1987).

[14] A. Loewenschuss, Y. Marcus, Chem. Rev. **84**, 89 (1984).

[15] Y. Marcus, J. Chem.Thermodyn. **48**, 70 (2012).

[16] Y. Marcus, Russ. J. Electrochem. **44**, 16 (2008).

[17] S. I. Drakin, A. V. Mikailov, Zh. Fiz. Khim. **36**, 1698 (1962).

[18] E. A. Mezhov, G. A. Reimarov, N. L. Khananashvili, Radiochemistry **34**, 5 (1992).

[19] V. I. Parfenyuk, Coll. J. Russ. Acad. Sci. **64**, 588 (2002).

[20] Y. Marcus, in D. V. Bostrelli, ed., Solution Chemistry Research Progress, Nova Science, Hauppauge, NY, 51–68 (2008).

[21] V. I. Parfenyuk, Russ. J. Electrochem. **44**, 50 (2008).

[22] H. F. Halliwell, S. W. C. Nyburg, Trans. Faraday Soc. **59**, 1126 (1963).

[23] Y. Marcus, J. Solution Chem. **15**, 291 (1986).

[24] R. Alexander, A. J. Parker, J. H. Sharpe, W. E. Waghorne, J. Am. Chem. Soc. **94**, 1148 (1972).

[25] S. Villermaux, J. J. Delpuech, J. Chem. Soc. Chem. Commun. 478 (1975).

[26] H. M. Koepp, H. Wendt, H. Strehlow, Z. Elektrochem. **64**, 483 (1960).

[27] G. Gritzner, Inorg. Chim. Acta **24**, 5 (1977).

[28] L. I. Krishtalik, N. M. Alpatova, E. V. Ovsyannikova, Electrochim. Acta **36**, 435 (1991).

[29] J.-I. Kim, Z. Phys. Chem. (Munchen) **113**, 129 (1978).

[30] M. H. Abraham, A. Nasehzadeh, Can. J. Chem. **57**, 71, 2004 (1979).

[31] Y. Marcus, Pure Appl. Chem. **55**, 977 (1983).

[32] Y. Marcus, M. J. Kamlet, R. W. Taft, J. Phys. Chem. **92**, 3613 (1988).

[33] Y. Marcus, Z. Naturfosch. A **50**, 51 (1995).

[34] C. Kalidas, G. Hefter, Y. Marcus, Chem. Rev. **100**, 819 (2000).

[35] Y. Marcus, Chem. Rev. **107**, 3880 (2007).

[36] Y. Marcus, Electrochim. Acta **44**, 91 (1998).

[37] G. Somsen, L. Weeda, J. Electroanal. Chem. Interface Electrochem. **29**, 375 (1971).

[38] E. M. Arnett, D. R. McKelvey, J. Am. Chem. Soc. **88**, 2598 (1966).

[39] H. L. Friedman, J. Phys. Chem. **71**, 1723 (1969).

[40] B. G. Cox, A. J. Parker, J. Am. Chem. Soc. **95**, 402 (1973).

[41] Y. Marcus, Pure Appl. Chem. **57**, 1103 (1985).

[42] G. Hefter, Y. Marcus, W. E. Waghorne, Chem. Rev. **102**, 2773 (2002).

[43] F. Hörzenberger, G. Gritzner, J. Chem. Soc. Faraday Trans. **89**, 3557 (1993).

[44] C. M. Criss, R. R. Held, E. Luksha, J. Phys. Chem. **72**, 2970 (1968).

[45] M. H. Abraham, Y. Marcus, J. Chem. Soc. Faraday Trans. 1 **82**, 3225 (1986).

[46] Y. Marcus, G. Hefter, J. Chem. Soc. Faraday Trans. **92**, 757 (1996).

[47] G. Hefter, Y. Marcus, J. Solution Chem. **26**, 249 (1997).

[48] Y. Marcus, G. Hefter, Chem. Rev. **104**, 3405 (2004).

[49] Y. Marcus, G. Hefter, T.-S. Pang, J. Chem. Soc. Faraday Trans. **90**, 1899 (1994).

[50] Y. Marcus, H. B. D. Jenkins, L. Glasser, J. Chem. Soc. Dalton Trans. 3795 (2002).

[51] Y. Marcus, J. Mol. Liq. **118**, 3 (2005).

[52] Y. Marcus, Introduction to Liquid State Chemistry, Wiley, Chichester (1977).

[53] H. Ohtaki, T. Radnai, Chem. Rev. **93**, 1157 (1993).

[54] Y. Marcus, J. Phys. Chem. B **113**, 10285 (2009).

[55] I. Howell, G. W. Neilson, J. Phys. Cond. Mat. **8**, 4455 (1996).

[56] T. S. Hofer, A. K. H. Weiss, B. R. Randolf, B. M. Rode, Chem. Phys. Lett. **512**, 130 (2011).

[57] B. M. Rode, C. F. Schwenk, T. S. Hofer, B. R. Randolf, Coord. Chem. Rev. **249**, 2993 (2005).

[58] A. Tongraar, S. Hannongbua, B. M. Rode, J. Phys. Chem. A **114**, 4334 (2010).

[59] H. H. Loeffler, B. M. Rode, J. Chem. Phys. **117**, 110 (2002).

[60] C. Kitayakornupong, V. Vchirawongkwin, B. M. Rode, J. Phys. Chen. B **114**, 12883 (2010).

[61] V. Vchirawongkwin, C. Kitayakornupong, A. Tongraar, B. M. Rode, J. Phys. Chen. B **115**, 12527 (2011).

[62] E. Hinteregger, A. B. Pribil, L. H. V. Lim, T. S. Hofer, B. R. Randolf, B. M. Rode, Inorg. Chem. **49**, 7964 (2010).

[63] V. Vchirawongkwin, T. S. Hofer, B. R. Randolf, B. M. Rode, J. Comput. Chem. **28**, 1006, 1057 (2007).

[64] V. Vchirawongkwin, A. B. Pribil, B. M. Rode, J. Comput. Chem. **31**, 249 (2010).

[65] S. S. Azam, T. S. Hofer, A. Bhattacharjee, L. H. V. Lim, A. B. Pribil, B. R. Randolf, B. M. Rode, J. Phys. Chem. B **113**, 9289 (2009).

[66] V. Vchirawongkwin, C. Kitayakornupong, B. M. Rode, J. Phys. Chen. B **114**, 11561 (2010).

[67] L. Eklund, T. S. Hofer, A. B. Pribil, B. M. Rode, I. Persson, Dalton Trans. **41**, 5209 (2012).

[68] V. Vchirawongkwin, B. M. Rode, I. Persson, J. Phys. Chen. B **111**, 4150 (2007).

[69] A. Bhattacharjee, A. B. Pribil, L. H. V. Lim, T. S. Hofer, B. R. Randolf, B. M. Rode, J. Phys. Chen. B **114**, 3921 (2010).

[70] A. B. Pribil, T. S. Hofer, B. R. Randolf, B. M. Rode, J. Comput. Chem. **29**, 2330 (2008).

[71] B. M. Rode, L. H. V. Lim, J. Mol. Liq. **157**, 29 (2010).

[72] S. A. A. Shah, T. S. Hofer, M. Q. Fatmi, B. R. Randolf, B. M. Rode, Chem. Phys. Lett. **426**, 311 (2006).

[73] T. S. Hofer, B. R. Randolf, B. M. Rode, I. Persson, Dalton Trans. 1512 (2009).

[74] T. S. Hofer, B. R. Randolf, B. M. Rode, Chem. Phys. **349**, 210 (2008).

[75] R. J. Frick, T. S. Hofer, A. B. Pribil, B. R. Randolf, B. M. Rode, J. Phys. Chem. A **113**, 12496 (2009).

[76] V. Vchirawongkwin, C. Kitayakornupong, A. Tongraar, B. M. Rode, Dalton Trans. **41**, 11889 (2012).

[77] O. M. D. Lutz, T. S. Hofer, B. R. Randolf, B. M. Rode, Chem. Phys. Lett. **536**, 50 (2012).

[78] P. Pedevilla, T. S. Hofer, B. R. Randolf, B. M. Rode, Aust. J. Chem. **65,** 1582 (2012).

[79] S. Durdagi, T. S. Hofer, B. R. Randolf, B. M. Rode, Chem. Phys. Lett. **406**, 20 (2005).

[80] C. B. Messner, T. S. Hofer, B. R. Randolf, B. M. Rode, Chem. Phys. Lett. **501**, 292 (2011).

[81] A. Payaka, A. Tongraar, B. M. Rode, J. Phys. Chem. A **113**, 3291 (2009).

[82] A. Payaka, A. Tongraar, B. M. Rode, J. Phys. Chem. A **114**, 10443 (2010).

[83] Y. Marcus, J. Chem. Phys. **137**, 154501 (2012).

[84] Y. Marcus, J. Phys. Chem. B **109**, 18541 (2005).

[85] Y. Marcus, G. Hefter, J. Solution Chem. **28**, 579 (1999).

[86] J. G. Matheison, B. E. Conway, J. Solution Chem. **3**, 455 (1974).

[87] H. Ulich, Z. Elektrochem, Angew. Phys. Chem. **36**, 497 (1930).

[88] Y. Marcus, J. Phys. Chem. B **116**, 7232 (2012).

[89] Y. Marcus, Ion Properties, Dekker, New York (1997).

[90] Y. Marcus, J. Solution Chem. **38**, 513 (2009).

[91] J. Padova, J. Chem. Phys. **40**, 691 (1964).

[92] Y. Marcus, J. Phys. Chem. B **118**, 10471 (2014).

[93] A. Einstein, Ann. Phys. **19**, 289 (1906).

[94] J. R. Schmidt, J. L., Skinner, J. Chem. Phys. **119**, 8062 (2003).

[95] P. C. F. Pau, J. O. Berg, W. G. McMillan, J. Phys. Chem. **94**, 2671 (1990).

[96] O. Ya. Samoilov, Disc. Faraday Soc. **24**, 141 (1957).

[97] T. J. Giligan, G. Atkinson, J. Phys. Chem. **84**, 208 (1980).

[98] N. Purdie, C. A. Vincent, Trans. Faraday Soc. **63**, 2745 (1967).

[99] J. Burgess, Metal Ion Solvation, Ellis Horwood, Chichester (1978).

[100] E. v. Goldhammer, in Modern Aspects of Electrochemistry, Plenum, New York, Vol. **10**, 77 (1975).

[101] J. Reuben, D. Fiat, J. Chem. Phys. **51**, 4918 (1969).

[102] H. Eyring, ed., Physical Chemistry: An Advanced Treatise, Acadermic, New York, Vol. **7**, 228 (1975).

[103] Y. Marcus, J. Phys. Chem. **95**, 8886 (1991).

[104] W. Linert, R. F. Jameson, G. Bauer, A. Taha, J. Coord. Chem. **42**, 211 (1997).

[105] W. Linert, R. F. Jameson, A. Taha, J. Chem. Soc. Dalton Trans. 3181 (1993).

5

MUTUAL EFFECTS OF IONS AND SOLVENTS

5.1 ION EFFECTS ON THE STRUCTURE OF SOLVENTS

Ions that are solvated in highly structured solvents (Section 3.3.1), such as light and heavy water, ethylene glycol, and formamide among few others, are expected to affect the structure of the solvents beyond the first (and second, if formed) solvation shell. However, there is very scant information regarding such effects in solvents other than H_2O. The effects of ions on the structure of water may be assessed by a variety of experimental methods as well as by computer simulations. Beyond the hydration shells of the ions and up to some distance away, some ions enhance the native structure of the water whereas other ions destroy it. The former are called cosmotropic ions or structure-makers and the latter kind of ions are chaotropic or structure-breakers. Recent reviews by Marcus [1, 2] may be consulted for further details.

5.1.1 Experimental Studies of Ion Effects on the Structure of Solvents

5.1.1.1 Self-diffusion of Water Molecules The rate of exchange of water molecules between the hydration shells of ions and bulk water, Section 4.5.2, indicates indirectly the effects of the ions on the water structure. According to Equation 4.46, the sign of the Gibbs energy of activation of this exchange, $\Delta G^{\neq}_{\text{exch}}$, permitted the division of ions into positively and negatively hydrated categories according to Samolilov [3] that correspond to structure-making and structure-breaking ones.

Ions in Solution and Their Solvation, First Edition. Yizhak Marcus.
© 2015 John Wiley & Sons, Inc. Published 2015 by John Wiley & Sons, Inc.

The self diffusion coefficients of water molecules in aqueous alkali halide solutions, $D_{W(E)}$, were obtained at 23°C by McCall and Douglass [4] and at 0°C by Endon et al. [5] from NMR measurements. More recent data, for 25°C, at moderate concentrations and including some divalent metal chlorides, are also available in the report by Müller and Hertz [6]. When both cation and anion are structure-breakers, that is, negatively hydrated according to $\Delta G^*_{\text{exch}} < 0$ as is the case for KX, RbX, and CsX, where X = Cl, Br, and I, then $(1 - D_{W(E)}/D_W^*)$ is negative. When at least one of the ions is strongly structure making, as is the case for LiX and NaX, where X = Cl and Br; and for LiI and also for MF, where M = K, Rb, and Cs; and M'Cl$_2$ for M' = Mg, Ca, and Zn, then $(1 - D_{W(E)}/D_W^*)$ is positive. When these tendencies are of the same magnitude and opposite, as for NaI, then $D_{W(E)} \sim D_W^*$ and $(1 - D_{W(E)}/D_W^*)$ is near zero. Data on aqueous NaClO$_4$, LiClO$_4$, and Mg(ClO$_4$)$_2$ at 25°C by Heil et al. [7] indicate that the structure-making properties of the cations predominate over those of the structure-breaking perchlorate anion so that $(1 - D_{W(E)}/D_W^*) > 0$. The self diffusion of water in aqueous Bu$_4$NCl was obtained at room temperature by Nowikow et al. [8] from quasi-elastic neutron scattering, yielding $(1 - D_{W(E)}/D_W^*) > 0$ and the salt is a net structure maker. On the basis of equating the values for K$^+$ and Cl$^-$, the ionic $(1 - D_{W(I)}/D_W^*)$ values are shown in Table 5.1.

The D/H isotope effect for the self diffusion coefficient of the water in solutions of CsCl in D$_2$O and H$_2$O was according to Sacco et al. [16] in agreement with the structure-breaking properties of both ions of this salt and with the more extensive ("stronger") hydrogen bonded network of the D$_2$O relative to H$_2$O, see Table 3.8.

5.1.1.2 Viscosity B-coefficients

Gurney appears to be the person who coined the terms "structure making" and "structure breaking" in the connection of the viscosities of aqueous ions [17], replacing subsequently those of positive and negative hydration. The quantitative measure of this effect is the Jones-Dole B-coefficient (Section 2.3.2.3). A critical compilation of B_η values for over 70 aqueous ions, based on the assumption that $B_\eta(\text{Rb}^+) = B_\eta(\text{Br}^-)$ which does not differ appreciably from $B_\eta(\text{K}^+) = B_\eta(\text{Cl}^-)$, was published by Jenkins and Marcus [9]. Structure-making ions have $B_\eta > 0$ and structure-breaking ones have $B_\eta < 0$. The values of the ionic B_η in aqueous solutions at room temperature are shown in Table 5.1.

The mechanism by which the ions affect the viscosity, given their assumed water structure–modifying behavior, is not really known. In a flowing aqueous electrolyte solution, the ions require void spaces to move into while breaking some of their hydration shells, so that both the sizes of the ions and the strength of their hydration play a role. The average intermolecular void space in water is $(V_W^* - V_{\text{vdW W}})/N_A = 0.0094 \, \text{nm}^3$, and it is created randomly by the thermal movement of the water molecules. The intrinsic volumes of ions that they take up in crystals were reported by Marcus et al. in Ref. 18. Large cations with volumes $(4\pi/3)r_I^3 > 0.010 \, \text{nm}^3$ are too large for moving into a randomly available hole near them. They need to destroy some of the hydrogen bonded structure of the water in order to create a cavity for their accommodation. They are "structure-breakers" and should accelerate the flow of the solution, accounting for their negative B_η values. Small cations with volumes $(4\pi/3)r_I^3 < 0.007 \, \text{nm}^3$ may fit into random cavities and

TABLE 5.1   **Relative Self Diffusion $(1 - D_{W(E)}/D_W)$ and Ionic B-coefficients of Viscosity and NMR Relaxation At 25°C**

	$1 - D_{W(I)}/D_W$	B_η	$B_{NMR(1H)}$	$B_{NMR(17O)}$ [a]	$(1 - \tau_{W(I)}/\tau_W)/10$	b
References	6	9	10	11	12	13
Li^+	0.113	0.146	0.14	0.120	0.141	−0.10
Na^+	0.077	0.085	0.06	0.053	0.053	0.34
K^+	−0.024	−0.009	−0.01	−0.017	−0.010	1.01
Rb^+	−0.041	−0.033	−0.04		−0.023	
Cs^+	−0.048	−0.047	−0.05	−0.014	−0.032	1.28
Ag^+		0.090	0.06			
NH_4^+		−0.008			−0.028	
$H(D)_3O^+$			0.06	0.036	0.062	
Me_4N^+		0.123	0.18	0.165 (0.172)	0.059	
Et_4N^+		0.385		0.444 (0.421)	0.096	
Pr_4N^+		0.916		0.889 (0.868)	0.137	
Bu_4N^+		1.275		1.33 (1.24)	0.180	
Ph_4P^+		1.073		0.831		
Be^{2+}	0.29 [4]	0.450				
Mg^{2+}	0.236	0.385	0.50			
Ca^{2+}	0.124	0.298	0.27			
Sr^{2+}		0.272	0.23			
Ba^{2+}	0.17 [4]	0.229	0.18			
Zn^{2+}	0.19 [4]	0.361				
Al^{3+}	0.41 [4]	0.750				
Th^{4+}	0.74 [4]	0.860				
F^-	0.086	0.127	0.14	0.120	0.161	
Cl^-	−0.024	−0.005	−0.01	−0.017	−0.010	0.56
Br^-	−0.046	−0.033	−0.04	−0.026	−0.027	0.46
I^-	−0.095	−0.073	−0.08	−0.055	−0.059	1.04
$OH(D)^-$	0.09 [4]	0.120	0.18	0.083	0.144	
CN^-		−0.024	−0.04	0.120		
SCN^-		−0.032 [14]	−0.07			0.75
N_3^-		−0.018	0.00			
ClO_3^-	−0.046	−0.024	−0.08			
ClO_4^-	−0.08 [4]	−0.058	−0.085			0.94
BrO_3^-		0.007	−0.06			
IO_3^-		0.138	0.02			
ReO_4^-		−0.055	−0.03			
NO_2^-		−0.024	−0.05			
NO_3^-	−0.08 [4]	−0.045	−0.05		−0.027	0.77
$CH_3CO_2^-$	0.13 [4]	0.236				
BPh_4^-		1.115		0.928		
CO_3^{2-}	0.19 [4]	0.278	0.25			−0.13
SO_3^{2-}	0.18 [4]	0.282	0.22			1.22
SO_4^{2-}		0.206	0.12			

[a] Values in parentheses are from Ref. 15.

enhance the hydrogen bonding through their electric fields. They are "structure-makers" and slow down the rate of flow having $B_\eta > 0$.

For anions, the border between structure-making and structure-breaking is at $(4\pi/3)r_I^3 \sim 0.020 |z_I| nm^3$, the magnitude of the charge also playing a role, as suggested by Marcus [19].

The scant data on electrolyte B-coefficients in solvents other than H_2O do show cases of negative values, denoting solvent structure breaking. For structure-breaking ions in D_2O, the B_η values are even more negative than those in light water, because D_2O is more structured than H_2O (Table 3.8) so that there is more structure to break. Thus, in D_2O $B_{\eta I}/M^{-1} = -0.02$ for K^+ and Cl^- (assumed equal), -0.05 for Br^-, -0.07 for Cs^+ and Et_4N^+, and -0.10 for I^-. In ethylene glycol, $B_{\eta E}/M^{-1} = +0.033$ for KI but -0.080 for CsI, and in glycerol, the $B_{\eta E}/M^{-1}$ for these two salts are -0.185 and -0.405, so that structure breaking by the three ions involved is indicated. Contrary to expectation, no cases of $B_{\eta E} < 0$ were reported by Notley and Spiro [20] for the highly structured formamide (Table 3.8).

5.1.1.3 NMR Signal Relaxation

The ^1H NMR longitudinal relaxation times of the water-proton, T_{1E}, was measured at 25°C in many aqueous electrolyte solutions by Engel and Hertz [10]. The results can be well expressed by:

$$[(1/T_{1E})/(1/T_{1W}) - 1] = B_{nmr}c_E + \cdots \tag{5.1}$$

which is analogous to the Jones-Dole expression for the viscosities (Section 2.3.2.3). The convention that $B_{nmr}(K^+) = B_{nmr}(Cl^-)$ was used to obtain the ionic values shown in Table 5.1. Structure-making ions or those classified as "positively hydrated" have $B_{nmr} > 0$ and structure-breaking or the "negatively hydrated" ones have $B_{nmr} < 0$. These ionic B_{nmr} values corresponded well with the ionic B_η values according to Engel and Hertz and also to Abraham et al. [10, 21]. The B_{nmr} values are limiting slopes, but the ratios of the NMR signal relaxation times, τ_I/τ_W, proportional to the ratios $(1/T_{1E})/(1/T_{1W})$, have the same signs up to large concentrations as reported by Chizhik [22].

Such ^1H NMR measurements of longitudinal relaxation times, T_{1E}, could be applied only to diamagnetic ions, and hence for (paramagnetic) transition metal cations, the ^{17}O NMR spin-lattice relaxation of D_2O molecules in aqueous salt solutions was measured and reported by Yoshida et al. [11]. Again, setting $B_{nmr}(K^+) = B_{nmr}(Cl^-)$ provided acceptable results in agreement with the ^1H B_{nmr} values [10] for ions studied by both methods. Also, the signs of $\Delta B_{nmr}/\Delta T$ from 0 to 25°C [10] and from 5 to 25°C [15, 21] agreed well for the two NMR methods. For most of the structure-breaking ions, these signs were opposite to those from viscosity, dB_η/dT, for a not well-understood reason.

5.1.1.4 Dielectric Relaxation

The so-called "slow water" epithet pertains to the cooperative reorientation time of water molecules, $\tau_W^* = 8.27 \pm 0.02$ ps in pure water (or 8.38 ps, with data at very high frequencies also included, see Section 5.2.1.2). The complex permittivities in aqueous alkali halide solutions measured as a function of the frequency by Kaatze and coworkers [12, 23] yielded the cooperative reorientation times τ_{WE} of water molecules in 1 M solutions at 25°C. The ionic values, τ_{WI}, were obtained from the relaxation data [12], setting the values for K^+ and Cl^- as equal. The function $\left(1 - \tau_{WI}/\tau_W^*\right)$ is shown in Table 5.1. (normalized by division by 10 to make it commensurate with the other values shown there). Structure-breaking ions have $\tau_{WI}/\tau_W^* < 1$ and structure-making ones have $\tau_{WI}/\tau_W^* > 1$. Another set of measurements of the cooperative reorientation time of bulk water, τ_{WE}, at salt concentrations $c_E < 1$M and at 25°C by Buchner and coworkers [24–31] yielded the coefficient b_E of Equation 5.2:

$$\tau_{WE} = \tau_W^* + a\left[\exp(-b_E c_E) - 1\right] \tag{5.2}$$

A comparison of these b_E coefficients with the viscosity B_η of the salts that were dealt with permitted Marcus [19] to split them into the ionic contributions b_I shown in Table 5.1. They are linear with the viscosity coefficients $b_I = (0.71 \pm 0.06) - (1.58 \pm 0.28)B_{\eta I}$, but with some scatter and a correlation coefficient of 0.897. The linearity of the b_I with the $B_{\eta I}$ means that the cooperative reorientation time of bulk water is a valid measure of the structure-making and structure-breaking properties of the ions. Values of b_I for ions not listed in Table 5.1. could also be derived from published data: malonate 2.76 [28], triflate 0.96, imide 1.47 [32], whereas others may be estimated if desired from the $B_{\eta I}$ [9]. The results for the larger tetraalkylammonium ions indicated that the dielectric relaxation times of water molecules near hydrophobic solutes are some three times slower than that of bulk water. Ice-like cages around the hydrophobic cations could cause hindered orientation of the water molecules according to Barthel et al. [25, 33], but shielding of water molecules located at the hydrophobic parts from "attack" by incoming water molecules could also be responsible for slowness of the exchange of water molecules [31, 34]. Numerical values of τ_{WE} for these salts have not been tabulated, but the slopes $d\tau_{WE}/dc_E > 0$ could be seen in figures and they contrast with those for salts with small ions, that is, $d\tau_{WE}/dc_E < 0$.

5.1.1.5 Vibrational Spectroscopy

Near-infrared spectroscopy was used by Choppin and Buijs [35] to study aqueous electrolytes by means of the resolved bands at 1.16, 1.20, and 1.25 μm. These bands were considered to correspond to water molecules with none, one, or two hydrogen bonds. Shifts to more hydrogen bonds per water molecule were related to water structure-making properties assigned to La^{3+}, Mg^{2+}, H^+, Ca^{2+}, OH^-, and F^-. Structure-breaking properties inferred from shifts to fewer hydrogen bonds were assigned to K^+, Na^+, Li^+, Cs^+, Ag^+, ClO_4^-, I^-, Br^-, NO_3^-, Cl^-, and SCN^-. However, the latter assignments for Na^+, Li^+, and Ag^+ do not agree with the results shown in Table 5.1, obtained from dynamic measures. The infrared absorbance contours of OD and OH stretching of solutions of HDO in H_2O and in D_2O were split in the presence of salts according to Kecki et al. [36]. The results led to the conclusion that the poorly hydrated ClO_4^- anion breaks down the water structure, contrary to the better hydrating anions: Cl^-, Br^-, and I^-. The 1.15 μm infrared band of 1 M aqueous electrolyte solutions consisted of two components as shown by

Bonner and Jumper [37], corresponding to hydrogen-bonded and non-bonded water groups. Cations increased the fraction of the hydrogen-bonded water relative to that in pure water whereas anions decreased it, but it is not clear how the observed changes were allocated to cations and anions.

FTIR spectroscopy was applied by Nickolov and Miller [38] to the O–D stretching vibration in 8 mass% HOD in H_2O in aqueous KF, CsF, NaI, KI, and CsCl with water-to-salt ratios optimally < 20. Water structure breaking was inferred from the narrowing and the blue-shifting of the peak of the 2380 cm^{-1} band and structure making from the opposite trends. The two fluoride salts were deemed to be structure makers and the others structure breakers. Attenuated total reflection infrared (ATR-IR) spectroscopy was recently applied by Kitadai et al. [39] to aqueous solutions of 22 electrolytes with ion concentrations up to 2 M. The O–H stretching band was monitored to yield the molar absorptivity of the water. The effects of individual cations were obtained using sulfate anions as a benchmark and those of the individual anions were obtained by subtraction of the effects of sodium and potassium ions from the salt values. The differences in area between the molar absorptivities at the lower (2600–3420 cm^{-1}) and the higher (3420–3800 cm^{-1}) frequency regions were well linearly correlated with other measures of the ionic effects on the structure of the water, namely the viscosity B_η (Section 5.1.1.2) and the NMR B_{NMR} (Section 5.1.1.3) coefficients and the structural entropy (see Section 5.1.1.7).

Raman spectra of alkali metal perchlorate solutions in H_2O and D_2O containing HOD were obtained according to Walrafen [40], resulting in pronounced splitting of the OD and OH stretching bands in the presence of ClO_4^-, in agreement with the infrared study of Kecki et al. [36]. This anion is a strong water structure-breaker, reducing the fraction of fully hydrogen bonded water molecules similarly to raising the temperature. The conclusion that the ClO_4^- anion is not hydrated is contrary to the results obtained from double pulse ultra-fast infrared spectroscopy by Omta et al. [41]. The Raman scattering intensities of the 6427 and 7062 cm^{-1} bands of 6 M HOD in D_2O were measured by Holba in the absence and presence of 0.5–2.0 M salts [42]. Such values of R_N, the ratio of the band intensity for 1 M salt solutions to that in the absence of salt, which are smaller than 1 denote structure-breaking effects, in the sense of decreasing the amounts of hydrogen-bonded water molecules, whereas $R_N > 1$ denote structure-making effects. The values of R_N are non-additive with respect to the constituting ions (at the 1 M concentrations employed): they are 0.944 for NaCl and 0.932 for KCl, but in reverse order, 0.876 for NaSCN and 0.878 for KSCN, and hence no splitting into ionic contributions can be made. The temperature dependence of the Raman spectra for the O–H stretching vibration of sodium halide salts in H_2O was used by Li et al. [43] who inferred that at 20°C F$^-$ does not affect the Raman spectrum appreciably, but Cl$^-$, Br$^-$, and I$^-$ ions do so by increasing water structure-breaking in this order as expected.

Pulsed two-color mid-infrared ultra-fast spectroscopy was used to study the effect of ions on the structural dynamics of their aqueous solutions by Bakker's group [44]. The pump pulse that excited the O–H (or O–D) stretch vibration to the first excited state was provided first, and then after a short delay, the probe pulse, which was red-shifted with respect to the first, probed the decay of this state. Solutions of 0.5 to

10 M of lithium, sodium, and magnesium halides and of KF, $NaClO_4$, $Mg(ClO_4)_2$, and Na_2SO_4 in 0.1 to 0.5 M HDO in D_2O were studied by this technique. The rotational anisotropy

$$R = \frac{\Delta\alpha_{\parallel} - \Delta\alpha_{\perp}}{\Delta\alpha_{\parallel} + 2\Delta\alpha_{\perp}} \tag{5.3}$$

was related to the reorientation dynamics of water molecules in the electrolyte solutions. Here $\Delta\alpha_{\parallel}$ and $\Delta\alpha_{\perp}$ are the change in the light absorption for the probe pulse being parallel and perpendicular to the pump pulse, and the orientational correlation times for bulk water molecules are $\tau_r = -t/\ln R(t)$, where t is the time delay. These τ_r times in 0.5–6 M $NaClO_4$, in 0.5 and 1 M $Mg(ClO_4)_2$, and in 1 M Na_2SO_4 were independent of the salt concentration, 3.4 ± 0.1 ps, the same as for pure water. These short times contrast with the cooperative orientation times, $\tau_w^* = 8.27 \pm 0.02$ ps, which was studied by dielectric relaxation that depends on the electrolyte concentration and on its nature (Section 5.1.1.4).

5.1.1.6 X-Ray Absorption and Scattering
The electronic structure of water and indirectly its hydrogen bonding structure can be probed by means of x-ray absorption spectroscopy (XAS) and x-ray Raman scattering (XRS) at the oxygen K edge. The x-ray absorption spectrum of pure water is characterized by a small pre-edge at 535 eV, a main peak around 538 eV and a postpeak (shoulder) near 541 eV. The observed results were interpreted in terms of changes in the electronic structure of the unoccupied orbitals of the water molecules as well as by the hydrogen bonding, namely in terms of free –OH groups characterized by the pre-edge and double hydrogen-bonded water molecules in the postedge.

The hydrogen-bonded network in aqueous salt solutions was investigated by Näslund et al. [45] by both XAS and XRS. The pre-edge, at 534–537 eV, is sensitive to the presence of solutes, but the main absorption or scattering peak around 538 eV is not. The 535 and 536.5 eV peaks were enhanced in 1 m aqueous NaCl and KCl compared with water, but on the contrary, the absorption at these energies in 2.7 m $AlCl_3$ was considerably smaller. These changes in the absorption were attributed to the effect of cations on the hydrogen bonding of the water, on the assumption that the chloride anion has little or no effect. Water structure breaking by K^+ and Na^+ was deduced from the increase in the fraction of single hydrogen bond donor water molecules, equivalent to a significant decrease in the fraction of tetrahedrally coordinated water, relative to the fractions that exist in pure water.

Aqueous 0.8 to 4.0 M solutions of NaCl, NaBr, and NaI were studied by XAS by Cappa et al. [46]. Little enhancement of the pre-edge at 535 eV was observed for NaCl, but NaBr and NaI exhibited appreciable enhancements of the 535 eV pre-edge, increasing in this order. The oxygen K edge XAS of aqueous 2 and 4 M chlorides of Li^+, Na^+, K^+, NH_4^+, $C(NH_2)_3^+$ (guanidinium), Mg^{2+}, and Ca^{2+} was then examined by Cappa et al. [47]. The spectra of the divalent cations were quite different from those of solutions of univalent ones, the latter not depending appreciably on the nature of the cation. The perturbation of the electronic structure of the water in the case of the univalent cation chlorides is thus due to the Cl⁻ anions, the XAS being mainly sensitive to donor hydrogen bonds of the water molecules. This is contrary to the assumption by Näslund et al. [45] that the chloride anion has little or no effect on the hydrogen bonding of the water. A possible explanation for the XAS of Mg^{2+} and Ca^{2+} chlorides differing from those of the univalent cation salts may be sought in the surface charge densities, $z_I/(r_I/nm)^2$ of Mg^{2+} (386) and Ca^{2+} (200) compared with those of the univalent cations Na^+ (96) and K^+ (53) (but not of Li^+ (210!)).

Contrary to the enhanced pre- and main-edge absorption intensities and decreased postedge intensities in 4 m NaCl solutions, the 2 and 4 m HCl spectra showed a diminution of the pre-edge intensity, no effect on the main-edge intensity, and an increase in the postedge intensity, linearly with the concentration reported by Cappa et al. [46]. The absorption intensity changes induced by the Cl⁻ anion were counteracted by the hydrated H_3O^+ cations. The oxygen K edge XAS of 4 and 6 M aqueous KOH solutions showed a new pre-pre-edge at 532.5 eV, as well as a strong enhancement of the pre-edge intensity, a strong diminution of the main-edge intensity, and an enhancement of the post-edge intensity with a blue shift of its energy. These phenomena were attributed by Cappa et al. to the OH⁻ anion [47], its behavior being fundamentally different from that of halide anions.

Altogether the XAS results yielded little *new* insight on the effects of ions on the water structure, having been obtained in rather concentrated solutions, that is, >1 M. Controversial interpretations [45, 46] do not contribute to the understanding.

5.1.1.7 *Structural Entropy*

A molar structural entropy of ions in solution, $\Delta_{str}S_I$, is obtained from the standard molar entropy of solvation of the ion, $\Delta_{solv}S_I^\infty$, when certain irrelevant quantities are subtracted from the latter. These include the compression entropy (change of available volume) on transfer from the gas to the solution and contributions from the formation of the ionic solvation shell and possible limitation of the ionic rotation of a polyatomic ion in the solution compared with the gas. The terms kosmotropes for water structure-making ions and chaotropes for water structure-breaking ions were introduced by Collins [48] (Section 5.1) according to whether $\Delta_{str}S < S_W^*$ or $\Delta_{str}S > S_W^*$, respectively. This view stressed the competition between the water–water interactions and the ion effects on the water structure.

However, it remains to specify explicitly the structural entropy of ions. The deduction from $\Delta_{solv}S_I^\infty$ of a "neutral entropy," equated with the value for a nonpolar solute of size (radius) similar to that of the ion was suggested by Abraham et al. [21, 49]:

$$\Delta S_n = 5.0 + 291(r_I/nm) \, J \cdot K^{-1} \cdot mol^{-1} \tag{5.4}$$

This expression took care of the compression entropy and the numerical coefficients pertain to ion hydration. The entropy change corresponding to the formation of the solvation shell was obtained from the temperature derivative of the Born expression for the Gibbs energy of solvation (Section 2.3.1.4):

$$\Delta S_{el} = \left(\frac{N_A e^2}{8\pi\varepsilon_0}\right) z^2 \left(r_I + d_S\right)^{-1} \varepsilon^{-1} \left(\frac{\partial \ln\varepsilon}{\partial T}\right)_P \tag{5.5}$$

It pertained to the solvent beyond the first solvation shell, the thickness of which was the molecular diameter of the solvent, d_S. Linear correlations of $\Delta_{\text{struc}}S = \Delta_{\text{solv}}S_I^\infty - \Delta S_n - \Delta S_{\text{el}}$ thus calculated for aqueous solutions of ions with their viscosity B_η coefficients and with the corresponding NMR B_{nmr} coefficients were noted by Abraham et al. [21].

A somewhat different approach by Marcus [50] specified the irrelevant entropy of compression as $\Delta_{\text{comp}}S = -26.7\,\text{J K}^{-1}\,\text{mol}^{-1}$, using the standard state of 0.1 MPa for the ideal gaseous ions and 1 M for the ions in solution. The same expression, Equation 5.5, for the electrostatic entropy of the solvent beyond the first solvation shell of the ion I was used for ΔS_{el}, but for the h_I solvent molecules within this shell that were translationally immobilized, the deduction of their entropy from $\Delta_{\text{solv}}S_I^\infty$ was:

$$\Delta_{\text{trim}}S(X^z) = \left(\frac{3}{2}\right)R\ln\left[\frac{M(\text{IS}_{hI})}{M(\text{I})}\right] - h_I S_{\text{trS}} \qquad (5.6)$$

The larger mass, M, of the solvated ion than that of the bare one yielded the first term (Section 4.4.3, with $h_I = 0.360\,|z_I|/(r_I/\text{nm})$ for aqueous solutions), and S_{trS} (26.0 J K^{-1} mol^{-1} for water) was the molar translational entropy of the liquid solvent that did not participate in the solvent structural effects. The supposition that sodium ions were indifferent with respect to the water structure making and structure breaking was then made [50]:

$$\Delta_{\text{str}}S(\text{Na}^+) = \Delta_{\text{hy}}S^\infty(\text{Na}^+) - \Delta_{\text{comp}}S - \Delta S_{\text{el}}(\text{Na}^+) - \Delta_{\text{trim}}S(\text{Na}^+) = 0 \qquad (5.7)$$

If $\Delta_{\text{str}}S < 0$, then the ions are assigned to structure-making category and if $\Delta_{\text{str}}S > 0$ to structure-breaking category.

Subsequently, $\Delta_{\text{str}}S$ was based by Marcus [51] on a model common for various thermodynamic functions of ion hydration used by him [52], based in turn on the width Δr_I of the electrostricted hydration shell. The volume of the fully compacted water molecules in this shell was $\pi d_W^3/6$ each (where $d_W = 0.276\,\text{nm}$ was the diameter of a water molecule), rather than V_W/N_A pertaining to bulk water. The hydration shell with h_I compacted water molecules had a volume

$$\left(\frac{4\pi}{3}\right)\left[(r_I + \Delta r_I)^3 - r_I^3\right] = \frac{h_I \pi d_W^3}{6} \qquad (5.8)$$

with $h_I = 0.360\,|z_I|/(r_I/\text{nm})$, as before, from which Δr_I could be deduced.

The entropic effects of the creation of a cavity in the water to accommodate the ion and of the compression from the gas to the solution were taken care of by a neutral term $\Delta S_{\text{nt}} = -3 - 600(r_I/\text{nm})\,\text{J}\cdot\text{K}^{-1}\cdot\text{mol}^{-1}$ [51]. The electrostatic entropy effects were calculated separately for the electrostricted hydration shell of width Δr_I and for the water near the ion but outside this shell. The permittivity and its temperature derivative in the former were assumed to have the infinitely large field values: $\varepsilon' = n_D^2 = 1.776$ and $(\partial\varepsilon'/\partial T)_P = 2(\partial n_D/\partial T)_P = -1\times10^{-4}\,\text{K}^{-1}$ (at 298.15 K), representing dielectric saturation. Then, in analogy with Equation 5.5, the electrostatic entropic effect in this shell was:

$$\Delta S_{\text{el1}} = \left(\frac{N_A e^2}{8\pi\varepsilon_0} \right) z^2 \left[\Delta r_{\text{I}} \left(r_{\text{I}} + \Delta r_{\text{I}} \right)^{-1} \right] \varepsilon'^{-2} \left(\frac{\partial \varepsilon'}{\partial T} \right)_P \tag{5.9}$$

and in the water layers beyond this shell:

$$\Delta S_{\text{el2}} = \left(\frac{N_A e^2}{8\pi\varepsilon_0} \right) z^2 \left(r_{\text{I}} + \Delta r_{\text{I}} \right)^{-1} \varepsilon_{\text{r}}^{-2} \left(\frac{\partial \varepsilon_{\text{r}}}{\partial T} \right)_P \tag{5.10}$$

The structural entropy was then obtained from:

$$\Delta_{\text{str}} S_{\text{I}} = \Delta_{\text{hyd}} S^{\infty} - \left[\Delta S_{\text{nt}} + \Delta S_{\text{el1}} + \Delta S_{\text{el2}} \right] \tag{5.11}$$

Due to cumulative errors in such calculations, only values of $\Delta_{\text{struc}} S(\text{I}^z) > 20\,\text{J K}^{-1}\,\text{mol}^{-1}$ indicated that the ion I^z was definitely water structure-breaking and only values $< -20\,\text{J K}^{-1}\,\text{mol}^{-1}$ indicated it to be structure making, whereas in-between values designated the ions to be borderline cases [51]. Applied to nearly 150 monatomic and polyatomic aqueous cations and anions, with charges $-4 \leq z_{\text{I}} \leq 4$, yielded the linear correlation with the viscosity B_{η}:

$$\Delta_{\text{str}} S_{\text{I}} / \text{J} \cdot \text{K}^{-1} \cdot \text{mol}^{-1} = 20 \left(z_{\text{I}}^2 + |z_{\text{I}}| \right) - 605 \left(B_{\eta} / \text{dm}^3 \text{mol}^{-1} \right) \tag{5.12}$$

(excluded were the tetraalkylammonium cations). The values of $\Delta_{\text{str}} S$, positive for large ions of low charge and negative for highly charged small ions, are shown in Table 5.2, adapted from the review by Marcus [1].

Positive values of $\Delta_{\text{str}} S_{\text{I}}$ at lower temperatures became negative at a characteristic limiting temperature when the temperature range was 15–65°C according to Krestov and coworkers [54, 55]. This decrease in the structure-breaking effect of the ions was explained by the diminishing inherent structure of the water as the temperature was raised (Section 3.3.1).

The structural heat capacity, $\Delta_{\text{str}} C_{\text{PI}}$, could also describe the effects of ions on the water structure as suggested by Marcus [51]. For this purpose, C_{p} replaced S, the factors $T(\partial^2 \varepsilon' / \partial T^2)_P$ and $T(\partial^2 \varepsilon_{\text{r}} / \partial T^2)_P$ replaced the corresponding factors in Equations 5.9 and 5.10, and $\Delta C_{\text{p nt}} = (175z - 48) + 1380(r/\text{nm})\,\text{J K}^{-1}\,\text{mol}^{-1}$. The resulting $\Delta_{\text{str}} C_{\text{PI}}$ values are shown in Table 5.2, with positive values for structure-making ions and negative ones for structure-breaking ones, with a wide borderline region of $\pm 60\,\text{J K}^{-1}\,\text{mol}^{-1}$.

5.1.1.8 Transfer from Light to Heavy Water

The effects of ions consisting of changing the average number of hydrogen bonds per water molecule that characterizes the water structure (Section 3.3.1), should, in principle, be the best estimate of their water structure affecting properties. The direct ion–water interactions in light and heavy water are very similar, because of the very similar solvating properties of molecules of these two kinds of water (Chapter 3). It was therefore assumed by

TABLE 5.2 Water Structural Entropy $\Delta_{str}S_I/\mathrm{J\,K^{-1}\,mol^{-1}}$ and Structural Heat Capacity $\Delta_{str}C_{PI}/\mathrm{J\,K^{-1}\,mol^{-1}}$ Effects [51] and the Changes of the Hydrogen Bond Geometrical Factors, $\Delta G_{HB(I)}$, of Representative Ions [53]

Ion	$\Delta_{str}S_I$	$\Delta_{str}C_{PI}$	$\Delta G_{HB(I)}$
Li^+	−52	147	−0.03
Na^+	−14	83	−0.22
K^+	47	0	−0.48
Rb^+	52	−38	−0.50
Cs^+	68	−83	−0.66
Ag^+	−15	47	−0.50
Tl^+	46	−49	−0.86
NH_4^+	5	28	−0.55
Me_4N^+	41	−30	−0.14
Et_4N^+	−5		
Pr_4N^+	−86		
Bu_4N^+	−144	62	1.02
Ph_4As^+	−34	153	−0.75
Mg^{2+}	−113	287	
Ca^{2+}	−59	215	
Sr^{2+}	−53	183	
Ba^{2+}	−18	132	−1.29
Mn^{2+}	−87	265	
Fe^{2+}	−152	242	
Co^{2+}	−123	268	
Ni^{2-}	−128	268	
Cu^{2+}	−103	280	
Zn^{2+}	−104	273	
Cd^{2+}	−89	244	−1.02
Hg^{2+}	−60	247	
Pb^{2+}	−21	156	
Al^{3+}	−38	646	
Y^{3+}	−111	415	
La^{3+}	−113	355	
Gd^{3+}	−99	392	
Lu^{3+}	−122	407	
Tl^{3+}	−59	404	
Bi^{3+}	−47		
Zr^{4+}	−44	867	
Th^{4+}	−119	653	
U^{4+}	−109	687	
OH^-	−51	−187	
F^-	−27	20	0.32
Cl^-	58	−62	−0.48
Br^-	81	−88	−0.84
I^-	117	−113	−1.13

TABLE 5.2   (Continued)

Ion	$\Delta_{str}S_I$	$\Delta_{str}C_{PI}$	$\Delta G_{HB(I)}$
CN^-	58		
N_3^-	58		
I_3^-	116	−179	
SCN^-	83	−33	
NO_2^-	47	−55	
NO_3^-	66	−59	
ClO_3^-	62	−58	
BrO_3^-	43	−80	−0.37
IO_3^-	−15	−51	
BF_4^-	93	−78	
ClO_4^-	107	−87	−0.40
MnO_4^-	100	−83	
HCO_2^-	23	−20	
$CH_3CO_2^-$	−32	−138	
BPh_4^-	−43	145	−0.75
CO_3^{2-}	−52	68	
SO_3^{2-}	−51	6	
SO_4^{2-}	8	−14	
SeO_4^{2-}	30		
CrO_4^{2-}	25	−15	
MoO_4^{2-}	−3	−53	
WO_4^{2-}	6	−47	
PO_4^{3-}	−131	103	
AsO_4^{3-}	−94		
$Fe(CN)_6^{3-}$	180	−472	
$Fe(CN)_6^{4-}$	108	131	

Ben-Naim [56] that the measurable difference in the standard chemical potentials $\Delta\mu_X^{\infty H-D}$ of a solute particle X introduced into H_2O and into D_2O depended solely on changes in the hydrogen bonding structure of the water. It should be represented by the product of the different energies of hydrogen bonding in the two kinds of water and the difference in the geometrical factor describing the average hydrogen bonding in them:

$$\Delta\mu_X^{\infty H-D} = \Delta^{H-D}e_{HB} \cdot \Delta G_{HB(X)} \qquad (5.13)$$

The difference in the strengths of the hydrogen bonds in D_2O and H_2O was taken by Marcus [50, 51, 53] to be $\Delta^{H-D}e_{HB} = -929\,J\,mol^{-1}$ at 25°C. The change, $\Delta G_{HB(X)}$, caused by the introduction of a particle of X, in the average geometrical factor $0 \le G_{HB(X)} \le 1$ specifying whether a hydrogen bond exists or not is

$$\Delta G_{HB(X)} = \left(\frac{2}{N}\right)\left[\left\langle\sum_N G_{HB}\right\rangle_{X_s} - \left\langle\sum_N G_{HB}\right\rangle_0\right] \qquad (5.14)$$

taken over all the configurations of the N water molecules of either kind present.

This approach was applied by Marcus and Ben-Naim [57] to ionic solutes $X = I^{\pm}$, but the experimentally measurable quantities, such as solubilities and EMF data, yielded rather unsatisfactory $\Delta\mu_I^{\infty HD}$ data for electrolytes [53]. Based on the assumption that the $\Delta\mu_I^{\infty H-D}$ value is equal for K^+ and Cl^-, the "best" available values at 25°C of $\Delta\mu_I^{\infty H-D}/J \cdot mol^{-1}$ ranged from −950 for Bu_4N^+ to +1200 for Ba^{2+} with probable errors of ±100. Incidentally, this assumption also equalizes the values for Ph_4As^+ and BPh_4^-, for which no preference for hydration by H_2O or D_2O is expected.

The effects of the ions on the structure of the water were then described by Marcus [51, 53] as the ratios $\Delta G_{HB(I)} = \Delta\mu_I^{\infty H-D}/\Delta^{H-D}e_{HB}$ according to Equation 5.13. The water structure effects of ions according to this approach are shown in Table 5.2 — structure makers having positive values and structure-breakers negative values. These results are unsatisfactory, due to the inaccuracy of the $\Delta\mu_I^{\infty H-D}$ data, making the divalent cations Ba^{2+} and Cd^{2+} appear as strong water-structure breakers and Li^+ as a mild structure breaker, contrary to all other information concerning these ions. The available data for the nine alkali metal and halide ions appear to be the most accurate, and their correlations with other quantities that describe the water structural effects of ions are:

$$\Delta G_{HB(I)} = -(0.54 \pm 0.11) + (4.75 \pm 1.39)\left(B_\eta/M^{-1}\right) \tag{5.15}$$

and

$$\Delta G_{HB(I)} = -(0.14 \pm 0.06) - (8.16 \pm 1.01) \times 10^{-3}\left(\Delta_{str}S_I/J \cdot K^{-1} \cdot mol^{-1}\right) \tag{5.16}$$

with standard errors of the fits of 0.2 units.

5.1.1.9   Internal Pressure   The internal pressure, $P_{int} \approx T\alpha_P/\kappa_T$, of water increases as the temperature is raised, $P_{int}/MPa = 63 + 0.352(T/K)$, contrary to most other liquids as shown in the review by Marcus [58]. This behavior was related to the breakdown of the large hydrogen bonded aggregates in water, and similar effects on the internal pressure of water in the presence of salts were ascribed by Dack [59] to their water structure-making and structure-breaking effects. The experimental P_{int} values for 1 M salt solutions at 25°C, accurate to 4 to 8 MPa, were compared with those of certain non-electrolytes (urea, formamide, acetonitrile, dioxane, and piperidine) that established the size (volume) effect of the solutes as:

$$P_{int\,vol}/MPa = 190 + 0.509V \tag{5.17}$$

where $V = M/\rho$ (molar mass divided by the density) is the molar volume of the solute, whether a non-electrolyte or a crystalline salt. The values of $\Delta P_{int} = P_{int}(1 M salt) - P_{int\,vol}$ were shown by Dack [59] in a figure to be either positive, for structure-breaking salts, or negative for a few structure-making salts. This was reasonable for (1:1) salts, but for multivalent salts at 1 M concentration, this calculation of ΔP_{int} overestimated it. For such salts, ΔP_{int} should be corrected by $[168 MPa - P_{int}(1 M\,salt)]/2$, where 168 MPa is the internal pressure of water at 25°C. The resulting values adopted from Dack and from Leyendekkers [59, 60] are shown in Table 5.3, but some questionable assignments to water structure breaking and structure making are noted.

TABLE 5.3 The internal Pressure Differences ΔP_{int} At 25°C of Aqueous 1 M (or 1 m) Electrolytes (values in parentheses corrected for valency, see the text), their Structural Temperature Differences $\Delta T = T_{str} - T$, and their Osmotic Coefficient Differences $\varphi(0.4\,m) - \varphi(0.2\,m)$

Electrolyte	ΔP_{int}/MPa	ΔP_{int}/MPa	ΔT/K(IR)	ΔT/K(NMR)	$\Delta\varphi$
References	59	60	60	61	62
LiCl	−13	−6	5		0.015
LiBr					0.016
LiI	−8	−6	8.0		0.029
NaOH	48				
NaF					−0.017
NaCl	31		3.9	5	−0.005
NaBr	28				0.001
NaI	15		6.5		0.009
NaSCN	5		8.5		0.004
NaClO$_4$	12		18.9	19	−0.008
KF					−0.005
KCl	21		4.6	3	−0.011
KBr	19		5.8	8	−0.010
KI	9		7^a	6	−0.005
KSCN	10		9.6		−0.011
KNO$_3$			8.0	8	−0.040
CsCl					−0.022
CsI					−0.025
Li$_2$SO$_4$	27(−10)	36(−1)			−0.009^b
Na$_2$CO$_3$	100(28)	88(14)	−4.0	−6	−0.034^b
Na$_2$SO$_4$	89(20)	75(6)	0.6		−0.048
(NH$_4$)$_2$SO$_4$	14(−23)		−1.8		−0.041
MgCl$_2$	13(−10)		−3		0.042
MgI$_2$					0.078
MgSO$_4$	35(−5)		−8.6	−8	−0.033
CaCl$_2$					0.032
SrClI$_2$					0.026
BaCl$_2$	64(7)	53(−4)	−6		0.016
BaI$_2$					0.058
ZnSO$_4$					−0.041
CuSO$_4$	45(−4)				−0.037
CdCl$_2$	44(0)				0.036

^{a}From Ref. 63.
^{b}From Ref. 64.

A multitude of additive terms make up the ionic P_{int} values according to Leyendekkers [65], only some of which are straight forward, the rest depending on ad-hoc assumed values of their parameters. Her work was discussed in detail in the review by Marcus [58]. It is rather questionable whether this treatment provides independent estimates of the effects of the ions on the structure of water.

5.1.1.10 Some Other Experimental Results The "structural temperature," T_{str}, of an electrolyte solution at a given actual temperature T is that temperature, at which pure water would have effectively the same inherent structure. Structural temperatures of electrolyte solutions are, however, only defined in terms of the method used for their determination.

The differences $\Delta T = T_{str} - T$ at $T = 25°C$ obtained by Leyendekkers [60] from infrared spectroscopy measuring the fraction of non-bonded OH groups absorbing light at the peak of free OH (that measured in water at 200°C) according to Luck [66, 67] are shown as $\Delta T(IR)$ in Table 5.3. The value of ΔT is positive when the structure-breaking properties of the ions making up a salt dominate over the structure-making ones but is negative otherwise. Raman spectroscopic data for fairly concentrated solutions of $NaClO_4$ reported by Walrafen [68], of NaBr and Bu_4NBr reported by Worley and Klotz [63], and infrared band shift data for the salts in the Me_4NBr to Bu_4NBr series reported by Bunzl [69] have also been interpreted in terms of the structural temperatures.

NMR data by Milovidova et al. [61] yielded $\Delta T(NMR)$ values of 1 M electrolytes in H_2O at 20°C that agree with the $\Delta T(IR)$ data and are shown in Table 5.3. However, the concentration and temperature dependences of ΔT do show differences between NMR and infrared results. In fact, Abrosimov [70] showed that ΔT changes sign from positive to negative for NaCl near 43°C and for KCl near 67°C. The balance between the structure-breaking and structure-making properties of the cation and anion, producing the sign and size of ΔT, is quite delicate. As seen from the dated references, the concept of structural temperature of aqueous electrolytes has more or less been abandoned in recent years.

A quite different approach by Marcus [62] compares the osmotic coefficient φ of aqueous electrolytes at two concentrations: $\Delta \varphi = \varphi(0.4\,m) - \varphi(0.2\,m)$ at 25°C (Table 5.3). Structure-making cations (Li^+, alkaline earth metal cations) yield $\Delta\varphi > 0$, being the larger for salts with the more structure-breaking anions. Structure-breaking cations (K^+, Cs^+) combined with any anions yields $\Delta\varphi < 0$. Sodium ions are borderline in this respect, yielding rather small positive or negative values of $\Delta\varphi$. Salts of a multivalent anion, CO_3^{2-} or SO_4^{2-}, have $\Delta\varphi < 0$ even for multivalent, structure-making cations such as Mg^{2+} and Zn^{2+}.

5.1.2 Computer Simulations of Ion Effects on the Structure of Solvents

The most extensive treatment of the structural effects of ions on the solvent surrounding them has been made by quantum-chemical treatment (charge field modified, in more recent years) of their first (or first and second) solvation shell combined with molecular-mechanical computer simulations of the solvent beyond the(se) solvation shell(s), the interface between these two regions being also carefully treated. Only small solvent molecules with few atoms, namely water (and for very few ions also ammonia), could be treated in this manner, because of the large expenditure of computer time required for the quantum chemical simulation.

The hydration numbers of ions that were obtained from computer simulations are already described in Section 4.4.2 and in Table 4.8. The ability of ions to affect the hydrogen bonded structure of the water, to enhance or diminish it, is closely related to the dynamics of the exchange of water molecules between the hydration shells of the ions and the bulk water, obtained by computer simulations and dealt with in the Section 5.2.1.

5.2 ION EFFECTS ON THE DYNAMICS OF THE SOLVENT

The information presented in Sections 5.1.1.1–5.1.1.4 and Table 5.1, although construed to pertain to the effects of ions on the structure of the solvent, in the sense of whether it is enhanced or loosened by the presence of ions, actually reflects the effects on the dynamics of the solvent in the immediate neighborhood of the ions. The mean residence times of water molecules in the vicinity of ions are indirectly measures of the effect of the ions on the structure of the water as described in Section 5.2.1. There are aspects of solvent dynamics that are not covered by these effects, such as the orientational relaxation rate and hydrogen-bond lifetimes. Two experimental methods have mainly been employed for obtaining such information: ultrafast mid-infrared and dielectric relaxation spectroscopy on the fs to ps time scales. Some slower processes were studied by NMR relaxation studies. Computer simulations added additional information, since it could be applied to individual ions rather than salts. As for the ion effects dealt with in the previous sections, the vast majority of the studies dealt with ions in aqueous solutions and only few ones considered ions in nonaqueous solvents

5.2.1 Mean Residence Times of Solvent Molecules Near Ions

The mean residence times, MRT, of water molecules in the immediate vicinity of ions were studied extensively by means of these quantum-mechanical combined with molecular-mechanical computer simulations as reviewed at the time by Hofer et al. [70]. The computational program employed has evolved over the years as was the minimal time t^*, above which a molecule is deemed to have left its position in the immediate vicinity of an ion, from 2 ps in the earlier studies to 0.5 ps used in the later ones. The MRT of water molecules in the bulk solvent, $\tau'^*_w = 1.7\,ps$, is only one-tenth of the time it takes the molecule to diffuse completely away. The *relative* mean residence times of water molecules in the second hydration shell to that in bulk water, $RMRT = \tau'_{w12}/\tau'^*_w$ (in %) at 25°C, are shown in Table 5.4. The MRT of water in the first hydration shells of multivalent ions are longer than could be studied by the computations. The $RMRT$s of water molecules near the ions are roughly proportional to the surface density of the charge on the ions, σ_I: $RMRT = 0.22 + 1.14(\sigma_I/C \cdot nm^{-2})$, but exceptions are noted.

The relative mean residence times, $RMRT$, of water molecules in the vicinity of ions indicate whether the ions are water structure making if they are >100% or structure breaking if they are <100% [91]. Note that the $RMRT$s for large univalent ions, both cations and anions, are <100%. This means that around such ions, the water molecules are more free to move than those bound in the hydrogen-bonded network

TABLE 5.4 Water Exchange Rates From the Vicinity of Ions: Rate Constants, $\log(k_r/s^{-1})$ for Exchange from the First Hydration Shell, and the Relative Mean Residence Times (RMRT in %) of Second Hydration Shell Molecules[a]

Ion	$\log k_r$	References	RMRT	References	Ion	RMRT	References
Li+	9.1	71	86	72	F-	~200	73
Na+	9.1	71	107	74	Cl-	110	73
K+	9.3	71	93	74	I-	65	75
Rb+	9.0	71	88	76	HS-	161, 142[a]	77
Cs+	9.0	71	76	78	NO3-	88	79
Ag+			80	80	ClO4-	88	81
Au+			127	82	HCO2-	132	83
Tl+			76	84	CH3CO2-	172	85
Be2+	3.5	86	320	74	HCO3-	54, 76[b]	87
Mg2+	5.2, 5.7	86, 88	280, 240[c]	89	HSO4-	83,167	90
Ca2+	8.5	86	260	91	SO3^2-	188	92
Sr2+	8.6	86	350	93	SO4^2-	153	81
Ba2+	8.9	86	320	91	CrO4^2-	135	81
V2+	1.9	94	440	91	HAsO4^2-	109	95, 96
Mn2+	7.4, 75	97	400 (24)	91	PO4^3-	229	81
Fe2+	6.5	97	320, 180[d]	98			

Ion	log k_r	References	RMRT	References	Ion	log k_r	References	RMRT	References
Co^{2+}	6.4,6.1	97	400(26)	99	Ru^{3+}	–6	88		
Ni^{2+}	4.5,4.4	97	?	100	Rh^{3+}	–7.5	88		
Cu^{2+}	9.3,8.3	88,97	450	91	In^{3+}	4.3	88		
Zn^{2+}	7.5	86	190	101	Sb^{3+}	8.3	94	140	102
Cd^{2+}	8.2	88	270(10)	91	La^{3+}	8.5	94	140	103
Hg^{2+}			88	91	Ce^{3+}		94,106	160	103
Hg_2^{2+}			180	104	Gd^{3+}	8.8,7.3	94		
Ge^{2+}			120	105	Lu^{3+}	7.9			
Sn^{2+}			110	107	Ir^{3+}			210	103
Pb^{2+}			140	108	Tl^{3+}	9.5	88	750	110
Pd^{2+}			270	109	Bi^{3+}	4	88	500	111
Pt^{2+}			?	109	Zr^{4+}			320	113
Al^{3+}	–0.8	86	1060	112	Ce^{4+}			350	103
Tl^{3+}	5.0,4.8	86,88	2180(37)	114	Hf^{4+}			910	113
V^{3+}	3.2	97	280	103	U^{4+}			480	113
Cr^{3+}	–6.3	86	440(22)	91	UO_2^{+}			250	115
Fe^{3+}	4.3,3.5	86,88	240,160	98	UO_2^{2+}			320	115
Co^{3+}			650(55)	91					
Ga^{3+}	3.3	88							

^{a}The $RMRT$ beyond d $t^* = 0.5$ ps (values in parentheses are the second shell MRT in ps for $t^* = 2.0$ ps).

^{b}$RMRT = 54\%$ near the H atom and 76% near the O atoms.

^{c}$RMRT = 83\%$ near three of the O atom, 167% near the fourth, and 324% near the H atoms.

^{d}$RMRT = 115\%$ near the S atom and 161% near the H atom.

of bulk water. According to this concept, the water-structure-breaking ions, those with <100% *RMRTs*, include K^+, Rb^+, Cs^+, Ag^+, Tl^+, I^-, and ClO_4^-, all the other ions studied to date being structure-makers, in particular multi-charged ions, as expected from the experimental studies in Sections 5.1.1. Vchirawongwin et al. [84] reported Tl^+ to be the strongest water structure-breaking ion, and according to Kritayakornupong et al. [114], Ti^{3+} appeared to be the most structure-making ion among the multivalent cations studied to date. Multivalent anions are only mildly structure making, because they are larger and have lower electrical fields that orient water molecules around them than cations of the same charge number.

The *RMRTs* of water molecules in the second hydration shell of cations (the first for univalent ones) are compared in Table 5.4. with $\log(k_r/s)$, the (logarithm of the) experimental (mainly from NMR measurements) rate constant of the first-order reaction of water molecules leaving the hydration shells of cations in exchange for incoming molecules. The larger the *RMRT* of the water molecules, the slower is the exchange as measured by $\log(k_r/s)$, but a definite proportionality or linear dependence could not be established.

Similar studies have been made on ions in liquid ammonia (at 240 K). The mean residence times of ammonia molecules in the second solvation shells of the ions studied are longer than for water molecules: 12.7 ps compared to 2.6 ps for Ag^+ [116], 28.5 ps compared with 6.5 ps for Co^{2+} [117], but shorter in the case of Cu^{2+} 3.2 ps [118] compared with 7.7 ps for water [91]. Molecular dynamics computer simulations of solutions of ions in liquid ammonia [119] yielded the self-diffusion coefficients of ammonia molecules, $D/10^{-9}$ $m^2 s^{-1}$, in the solvation shells of K^+ 6.1 and of I^- 7.4, shorter than the value for ammonia molecules in the bulk liquid, 11.5 ± 1.5. These studies thus indicate that K^+ and I^- are structure breakers and Ag^+ and Co^{2+} are structure makers regarding the inherent structure of liquid ammonia.

5.2.2 Experimental Studies of Ion Effects on the Solvent Orientation Dynamics

5.2.2.1 Ultrafast Infrared Spectroscopy

Pulsed two-frequency (ultrafast, femtosecond) polarization-resolved mid-infrared spectroscopy was used in a series of papers by Bakker and coworkers to study the effect of ions on the structural dynamics of their aqueous solutions [44, 120]. The solvent consisted of mixtures of H_2O and D_2O (generally 0.1 M HDO in D_2O) and the first, the pump, pulse excited the O–H or O–D stretch vibration to the first excited state that then relaxed at a measurable rate. The second, the probe, pulse was red-shifted with respect to the first and probed the decay of this excited state. Generally, fairly concentrated electrolyte solutions were required for the application of this technique, in the range 0.5 to 10 M. The rotational anisotropy is as follows:

$$R = \left(\frac{\Delta\alpha_{\parallel} - \Delta\alpha_{\perp}}{\Delta\alpha_{\parallel} + 2\Delta\alpha_{\perp}} \right) \tag{5.18}$$

was the primary result of the method, where $\Delta\alpha_{\parallel}$ and $\Delta\alpha_{\perp}$ were the light absorption changes for the probe pulse, parallel and perpendicular to the pump pulse. The reorientation dynamics of water molecules in the electrolyte solutions is then obtained from

$$\tau_{or} = \frac{-t}{\ln R(t)} \tag{5.19}$$

after a delay of 3 ps between pump and probe pulses, in order to remove vibrational relaxation and spectral diffusion effects that occur much faster than the molecular reorientation. In subsequent studies, R was expressed as a sum of two exponential functions:

$$R_{OD\cdots X}(t) = A_f \exp\left(\frac{-t}{\tau_f}\right) + A_s \exp\left(\frac{-t}{\tau_s}\right) \tag{5.20}$$

where subscript f denotes fast relaxation for OD groups hydrogen bonded to water having bulk relaxation dynamics and subscript s pertains to a slow process for OD groups hydrogen bonded to the anion or the water molecules hydrating it.

In pure water, the exponential relaxation rate is described by $\tau_{or} = 2.5 \pm 0.2$ ps for the O–D stretch vibration in H_2O, somewhat shorter than the 3.0 ± 0.2 ps for the O–H relaxation in D_2O. This is because the relaxation depends on the collective motion of neighboring water molecules and the somewhat larger viscosity of the heavier isotopic form of water [120].

The presence of ions affects the reorientation times of dynamics of the water molecules in their vicinity, resulting in the so-called "slow water" being formed. In 3 M NaCl at 27°C τ_{or} of water molecules, hydrogen bonded to the chloride anion was 9.6 ps, diminishing to 4.2 ps at 106°C [44]. The values of τ_{or} became shorter in the presence of iodide but longer in the presence of bromide anions. This indicated hydrodynamic radii of the anions of 0.213 nm for Cl⁻, 0.237 nm for Br⁻, but only 0.205 nm for iodide (smaller than the crystal ionic radius, 220 nm!), as obtained from the Stokes–Einstein expression. In a 6.0 M $NaClO_4$ solution, the orientational relaxation of the O–H$\cdots$OClO$_3^-$ group is $\tau_{or} = 7.6 \pm 0.3$ ps at room temperature. It is noteworthy that the water reorientation times in aqueous $Mg(ClO_4)_2$ and Na_2SO_4 are the same as those of pure water, so that the ions do not seem to affect these times outside the hydration shells according to Bakker [44].

As the ion concentration increases, it is not possible to disentangle the effects of cations and of anions on the water reorientation dynamics because of the formation of solvent-shared ion pairs. Nevertheless, van der Post and Bakker [121] showed that sodium ions at concentrations up to 6 m slow down the reorientation of water molecules in aqueous NaCl and NaI, compared with CsCl and KI at the same concentration. Small effects are shown as the concentration of LiI increases up to 2 m, but gradual slowing down of τ_{or} is seen in aqueous Cs_2SO_4 and $Mg(ClO_4)_2$ and much more so in aqueous Na_2SO_4 and $MgSO_4$, but the effects diminish as the temperature increases from 22 to 70°C as found by Tielrooij et al. [122]. In 6 m NaOH solutions, the reorientation time of the OH⁻ hydration complex is $\tau_{or} = 12 \pm 2$ ps, that is, much slowed down relative to bulk water, because it is a large hydrogen-bonded structure that reorients as a whole according to Liu et al. [123].

The mutual effects of cations and anions on the water reorientation rates have also been studied. A comparison between solutions of 4 m aqueous LiCl, CsI, and CsF shows that there is a considerably larger amount of "slow water" in the CsF solutions, but hardly any effects in the former two salt solutions. The combination of a strongly hydrated ion (F^-) with a weakly hydrated one (Cs^+) is responsible for this effect according to Tielrooij et al. [124]. In aqueous alkali metal formate solutions, the time constant for the "slow water" was estimated as $\tau_{or} \approx 20$ ps, and its fraction increased in the order $Cs^+ < K^+ < NH_4^+ < Li^+ < Na^+$ (note the out-of-order position of Na^+). This, again, emphasizes the cooperative effects of cations and anions as suggested by Pastorczak et al. [125]. In up to 4 m of aqueous R_4NBr (R = methyl to butyl), a very strong slowing down of the water reorientation was found, the fraction of "slow water" increasing with the size of R, a phenomenon that is associated with the hydration of the hydrophobic alkyl groups. The values of τ_{or} at low concentrations increase from 15 ps for Me_4NBr (independent of the concentration) to 20 ps for Et_4NBr to >40 ps for Pr_4NBr and Bu_4NBr, for the latter three salts increasing with the concentration according to van der Post et al. [126, 127]. A comparison of the relaxation processes in aqueous guanidinium chloride and tetramethyl-guanidinium chloride again brings out the effect of the hydrophobic groups. Whereas the former salt has very little effect on the reorientation dynamics of the water molecules, the latter salt has a considerable fraction of "slow water" around its ions that acts in a concerted manner [127].

As a generalization, ultrafast infrared spectroscopy in dilute aqueous salt solutions is rather insensitive to the nature of the cation (unless large and hydrophobic) but does respond to the anion by its hydrogen bonding to the O–D probe. At larger concentrations, solvent-shared ion pairs show a cooperative cation–anion effect on the reorientation rate of the water molecules.

A different ultrafast spectroscopic method that was applied to electrolyte solutions is the pulsed optical Kerr effect, in which a short high intensity laser beam changes transiently the refractive index of the solution. Its application to alkali halide solutions in formamide by Palombo and Meech [128] resulted in a rotational orientation time for the formamide molecules of 12.7 ps, which is increased in the salt solutions, linearly with the viscosity (see Equation 5.22). However, the slope of the viscosity dependence is affected by the nature of the cation, being larger for NaI than for KI solution. It must be conceded that the obtained relaxation time constants do not pertain to the individual molecular rotations in the case of formamide, as appears to be the case for other highly hydrogen bonded solvents, but to the collective breaking of the hydrogen bonds, as concluded earlier by Barthel et al. [129].

5.2.2.2 High-frequency Dielectric Relaxation Spectroscopy

Earlier studies of dielectric relaxation times in aqueous alkali halide solutions by Kaatze and coworkers [12, 130] were obtained from the complex permittivities as a function of the frequency, but at frequencies up to about 100 GHz only as noted by Buchner [131]. The complex permittivity of conducting solutions has to be corrected for the conductance. The remainder can be expressed as Debye equations:

$$\varepsilon(v) = \sum S_j \left(1 - 2\pi i v \tau_j\right)^{-1} \qquad (5.21)$$

where the summation extends over several relaxation processes with amplitudes S_j and time constants τ_j ($i = -1^{1/2}$). The advantage of dielectric relaxation spectroscopy is its applicability to a study of molecular solvent orientation processes of dipolar non-hydrogen-bonding solvents, in contrast with the cooperative effects in hydrogen-bonded solvents, and the effects of ions on these processes, because the dielectric relaxation measures the solvent dipole orientation times (when no ion pairing interferes). The cooperative dipole reorientation times of water molecules of pure water at 25°C was $\tau_{1W} = 8.38 \pm 0.02$ ps, but accompanied with a faster individual molecular process with $\tau_{2W} \sim 1.1$ ps according to Buchner et al. [132], ascribed to rotation of water molecules with at most one hydrogen bond [25]. More recently developed technology permitted the study of the dielectric relaxation processes at frequencies in the terahertz range, up to 18 THz in water and 5 THz in methanol. The high frequency process time constant in water was reduced to $\tau_{2W} = 0.42$ ps with four additional damped harmonic oscillations with time constants of 0.30, 0.176, 0.071, and 0.048 ps. The corresponding relaxation time constants in methanol were $\tau_{1MeOH} = 51.8$ ps, $\tau_{2MeOH} = 8.04$ ps, and $\tau_{3MeOH} = 0.89$ ps with two oscillators, and in ethylene glycol (measured up to 89 GHz), these were $\tau_{1EG} = 122$ ps, $\tau_{2EG} = 21.4$ ps, and $\tau_{3EG} = 2.88$ ps, determined by Fukasawa et al. [133].

Sodium salt solutions with various anions, NaOH, NaCl, NaBr, NaI, $NaNO_3$, $NaClO_4$, NaSCN, Na_2CO_3, and Na_2SO_4 showed little dependence of the a coefficients of Equation 5.2 for the water relaxation on the nature of the anions ($a = 1.46 \pm 0.22$ ps, with $NaNO_3$ being somewhat lower, 1.06 ps). This corresponds to the fast water molecule rotation process. The strong water-structure-making anions do not conform: NaOH having $a = 0 \pm 0.15$ ps and Na_2CO_3 having $a = 3.52 \pm 0.4$ ps) according to Wachter et al. [28]. Tetraalkyl-ammonium bromides (with propyl, butyl, and pentyl groups) have also a "slow water" relaxation component, with $\tau_{sW} = 18.3 \pm 0.9$ ps extrapolated to zero concentration, absent in the smaller chain salts (methyl and ethyl) as found by Buchner et al. [134]. This relaxation, about 2.5 times slower than that of bulk water, signifies an increase in the average hydrogen bonds per water molecule in the vicinity of the hydrophobic groups.

The fraction of "slow water" was studied by pulsed terahertz dielectric relaxation spectroscopy in aqueous salt solutions: it was very small for CsF, CsCl, CsI, and Cs_2SO_4, larger for NaCl, $MgCl_2$, and LiCl (up to 5, 10, and 20% at 1 m) and up to 35% at 1 m for $MgCl_2$ and $MgSO_4$ according to Bakker and coworkers [122, 123, 127]. Salts comprising ions that have themselves dipoles, such as formate, acetate, and trifluoroacetate, complicate the interpretation of the dielectric relaxation spectra; these aqueous sodium salts have a composite relaxation of water and anion with a time constant of ~ 15 ps according to Rahmen et al. [135, 136]. Large alkyl groups attached to a carboxylate one in sodium propanoate and butyrate show "slow water" in the dielectric relaxation spectra due to the hydrophobic moiety [137]. For the larger tetraalkylammonium ions, the slowing down of the relaxation of the water molecules surrounding the hydrophobic parts of the ions was attributed by Buchner and Hefter [31] to their being shielded by these parts from "attack" by incoming water molecules.

As a generalization, high-frequency dielectric relaxation spectroscopy in dilute aqueous salt solutions is rather insensitive to the nature of the anion, provided it is not dipolar itself, but does respond to the cation that orients the water dipoles around it through its electric field, and affects the water reorientation times, manifested by the appearance of the so-called "slow water."

The application of dielectric relaxation studies to electrolytes in nonaqueous solvents has been rather sparse. The molecular rotational correlation time τ'_{sl} is related to the solvent relaxation time τ_{sl} according to Barthel et al. [138] as:

$$\tau'_{sl} = \left[\frac{2\varepsilon_s + \varepsilon_\infty}{3\varepsilon_s}\right]\left(\frac{g'}{g_K}\right)\tau_{sl} = \left(\frac{3v_S C f_\perp}{k_B T}\right)\eta_S \tag{5.22}$$

In the first equality, ε_s is the static permittivity of the solvent, ε_∞ its infinite frequency permittivity, (g'/g_K) is the ratio of the dynamic coupling and the Kirkwood dipole orientation correlation parameters (set equal to unity for lack of detailed information). The second equality relates τ'_{sl} to the solvent viscosity η_S via the solvent molecular rotational volume, where $v_S = (4\pi/3)a^2 b$ is its molecular volume (a and b are the half-axis lengths of the ellipsoid of rotation representing the solvent molecule), $C = 1$ for stick boundary conditions, and $f_\perp$ is a factor describing the deviation of the molecular shape from sphericity (the measure by how much $a \neq b$).

The methodology was applied by Barthel et al. [138] to solutions of NaI and Bu$_4$NBr in acetonitrile, yielding solvent molecular rotational correlation times $\tau'_{sl} = 2.5$ to $3.0\,ps$ for the pure solvent, according to various assumptions regarding (g'/g_K), that increase with electrolyte concentration due to the increased viscosity. It was applied by Wurm et al. [139] to LiClO$_4$, NaClO$_4$, and Bu$_4$NClO$_4$ in DMF and DMA. For the pure solvents DMF and DMA, $\tau'_{sl} = 6.6$ and $8.9\,ps$, respectively, corresponding to rotation volumes $v_S C f_\perp/10^{-30}\,m^3 = 11.3$ and 13.8. These volumes diminish by a factor of about 4 in the presence of the electrolytes, but the reasons for this were not provided. Application of the method to NaCF$_3$CO$_2$, Mg(CF$_3$CO$_2$)$_2$, and Ba(ClO$_4$)$_2$ in DMF by Placzek et al. [140] yielded $\tau'_{sl} = 7.2\,ps$ for pure DMF, somewhat different from the earlier value shown above. The effects of the salts on this rotation time were not discussed.

5.2.2.3 NMR Relaxation Times

The inverse of the T_1 NMR relaxation time in pure solvents and solutions is proportional to the solvent molecular rotation time. For quadrupolar nuclei such as ^{2}H and ^{14}N with spin number $I = 1$ (see McConnell [141]) under narrow conditions:

$$\frac{1}{T_1} = \left(\frac{3\pi^2}{10}\right)\left[\frac{(2I+3)}{I^2(2I-1)}\right]\left(1+\xi^2/3\right)\left(\frac{e^2 qQ141}{h}\right)^2 \tau_{or} \tag{5.23}$$

where ξ is the asymmetric parameter for the electric field gradient (generally negligible) and $(e^2 qQ/h)$ is the quadrupole coupling constant. It is not necessary to involve the latter quantity directly when the effect of ions on the solvent molecular rotation time is required, because then the B_{NMR} coefficients obtained according to

Equation 5.1 are related to the solvent orientation times as noted by Engel and Hertz [10]:

$$\frac{\tau_{S(I)or}}{\tau_{Sor}} = 1 + B_{NMR}c_S/h_{I(S)} \tag{5.24}$$

Here $\tau_{S(I)or}$ is the orientation time of the solvent S in the presence of the ion I, τ_{Sor} is that time for the pure solvent, c_S is the molar concentration of the solvent, and $h_{I(S)}$ is the coordination number of the solvent in the solvation shell of the ion.

NMR values of $\tau_{S(I)or}/\tau_{Sor}$ for aqueous ions at 22°C from Chizhik [22] and 25°C from Engel and Hertz [10] and in a few nonaqueous solvents at 25°C from Engel and Hertz [10] and from Sacco et al. [142] for the specified coordination numbers $h_{I(S)}$ are shown in Table 5.5. The splitting of the ionic values is done at the level of the B_{NMR} values, generally equating those for K$^+$ and Cl$^-$. Values of $\tau_{S(I)or}/\tau_{Sor} < 1$ signify that the ion is breaking the structure of the solvent, whereas when this ratio is > 1 the structure in the solvation shell is enhanced.

TABLE 5.5 The Molecular Reorientation Time Constants in the Presence of Ions Relative to the Time Constants of the Pure Solvents, $\tau_{S(I)or}/\tau_{Sor}$, in Various Solvents for the Specified Coordination Numbers $h_{I(S)}$, Obtained by NMR

Ion	$h_{I(S)}$	Water		MeOH	EG	FA	NMF	DMSO
References	22	22	10	10	10	10	10	142
Li$^+$	4	2.3	2.5^a	4.1			2.4	
Na$^+$	6	1.6		2.0	2.0	2.4	1.7	1.9
K$^+$	8	1.0	0.9	1.6	1.1	1.5	1.5	1.5
Rb$^+$	8 [10]				1.0	1.4	1.4	1.5
Cs$^+$	8 [10]				0.9	1.3	1.3	1.4
NH$_4^+$	4	0.8						
H$_3$O$^+$	3	1.9						
Mg^{2+}	6	4.5	5.2					
Ca^{2+}	6	3.3						
Sr^{2+}	8	2.7						
Ba^{2+}	12	2.1						
F$^-$	4	3.3						
Cl$^-$	4	1.3			1.2	1.7	2.0	
Br$^-$	4	0.9		1.6^a	1.0	1.5^a	2.0	2.2
I$^-$	4	0.8	0.2^a	1.4^a	0.8	1.1^a	1.6^a	1.8
CN$^-$	6 [10]		0.6					
N$_3^-$	6	0.8						
SCN$^-$	4		1.4^a					
NO$_3^-$	6	0.9						
ClO$_4^-$	8 [10]		0.4	1.1				
CO$_3^{2-}$	9	2.7	2.6					
SO$_4^{2-}$	8	3.2	1.9					

^{a}Extrapolated from data for varying $h_{I(S)}$ to the value specified in column 2.

5.2.3   Computer Simulations of Reorientation Times

The most well-documented effects of ions on the dynamics of the solvent obtained by computer simulations pertain to the residence times of the water molecules in the hydration shells of the ions as listed already in Table 5.4. Much less information resulted from computer simulations regarding the reorientation dynamics of the water molecules for comparison with the experimental results in Table 5.5, and hardly anything regarding the dynamics of nonaqueous solvent molecules in the solvate shells of ions.

The reorientation times in the first hydration shell relative to those in bulk water, $\tau_{W(I)or}/\tau_{Wor}$, were obtained by Balbuena et al. [143] from semi-continuum molecular dynamics by means of the SPC/E water model (500 water molecules per ion). On the assumption of a coordination number of $N_{co} = h_{I(W)} = 6$, the $\tau_{W(I)or}/\tau_{Wor}$ are 5.0 for Na$^+$, 2.1 for K$^+$, 1.5 for Rb$^+$, and 1.1 for Cl$^-$ at 25°C, showing faster reorientation as the water binding weakens. For tetramethylammonium ions and the same water model (215 water molecules per ion) Carcia-Tarres and Guardia, [144] found that the ratio $\tau_{W(I)or}/\tau_{Wor} = 1.55$ is the average over the three principal rotation vectors. According to Rode and coworkers, for both Rb$^+$ [76] and Cs$^+$ [78], the corresponding average reorientation time ratios are 0.20, obtained from combined quantum mechanical and molecular dynamics simulations. Note the much smaller ratio for Rb$^+$ obtained by this method compared with the purely molecular dynamics result shown above. The Tl$^+$ ion, deemed the "strongest water structure-breaking metal ion" [84] has an even smaller average ratio $\tau_{W(I)or}/\tau_{Wor} = 0.18$.

Divalent cations, on the other hand, have $\tau_{W(I)or}/\tau_{Wor}$ ratios larger than 1, but there are considerable differences between the rotation constants around the three principal axes. According to Rode and coworkers, the reorientation constants of water molecules in the first hydration shell relative to that in the bulk solvent is largest around the y-axis and are as follows: Ba^{2+}, 1.14 [145]; Pb^{2+}, 3.5 [146]; and Zn^{2+}, 14.7 [147], increasing for diminishing ion sizes, but it is largest around the z-axis for Mg^{2+}: 6.5 [148]. For the trivalent La^{3+}, this ratio is again largest for water molecule rotation around the y-axis: 6.8 [149].

5.3   SOLVENT EFFECTS ON THE PROPERTIES OF IONS IN SOLUTION

5.3.1   Bulk Properties

The properties of ions in solution depend, of course, on the solvent in which they are dissolved. Many properties of ions in water are described in Chapters 2 and 4, including thermodynamic, transport, and some other properties. The thermodynamic properties are mainly for 25°C and include the standard partial molar heat capacities and entropies (Table 2.8) and standard molar volumes, electrostriction volumes, expansibilities, and compressibilities (Table 2.9), the standard molar enthalpies and Gibbs energies of formation (Table 2.8) and of hydration (Table 4.1), the standard molar entropies of hydration (Table 4.1), and the molar surface tension increments (Table 2.11). The transport properties of aqueous ions include the limiting molar conductivities and diffusion coefficients (Table 2.10) as well as the B-coefficients obtained from viscosities and NMR data (Table 2.10). Some other properties of aqueous ions are also presented: the molar static dielectric decrement in Table 2.11 and the hydration numbers obtained from computer simulations and some experimental methods in Tables 4.8 and 4.9, respectively.

Thermodynamic properties of ions in nonaqueous solvents are described in terms of the transfer from water as the source solvent to nonaqueous solvents as the targets of this transfer. These properties include the standard molar Gibbs energies of transfer (Table 4.2), enthalpies of transfer (Table 4.3), entropies of transfer (Table 4.4) and heat capacities of transfer (Table 4.5) as well as the standard partial molar volumes (Table 4.6) and the solvation numbers of the ions in non-aqueous solvents (Table 4.10). The transfer properties together with the properties of the aqueous ions yield the corresponding properties of ions in the nonaqueous solvents.

It remains to present other properties of ions in nonaqueous solvents, those of paramount importance being their transport properties. Values of limiting conductivities λ_I^∞ of univalent ions in several nonaqueous solvents at 25°C are available in the report by Krumgalz [150], reproduced in part in Table 5.6, which also shows the viscosities of these solvents. The values of the λ_I^∞ for the tetraalkylammonium ions have been related by Marcus [151] to the hydrodynamic radii r_{St} as $\lambda_I^\infty = 8.201/\eta_S r_{St}$, the numerical coefficient being $F^2/6\pi N_A$. There are essentially very sparse conductivity data regarding multivalent ions in the nonaqueous solvents because of the low solubilities of salts of such ions in them, as is demonstrated by the large positive standard molar Gibbs energies of transfer of such ions from water into nonaqueous solvents, Table 4.2.

It should be noted that the hydrodynamic radii r_{St} of the ions decrease, hence the limiting conductivities λ_I^∞ increase, as the ionic radius r_I increases in the series of the alkali metal cations and the halide anions for any solvent. This is due to the stronger solvation of the smaller ions that is manifested by larger solvation numbers, and hence larger hydrodynamic volumes and radii leading to slower mobilities. The tetraalkylammonium ions are poorly solvated in any case, and hence as their hydrodynamic radii increase with the size of the alkyl chain, their conductivities diminish. In the discussion of the diffusivities of ions in Chapter 4, a consequence of Equation 4.37 for ions I in solvents S is that the product of the ionic diffusion coefficient and the viscosity of the solvent, $D_I \eta_S$, for a given ion (with a fixed radius) are independent of the nature of the solvent. The ionic self-diffusion coefficient at infinite dilution, D_I^∞, is proportional to the limiting molar ionic conductivity, λ_I^∞, Equation 2.28. Hence, also the product $\lambda_I^\infty \eta_S$ should be independent for a given ion of the nature of the solvent. This is the well-known *Walden rule*, which should hold, provided the hydrodynamic ionic radius is independent of the solvent. This premise was demonstrated for large ions, such as the tetraalkylammonium ones by Krumgalz and by Marcus [150, 151] for many solvents, but the agreement is not as good for those solvents that have a hydrogen-bonded network, such as water, ethylene glycol, and formamide.

Another quantity of interest is the ionic viscosity B-coefficient, Equation 2.29, that in aqueous solutions describes the effect of the ion on the structure of the solvent water. Some values of $B\eta_I$ in nonaqueous solvents have been compiled by Jenkins and Marcus [9] and are reproduced in Table 5.7. It should be noted that practically in all the solvents (except light and heavy water), all the $B\eta_I$ values are positive and the ions appear to enhance the structure of the solvent. However, the splitting of the

TABLE 5.6 Limiting Molar Ionic Conductivities, λ_1^∞/S·cm²·mol⁻¹, in Nonaqueous Solvents at 25°C [150, 151]

Ion	MeOH	EtOH	PrOH	EG	Me_2CO	PC	FA	NMF	DMF	TMU	HMPT	MeCN	DMSO	$MeNO_2$	$PhNO_2$	Py
η_s	0.551	1.083	1.943	16.34	0.303	2.33	3.30	1.65	0.802	1.395	3.11	0.96	1.99	0.644	1.784	0.884
H^+	146.1	62.7	45.7				10.4		35.0				15.5	64.5		
Li^+	39.6	17.1	8.2	1.91	69.2	7.1	8.3	10.1	26.1	14.5	5.6		11.8	53.9	16.6	25.9
Na^+	45.2	20.3	10.2	2.88	70.2	9.1	9.9	16.0	30.0	16.1	6.2	76.8	13.9	56.8	17.8	26.6
K^+	52.4	23.5	11.8	4.42	81.1	11.1	12.4	16.5	31.6	15.7	6.4	83.7	14.7	58.1	18.6	31.8
Rb^+	56.2	24.9	13.2	4.61		11.7	12.8	17.9	33.2	16.0	6.9	85.8	15.0		19.9	
Cs^+	61.5	26.3	14.0	4.53	84.1	12.2	13.4	18.5	35.4	16.9	7.3	87.4	16.2			
Ag^+	50.1	20.6		4.05		12.0			35.7		6.2	86.0	*16.1*	50.8		34.4
Tl^+				5.08		13.1	15.6		39.5			91.1		58.8		
NH_4^+	57.6	22.1			89.5		14.9	24.6	39.4		6.5	97.1		62.8	17.9	46.6
Me_4N^+	68.0	29.9	15.2	3.04	101.8	14.2	12.8	35.3	39.0	25.1		94.6	17.7	55.9	20.1	
Et_4N^+	60.5	29.1	15.2	2.20	94.0	13.2	10.4	26.0	36.0	23.0		85.1	16.2	48.7	14.9	
Pr_4N^+	45.7	22.8	12.5	1.74	76.7	10.5	7.9	13.9	29.4			70.3		40.0	12.3	
Bu_4N^+	38.8	19.7	11.0	1.51	69.2	9.0	6.5	10.2	26.4	15.3		61.8	10.8	34.8		
Cl^-	52.4	21.9	10.2	5.30	109.3	18.8	17.5	25.6	53.8		19.5	100.4	23.4	62.5	22.6	51.4
Br^-	56.3	24.5	12.0	5.21	113.9	18.4	17.5	28.2	53.4	30.1	17.5	100.7	23.8	62.8	22.0	51.2
I^-	62.6	27.0	13.7	4.79	116.2	18.8	16.9	28.4	51.1	18.8	16.3	102.6	23.6	63.6	21.3	19.0
SCN^-	61.9	27.5	14.65		121.3	22.2	17.5	28.7	59.2	33.1	19.7	113.3	28.9	72.1	23.0	
NO_3^-	61.0	24.8	11.2	5.04	127.2	20.9	17.7		57.1	32.4	19.3	106.2	26.8	66.7		52.5
ClO_4^-	70.8	30.5	16.2	4.45	115.8	18.9	16.6	27.4	51.6	28.4	15.0		24.4	65.8	21.6	47.5
$CH_3CO_2^-$		23.0					12.1								24.0	
$Picrate^-$	46.9	25.4	12.3		86.1	13.1		19.8	37.1		10.8	77.3	17.1	45.4	16.5	33.6

TABLE 5.7 The Ionic Viscosity B-coefficients in Nonaqueous Solvents at 25°C

	D₂O	MeOH	BuOH	MeOEtOH	Me₂CO	EC	PC	FA	NMF	DMF	TMU	MeCN	HMPT	DMSO	TMS	DCIE
Li⁺	0.15	0.30		0.18		0.63	0.79	0.35	0.07		0.87	0.51	1.13	0.61	1.07	
Na⁺	0.08	0.31	1.13	0.32	0.63	0.71		0.37	0.09	0.72	0.78	0.49	1.17	0.53	1.30	
K⁺	-0.02	0.27	0.96	0.17	0.74	0.58	0.60	0.19	0.11	0.77	0.59	0.51	0.89	0.54	1.11	
Rb⁺				0.13		0.56		0.23			0.56		0.88	0.52	1.04	
Cs⁺	-0.07	0.03		0.08		0.56		0.13	0.15		0.39		0.87	0.49	0.91	
Ag⁺		0.57			0.56					0.70	0.71	0.52		0.55		
NH₄⁺			0.92					0.12	-0.07							
Me₄N⁺	0.13	0.10	0.15	0.07	0.36		0.26	0.10	0.04			0.32	0.45	0.41	0.45	0.40
Et₄N⁺	0.36	0.02	0.31	0.11	0.46		0.40	0.13	0.23	-0.17		0.39	0.55	0.48	0.55	0.50
Pr₄N⁺	0.84	0.19		0.21					0.37	0.02		0.48	0.64	0.56	0.64	
Bu₄N⁺	1.31	0.34	0.72	0.30	0.70	0.46	0.59	0.51		0.12	0.80	0.62	0.99	0.61	0.79	0.69
Ph₄P⁺		0.91			1.06					0.93	1.00	0.81	1.90			
F⁻																
Cl⁻	-0.02	0.54	0.39		0.34	0.10	0.46	0.19	0.52	0.39	0.96	0.35	0.74	0.26	-0.01	0.33
Br⁻	-0.05	0.50	0.33	0.12		0.11		0.14	0.47			0.35	0.74	0.30	0.06	0.29
I⁻	-0.10	0.46	0.30		0.10	0.08	0.30	0.11	0.45	1.21		0.28	0.60	0.274	0.04	0.23
SCN⁻		0.29	0.29		0.10			0.06						0.26	0.23	0.23
NO₃⁻								0.21				0.25		0.27		
ClO₄⁻		0.28		0.08	0.22	0.01	0.27				0.70	0.23		0.26	-0.07	0.13
BF₄⁻				0.09			0.41									
BPh₄⁻		0.79	0.70	0.37	0.86	0.59	0.72			0.90	1.12	0.74	1.84	0.71	0.95	0.75

measured $B_{\eta E}$ of the salts into the ionic contributions is not generally the same as for aqueous solutions and varies from solvent to solvent. The splitting according to $B(K^+) = B(Cl^-)$, which is very similar to $B(Rb^+) = B(Br^-)$ used for aqueous solutions [9], is confined to heavy water and formamide. For most of the other solvents, it is according to the ratio of the cubes of the van der Waals radii of two selected large ions, most often Bu_4N^+ and BPh_4^-. The result is that the B-coefficients of the small cations are considerably larger than those of anions, except for NMF, where the reverse relationship holds. There the splitting is according to the ratio of the cubes of the van der Waals radii of two selected large cations, Bu_4N^+ and Pe_4N^+.

There are some data available for the (static) dielectric decrements of salts in non-aqueous solvents in the compilations by Barthel et al. [33, 152]. Values of the relative limiting decrements, that is, $-\delta/\varepsilon_s(c_E = 0)$, in nonaqueous solvents are shown in Table 5.8, where the corresponding values for aqueous salts are also shown for comparison. Note that the linearity of the $\varepsilon_s = f(c)$ curves breaks down at considerably lower concentrations than in water in solvents of relatively low permittivity, in which ion pairing is expected. This may explain the discrepancies in the values obtained in methanol between the entries in Refs. [152] and [33].

The data are too few to evaluate single ion contributions to these decrements, although additivity of such values is expected. Trends seen in aqueous solutions, that the decrements increase in absolute values as the size of the anion increases, are also

TABLE 5.8   The Relative Dielectric Decrement $-\delta/\varepsilon_s(c_E=0)/dm^3 \cdot mol^{-1}$ of Salts in Nonaqueous Solvents and Water at 25°C [a,b]

Salt	H$_2$O	MeOH	PC	FA	NMF	DMF	DMA	MeCN	DMSO
LiCl	0.19	(0.88)							
LiBr	0.21							0.15	
LiI	0.19								
LiSCN			0.34						0.36
LiNO$_3$	0.22	0.92		0.23	(0.53)	(0.13)			0.30
		(0.66)		(0.24)					(0.23)
LiClO$_4$	0.22			0.26	0.80	0.44	0.93	0.21	
NaCl	0.16	1.72							
NaBr	0.17	1.29							
		(0.86)							
NaI	0.18	1.20	0.57			0.21		0.00	0.34
		(0.92)							
NaClO$_4$	0.22	1.01	0.62	0.29	0.83	0.46	0.90	0.22	
KI		(0.83)							
NH$_4$Br	0.13	1.01							
Bu$_4$NCl	0.14	1.29							
Bu$_4$NBr	0.15	1.01						0.38	
Bu$_4$NI	0.17	0.92	0.60						
Bu$_4$NClO$_4$	0.20	0.89	0.62	0.44	0.89		0.93		

[a]from Ref. 33.
[b]Values in parentheses are from Ref. 152.

followed in the nonaqueous solutions, but the effects for the cations are too obscure for a trend to be seen. The trends among the solvents are not clear either, they certainly cannot distinguish between protic (hydrogen bonded) and aprotic solvents. The values of $-\delta/\varepsilon_s(c_E = 0)$ are relatively small for H_2O, FA, and MeCN, they are intermediate for DMF and DMSO, and are large for MeOH, PC, NMF, and DMA.

Data are also available for the molar surface tension increments, $\Delta\sigma/c_E$, where $\Delta\sigma = \sigma - \sigma(c_E = 0)$ for a few salts in a few nonaqueous solvents in several publications [153–157]. The values of $\sigma/\sigma_S = f(c_E)$ for 35°C in Ref. 155 for formamide and N-methylformamide were read from figures and converted to $\Delta\sigma/c_E$ using σ_S at 35°C from Refs. 158 and 159, respectively. The values of $\Delta\sigma = (100/\sigma_S)(\sigma - \sigma_S)$ as functions of $100x_E$ reported in Ref. 153 and of $d\sigma/dm$ reported in Ref. 156 were converted to $\Delta\sigma/c_E$ as appropriate, using the solvent density, ρ_S, because only quite dilute solutions were considered. The resulting values of $\Delta\sigma/c_E$ are shown in Table 5.9. There are some inconsistencies between the data reported by different authors, but the trends with respect to the sizes of the anions are clear: the surface tension increments for the alkali halide salts in the nonaqueous solvents *increase* with the sizes of the anions $Cl^- < Br^- < I^-$, contrary to the case in aqueous solutions, in which they *decrease* in this order, Table 2.12. The trends among salts with different cations, on the other hand, are difficult to discern. The ions of the salts considered are all depleted from the surface layer of the solution that has a thickness $\delta = (2RT)^{-1}\Delta\sigma/c_E$ according to Aveyard and Thompson [156].

TABLE 5.9   The Molar Surface Tension increments, $\Delta\varepsilon/c_E$, of Salts in Nonaqueous Solvents and Water at 25°C

Salt	H_2O	MeOH	EtOH	Me_2CO	PC	FA	NMF	DMF	DMSO	Py
LiCl	1.85	1.55[a]	0.82[b]	1.00[c]		1.42[a]	1.10[b]	1.50[a]	1.84[a]	0.69[b]
		1.40[b]						1.64[c]		
LiBr	1.60	3.06[b]	1.19[b]	1.11[b]					2.56[a]	2.56[b]
LiI	1.00	2.89[b]	2.28[b]	1.64[b]	0.87[a]				2.17[a]	2.76[b]
				1.51[d]						2.59[d]
NaCl	2.10	2.86[d]				1.41[e]	1.13[e]	1.18[e]		
NaBr	1.85	1.91[b]				2.60[e]	2.73[e]	2.15[d]		
		1.94[d]						2.84[e]		
NaI	1.25	2.01[b]	1.98[b]	1.40[b]	1.23[a]	3.64[e]	2.98[e]	3.10[e]		2.14[b]
		2.34[d]		2.54[d]						
KCl	2.00	2.58[c]				1.52[e]	1.38[b]	1.44[e]		
KBr	1.75	1.71[c]				3.59[e]	2.80[e]	2.91[e]		
KI	1.15	2.42[b]			1.00[a]	4.56[e]	3.47[e]	3.61[e]		
		2.05[c]						2.30[c]		
CsCl	1.70	2.04[c]								

[a] From Ref. 156.
[b] From Ref. 153.
[c] From Ref. 154.
[d] From Ref. 157.
[e] From Ref. 155 at 35°C.

5.3.2 Molecular Properties

The solvation shells of ions may not be spherically symmetrical, a phenomenon that occurs at the surfaces of solutions. The solvent molecules may then polarize the ions, even if they are monatomic, and thus affect their properties. However, in the bulk of the solutions, monatomic ions reside in a spherically symmetrical field and are themselves little affected by the solvent. This is not necessarily the case for poly-atomic ions, even for those that have a globular shape. This effect is most noted in the vibrational and rotational relaxation times of polyatomic ions. The relaxation is measurable by the pump/probe technique, where the laser beam at the wave-number of the ground state is followed after a short interval by a probe beam at the level of the excited state, following its exponential decay.

In aqueous 2.3 M NaCN, the stretching vibration relaxation time is $\tau_1 = 6.7$ ps according to Heilweil et al. [160] compared with the relaxation time $\tau_1 = 1.2$ ps of the asymmetrical stretch of N_3^- in H_2O and $\tau_1 = 2.4$ ps in D_2O and $\tau_1 = 18.3$ ps for SCN^- in D_2O. For the latter vibration in MeOH, the time is shorter, $\tau_1 = 11.0$ ps, and in the same solvent for OCN^- it is only 2.9 ps and for N_3^- only 2.4 ps, but for the latter anion in HMPA, it is 14.8 ps as reported by Li et al. [161].

The corresponding reorientational relaxation times for these linear anions were also studied by the infrared pump/probe technique by Li et al. [162]. However, NMR spin-lattice relaxation times have been employed for other polyatomic ions, using the nuclei ^{14}N for NO_3^- [162], ^{17}O for ClO_4^- [163], ^{15}N for NH_4^+ [164], and ^{13}C for several tetraalkylammonium cations [165] according to Masuda and coworkers. The results are shown in Table 5.10. It was concluded that these times do not follow the expected (Stokes–Einstein–Debye) hydrodynamic solvent sequences according to the solvent

TABLE 5.10 Rotational orientation Correlation Times τ_r/ps of Several Polyatomic Ions in Various Solvents

Solvent	N_3^-	OCN^-	SCN^-	NO_3^-	ClO_4^-	NH_4^+	Et_4N^+	Pr_4N^+	Bu_4N^+
Water	7.1[a]		4.7[a]	1.14	6.16	0.93	6.7	49	115
MeOH	12.9	6.7	8.7	3.38	5.85	3.4	7.3	41	89
EtOH				6.12	6.86	5.8			
1-PrOH				11.6	8.22				
EG					13.9	12.4			
Me$_2$CO					4.70	0.87	3.1	11	24
PC					7.34	1.03			
MeCN			1.08		5.06	0.53	2.9	13	29
PhCN					8.45				
MeNO$_2$					5.87	0.3	4.6	23	51
PhNO$_2$					9.36				
DMF					5.41	3.4	7.5	32	69
TMU					8.12				
HMPA	5.6				7.01	3.1	38	150	290
DMSO					6.51	6.3	14	75	160

[a] In D_2O.

viscosities nor the expected (Hubbard–Onsager) dielectric friction sequences (except for the nitrate rotation that follows the latter for the limited list of solvents studied by Nakahara and Emi [172]). Some correlation of the τ_r times with the (electron pair) donor properties of the solvents was however found by Masuda [164].

The rotational correlation times τ_r were also measured by the NMR technique with ^{13}C nuclei of the para-carbon of the phenyl rings for Ph_4As^+ and Bph_4^- in two solvents by Masuda and Muramoto [165]. In MeCN, these times were practically the same, 27 and 28 ps, about the same as for Bu_4N^+, 29 ps. However, in water they differed, being 92 and 69 ps, respectively, smaller than for Bu_4N^+, 115 ps. Specific interactions of these two ions (Bph_4^- and Ph_4P^+) that are for many purposes quite similar have also been found for the ^1H chemical shifts of the phenyl groups, being more high field for Bph_4^- than for Ph_4As^+ and Ph_4P^+ in this order in the solvents compared: water, ethanol, acetonitrile, and DMSO as found by Coetzee and Sharpe [166]. What are the consequences of these specific interactions on the energetics of the interactions of these three ions with solvents in their dilute solutions difficult to discern?

Ions at solution/vapor surfaces cannot be said to have bulk properties, so that the influence of the nature of the solvent on their effects on bubble coalescence [167] and thin film rupture [168] studied by Henry and coworkers may be briefly mentioned here. Bubbles in a liquid should be unstable and when bubbles collide they coalesce in order to minimize the interfacial area and the energy of the system. Some electrolytes are known to inhibit this coalescence and stabilize the bubbles, as evidenced by the white foam of breaking sea waves. In aqueous solutions, ions can be assigned to classes α and β: if the cations and anions belong to the same class ($\alpha\alpha$ or $\beta\beta$), they inhibit the coalescence, but if they belong to different classes ($\alpha\beta$ or $\beta\alpha$), they have no effect. Li^+, Na^+, K^+, NH_4^+, Mg^{2+}, and Ca^{2+} among the cations and Cl^-, Br^-, I^-, NO_3^-, and SO_4^{2-} among the anions belong to class α in aqueous solutions, whereas H^+, Me_4N^+, SCN^-, ClO_3^-, ClO_4^-, and $MeCO_2^-$ belong to class β. Of the latter, only H^+, Me_4N^+, and ClO_4^- diminish the surface tension of water, Table 2.12, and are enriched at the water/vapor interface, whereas all the class α ions as well as SCN^-, ClO_3^-, and $MeCO_2^-$ increase the surface tension and are repulsed from the interface (chlorate and acetate only mildly). These assignments of ions to the two classes are not strictly retained in nonaqueous solutions: although in formamide the same classification as in water prevails, in propylene carbonate SCN^- and ClO_4^- are assigned to class α rather than to β and so is ClO_4^- in DMSO solutions as determined by Henry and Craig [167].

REFERENCES

[1] Y. Marcus, Chem. Rev. **109**, 1346 (2009).

[2] Y. Marcus, Pure Appl. Chem. **82**, 1889 (2010).

[3] O. Ya. Samoilov, Faraday Disc. **24**, 141 (1957).

[4] D. W. McCall, D. C. Douglass, J. Phys. Chem. **69**, 2001 (1965).

[5] L. Endon, H. G. Hertz, B. Thuel, M. D. Zeidler, Ber. Bunsengesell. Phys. Chem. **71**, 1008 (1967).

[6]   K. J. Müller, H. G. Hertz, J. Phys. Chem. **100**, 1256 (1996).

[7]   S. R. Heil, M. Holz, T. M. Kastner, H. Weingärtner, J. Chem. Soc., Faraday Trans. **91**, 1877 (1995).

[8]   A. Nowikow, M. Rodnikova, J. Barthel, O. Sobolev, J. Mol. Liq. **79**, 203 (1999).

[9]   H. B. D. Jenkins, Y. Marcus, Chem. Rev. **95**, 2695 (1995).

[10] G. Engel, H. G. Hertz, Ber. Bunsengesell. Phys. Chem **72**, 808 (1968).

[11] K. Yoshida, K. Ibuki, M. Ueno J. Solution Chem. **25**, 435 (1996).

[12] U. Kaatze, J. Solution Chem. **26**, 1049 (1997).

[13] Y. Marcus, Chem. Rev., **113**, 6536 (2013).

[14] Y. Marcus, J. Chem. Eng. Data **57**, 617 (2012).

[15] K. Fumino, K. Yukiyasu, A. Shimizu, Y. Taniguchi, J. Mol. Liq. **75**, 1 (1998).

[16] A. Sacco, H. Weingärtner, B. M. Braun, M, Holz, J. Chem. Soc., Faraday Trans. **90**, 849 (1994).

[17] R.W. Gurney, Ionic Processes in Solution, McGraw-Hill, New York (1953).

[18] Y. Marcus, H. B. D. Jenkins, L. Glasser, J. Chem. Soc., Dalton Trans., 3795 (2002).

[19] Y. Marcus, Langmuir **29**, 2881 (2013).

[20] J. M. Notley, M. Spiro, J. Phys. Chem. **70**, 1502 (1966).

[21] M. H. Abraham, J. Liszi, E. Papp, J. Chem. Soc., Faraday Trans. **78**, 197 (1982).

[22] V. I. Chizhik, Mol. Phys. **90**, 653 (1997).

[23] W.-Y. Wen, U. Kaatze, J. Phys. Chem. **81**, 177 (1977).

[24] R. Buchner, G. Hefter, J. Barthel, J. Chem. Soc. Faraday Trans. **90**, 2475 (1994).

[25] J. Barthel, R. Buchner, B. Wurm. J. Mol. Liq. **98**, 51 (2002).

[26] T. Chen, G. Hefter, R. Buchner, J. Phys. Chem. A **107**, 4025 (2003).

[27] A. Tromans, P. M. May, G. Hefter, T. Sato, R. Buchner, J. Phys. Chem. B **108**, 13789 (2004).

[28] W. Wachter, W. Kunz, R. Buchner, G. Hefter, J. Phys. Chem. A **109**, 8675 (2005).

[29] W. Wachter, R. Buchner, G. Hefter, J. Phys. Chem. B **110**, 5147 (2006).

[30] W. Wachter, S. Fernandez, R. Buchner, G. Hefter, J. Phys. Chem. B **111**, 9010 (2007).

[31] R. Buchner, G. Hefter, Phys. Chem. Chem. Phys. **11**, 8984 (2009).

[32] M. L. T. Asaki, A. Redondo, T. A. Zawodzinski, A. J. Taylor, J. Chem. Phys. **116**, 8469 (2002).

[33] J. Barthel, R. Buchner, P.-N. Eberspächer, M. Münsterer, J. Stauber. B. Wurm, J. Mol. Liq. **78**, 83 (1998).

[34] M. F. Kropman, H.-K. Nienhuys, H. J. Bakker. Phys. Rev. Lett **88**, (2002) 77601 (1998).

[35] G. R. Choppin, K. J. Buijs, Chem. Phys. **39**, 2042 (1963).

[36] Z. Kecki. P. Dryjanski, E. Kozlowska, Rocz. Chem. **42**, 1749 (1968).

[37] O. D. Bonner, C. F. Jumper, Infrared Phys. **13**, 233 (1973).

[38] Z. S. Nickolov, J. D. Miller, J. Coll. Interf. Sci. **287**, 572 (2005).

[39] N. Kitadai, T. Sawai, R. Tonoue, S. Nakashima, M. Katsura, K. Fukushi, J. Solution Chem. **43**, 1055 (2014).

[40] G. E. Walrafen, J. Chem. Phys. **52**, 4176 (1970).

[41] A. W. Omta, M. F. Kropman, S. Woutersen, H. J. Bakker, J. Chem. Phys. **119**, 12457 (2003).

[42] V. Holba, Coll. Czech. Chem. Comm. **47**, 2484 (1982).

[43] R. Li, Zh. Jiang, F. Chen, H. Yang, Y. Guan, J. Mol. Struct. **707**, 83 (2004).

[44] H. J. Bakker, Chem. Rev. **108**, 1456 (2008).

[45] L.-Å. Näslund, D. C. Edwards, P. Wernet, U. Bergmann, H. Ogasawara, L. G. M. Pettersson, S. Myneni, A. Nilsson, J. Phys. Chem. A **109**, 5995 (2005).

[46] C. D. Cappa, J. D. Smith, K. R. Wilson, B. M. Messer, M. K. Gilles, R. C. Cohen, R. J. Saykally, J. Phys. Chem. B **109**, 7046 (2005).

[47] C. D. Cappa, J. D. Smith, B. M. Messer, R. C. Cohen, R. J. Saykally, J. Phys. Chem. B **110**, 5301 (2006); **111**, 4776 (2007).

[48] K. D. Collins, Biophys. J. **72**, 65 (1997).

[49] M. H. Abraham, J. Liszi, J. Chem. Soc., Faraday Trans. 1 **74**, 2858 (1978).

[50] Y. Marcus, J. Chem. Soc., Faraday Trans. 1 **82**, 233 (1986).

[51] Y. Marcus, J. Solution Chem. **23**, 831 (1994).

[52] Y. Marcus, Pure Appl. Chem. **59**, 1093 (1987).

[53] Y. Marcus, Russ. J. Electrochem. **44**, 16 (2008).

[54] G. A. Krestov, V. K. Abrosimov, Zh. Strukt. Khim. **5**, 510 (1964).

[55] G. A. Krestov, Thermodynamics of Solvation, Ellis Horwood, New York (1991).

[56] A. Ben-Naim, J. Phys. Chem. **79**, 1268 (1975).

[57] Y. Marcus, A. Ben-Naim, J. Chem. Phys. **83**, 4744 (1985).

[58] Y. Marcus, Chem. Rev. **113**, 6536 (2013).

[59] M. R. Dack, Aust. J. Chem. **29**, 771 (1976).

[60] J. V. Leyendekkers, J. Chem. Soc., Faraday Trans. 1 **79**, 1109 (1983).

[61] N. D. Milovidova, B. M. Moiseev, L. I. Fedorov, Zh. Strukt. Khim. **11**, 136 (1970); Russ. J. Struct. Chem. 11, 121 (1970).

[62] Y. Marcus, unpublished results (2011).

[63] J. D. Worley, I. M. Klotz, J. Chem. Phys. **45**, 2868 (1966).

[64] R. N. Goldberg, J. Phys. Chem. Ref. Data **10**, 671 (1981).

[65] J. V. Leyendekkers, J. Chem. Soc., Faraday Trans. 1 **79**, 1123 (1983).

[66] W. Luck, Ber. Bunsenges. Phys. Chem. **69**, 69 (1965).

[67] W. A. P. Luck, in F. Franks, ed., Water, A Comprehensive Treatise, Plenum, New York, Vol. **2**, Ch. 4 (1975).

[68] G. E. Walrafen, J. Chem. Phys. **52**, 4176 (1970).

[69] K. W. Bunzl, J. Phys. Chem. **71**, 1358 (1967).

[70] V. K. Abrosimov, Zh. Strukt. Khim. **14**, 154 (1973); Russ. J. Struct. Chem. 14, 133 (1973).

[71] T. J. Giligan, G. Atkinson, J. Phys. Chem. **84**, 208 (1980).

[72] G. F. Bene, T. S. Hofer, B. R. Rudolf, B. M. Rode, Chem. Phys. Lett **521**, 74 (2012).

[73] A. Tongraar, B. M. Rode, Chem. Phys. Lett. **403**, 314 (2005).

[74] S. S. Azam, T. S. Hofer, A. Bhattacharjee, L. H. V. Lim, A. B. Pribil, B. R. Randolf, B. M. Rode, J. Phys. Chem. B **113**, 9289 (2009).

[75] A. Tongraar, S. Hannongbua, B. M. Rode, J. Phys. Chem. A **114**, 4334 (2010).

[76] T. S. Hofer, B. R. Randolf, B. M. Rode, J. Comput. Chem. **26**, 949 (2005).

[77] C. Kritayakornupong, V. Vchirawongkwin, B. M. Rode, J. Phys. Chem. B **114**, 12883 (2010).

[78] C. Schwenk, T. S. Hofer, B. M. Rode, J. Phys. Chem. A **108**, 1599 (2004).

[79] A. Tongraar, P. Tangkawanwanit. B. M. Rode, J. Phys. Chem. A **110**, 12918 (2006).

[80] M. Blauth,A. B. Pribil, B. R. Randolf, B. M. Rode, T. S. Hofer, Chem. Phys. Lett. **500**, 251 (2010).

[81] A. B. Pribil, T. S. Hofer, V. Vchirawongkwin, B. R. Randolf, B. M. Rode, Chem. Phys. **346**, 182 (2008).

[82] L. Liu, J. Hunger, H. J. Bakker, J, Phys. Chem. A **115**, 14593 (2011).

[83] A. Payaka, A. Tongraar, B. Michael Rode, J. Phys. Chem. A **113**, 3291 (2009).

[84] V. Vchirawongwin, T. S. Hofer, B. R. Randolf, B. M. Rode, J. Comput. Chem. **28**, 1006 (2007).

[85] A. Payaka, A. Tongraar, B. M. Rode, J. Phys. Chem. A **114**, 10443 (2010).

[86] R. G. Pearson, P. C. Ellgen, in H. Eyring, ed., Physical Chemistry. An Advanced Treatise, Acadermic Press, New York, **7**, 228 (1975).

[87] V. Vchirawongkwin, A. B. Pribil, B. M. Rode, J. Comput. Chem. **31**, 249 (2010).

[88] J. Burgess, Metal Ion Solvation, Ellis Horwood, Chichester (1978).

[89] A. Tongraar, B. M. Rode, Chem. Phys. Lett. **409**, 304 (2005).

[90] V. Vchirawongkwin, C. Pornpiganon, C. Kritayakornupong, A. Tongraar, B. M. Rode, J. Phys. Chem. B **116**, 11498 (2012).

[91] T. S. Hofer, H. T. Tran, C. F. Schwenk, B. M. Rode, J. Comp. Chem. **25**, 211 (2004).

[92] L. Eklund, T. S. Hofer, A. B. Pribil, B. M. Rode, I. Persson, Dalton Trans. **41**, 5209 (2012).

[93] T. S. Hofer, B. R. Randolf, B. M. Rode, J. Phys. Chem. B **110**, 20409 (2006).

[94] N. Purdie. C. A. Vincent, Trans. Faraday Soc. **63**, 2745 (1967).

[95] T. Sakwarathorn, S. Pongstabodee, V. Vchirawongkwin, L. R. Canaval, A. O. Tirler, T. S. Hofer, Chem. Phys. Lett. 595–596, 226–229 (2014).

[96] A. Bhattacharya, A. B. Pribil, T. S. Hofer, R. M Randolf, B. M. Rode, J. Phys. Chem. B **114**, 3921 (2010).

[97] E. v. Goldhammer, in Modern Aspects of Electrochemistry, Plenum Press, New York **10**, 77 (1975).

[98] S. T. Moin, T. S. Hofer, A. B. Pribil, B. R. Randolf, B. M. Rode, Inorg. Chem. **49**, 5101 (2010).

[99] C. Kritayakornupong, K. Plankensteiner, B. M. Rode, J. Chem. Phys. **119**, 6068 (2003).

[100] Y. Inada, A. M. Mohammed, H. H. Loeffler, B. M. Rode, Phys Chem A **106**, 6783 (2002).

[101] M. Q. Fatmi, T. S. Hofer, B. R. Randolf, B. M. Rode, J. Chem. Phys **123**, 054514 (2005).

[102] L. H. V. Lin, A. Bhattacharjee, S. S. Asam, T. S. Hofer, B. R. Randolf, B. M. Rode, Inorg. Chem. **49**, 2132 (2010).

[103] T. S. Hofer, A. K. H. Weiss, B. R. Randolf, B. M. Rode, Chem. Phys. Lett. **512**, 139 (2011).

[104] T. S. Hofer, B. R. Randolf, B. M. Rode, Chem. Phys. **349**, (2008) 210.

[105] S. S. Azam, L. H. V. Lin, T. S. Hofer, B. R. Randolf, B. M. Rode, J. Comput. Chem. **31**, 278 (2010).

[106] J. Reuben, D. Fiat, J. Chem. Phys. **51**, 4918 (1969).

[107] L. H. V. Lim, T. S. Hofer, A. B. Pribil, B. M. Rode, J. Phys. Chem. B **113**, 4372 (2009).

[108] A. Bhattacharjee, T. S. Hofer, A. B. Pribil, B. R. Randolf, L. H. V. Lin, A. F. Lichtenberger, B. M. Rode, J. Phys. Chem. B **113**, 13007 (2009).

[109] T. S. Hofer, B. R. Randolf, B. M. Rode, I. Persson, Dalton Trans. 1512 (2009).

[110] V. Vchirawongwin, T. S. Hofer, B. R. Randolf, B. M. Rode, J. Comput. Chem. **28**, 105 (2007).

[111] S. Durdagi, T. S. Hofer, B. R. Randolf, B. M. Rode, Chem. Phys. Lett. **406**, 20 (2005).

[112] T. S. Hofer, B. R. Randolf, B. M. Rode, J. Phys. Chem. B **112**, 11726 (2008).

[113] C. B. Messner, T. S. Hofer, B. R. Randolf, B. M. Rode Chem. Phys. Lett. **501**, 292 (2011).

[114] C. Kritayakornupong, K. Plankensteiner, B. M. Rode, ChemPhysChem **5**, 499 (2004).

[115] R. J. Frick, T. S. Hofer, A. B. Pribil, B. R. Randolf, B. M. Rode, Phys. Chem. Chem. Phys. **12**, 11736 (2010).

[116] R. Armunanto, C. F. Schwenk, B. R. Randolf, B. M. Rode, Chem. Phys. Lett. **388**, 395 (2004).

[117] R. Armunanto, C. F. Schwenk, B. R. Randolf, B. M. Rode, Chem. Phys. **305**, 135 (2004).

[118] C. F. Schwenk, B. M. Rode, ChemPhysChem **5**, 342 (2004).

[119] A. Tongraar, S. Hannongbua, B. M. Rode, Chem. Phys. **219**, 279 (1997).

[120] H. J. Bakker, J. L. Skinner, Chem. Rev. **110**, 1498 (2010).

[121] S. T. van der Post, H. J. Bakker, Phys. Chem. Chem. Phys. **14**, 6280(2012).

[122] K. J. Tielrooij, N. Garcia-Araez, M. Bon, H. J. Bakker, Science **326**, 1003 (2010).

[123] L. Liu, J. Hunger, H. J. Bakker, J. Phys. Chem. A **115**, 14593 (2011).

[124] K. J. Tielrooij, S. T. van der Post, J. Hunger, M. Bon, H. J. Bakker, J. Phys. Chem. B **115**, 12638 (2011).

[125] M. Pastorczak, S. T. van der Post, H. J. Bakker, Phys. Chem. Chem. Phys. **15**, 17767 (2013).

[126] S. T. van der Post, S. Scheidelaar, H. J. Bakker, J. Mol. Liq. **176**, 22 (2012).

[127] S. T. van der Post, K.-J. Tielrooji, J. Hunger, E. H. G. Backus, H. J. Bakker, Faraday Disc. **160** 171 (2013).

[128] F. Palombo, S. R. Meech, J. Phys. Chem. B **116**, 13481 (2012).

[129] J. Barthel, R. Buchner, B. Wurm, J. Mol. Liq. **98–99**, 51 (2002).

[130] K. Giese, U. Kaatze, R. J. Pottel, J. Phys. Chem. **74**, 3718 (1970).

[131] R. Bucher, Pure Appl. Chem. **80**, 1239 (2008).

[132] R. Buchner, J. Barthel, J. B. Gill, Phys. Chem. Chem. Phys. **1**, 105 (1999).

[133] T. Fukasawa, T. Sato, J. Watanabe, Y. Hama, W. Kunz, R. Buchner, Phys. Rev. Lett. **95**, 197802 (2005).

[134] R. Buchner, C. Hölzl, J. Stauber, J. Barthel, Phys. Chem. Chem. Phys. **4**, 2169 (2002).

[135] H. M. A. Rahman, G. Hefter, R. Buchner, J. Phys. Chem. B. **116**, 314 (2012).

[136] H. M. A. Rahman, R. Buchner, J. Mol. Liq. **176**, 93 (2012).

[137] H. M. A. Rahman, G. Hefter, R. Buchner, J. Phys. Chem. B. **117**, 2142 (2013).

[138] J. Barthel, M. Kleebauer, R. Buchner, J. Solution Chem. **24**, 1 (1995).

[139] B. Wurm, M. Münsterer, J. Richardi, R. Buchner, J. Barthel, J. Mol. Liq. **119**, 97 (2005).

[140] A. Placzek, G. Hefter, H. M. A. Rahman, R. Buchner, J. Phys. Chem. B **115**, 2234 (2011).

[141] J. McConnell, The Theory of Nuclear Magnetic Relaxation in Liquids, Cambridge University Press, London, Chap. 9 (1987).

[142] A. Sacco, M. Carbonara, M. Holz, J. Chem. Soc., Faraday Trans 1 **85**, 1257 (1989).

[143] P. B. Balbuena, K. P. Johnston, P. J. Rossky, J.-K. Hyun, J. Phys. Chem. B **102**, 3806 (1998).

[144] L. Carcia-Tarres, E. Guardia, J. Phys. Chem. B **102**, 7448 (1998).

[145] T. S. Hofer, B. M. Rode, B. R. Randolf, Chem. Phys. **312**, 81 (2005).

[146] T. S. Hofer, B. R. Randolf, B. M. Rode, Chem. Phys. **323**, 473 (2006).

[147] M. Q. Fatmi, T. S. Hofer, B. R. Randolf, B. M. Rode, J. Phys. Chem. **110**, 616 (2006).

[148] A. Tongraar, B. M. Rode, Chem. Phys. Lett. **409**, 304 (2005).

[149] T. S. Hofer, H. Scharnagl, B. R. Randolf, B. M. Rode, Chem. Phys. **327**, 31 (2006).

[150] B. S. Krumgalz, J. Chem. Soc., Faraday Trans. 1 79, 571 (1983).

[151] Y. Marcus, in D. V. Bostrelli, ed., Solution Chemistry Research Progress, Nova Science, Hauppauge, NY. 51 (2008).

[152] J. M. G. Barthel, H. Krienke, W. Kunz, in H. Baumgärtel, E. U. Franck, W. Grünbein, eds., Physical Chemistry of Electrolyte Solutions, Steinkopf-Springer, Darmstadt-New York (1998).

[153] P. P. Kosakewitsch, Z. Phys. Chem. **136**, 195 (1928).

[154] R. A. Stairs, W. T. Rispin, R. C. Makhija, Can. J. Chem. **48**, 2755 (1970).

[155] D. Chandra, R. Gopal, Indian J. Chem. A **14**, 391 (1976).

[156] R. Aveyard, Y. Thompson, Can. J. Chem. **57**, 856 (1979).

[157] A. A. Abramzon, R. D. Gaukhberg, Russ. J. Appl. Chem. **66**, 1985 (1993).

[158] D. England, B. Midgley, I. S. Hughes, J. Coll. Interf. Sci. **136**, 327 (1990).

[159] R. Gopal, S. A. Rizvi, J. Indian Chem. Soc. **43**, 179 (1966).

[160] E. J. Aveyard, Y. Thompson, F. E. Doany, R. Moore, R. M. Hochstrasser, J. Chem. Phys. **76**, 5632 (1982).

[161] M. Li, J. Owrutsky, M. Sarisdky, J. P. Cluver, A. Yodh, R. M. Hochstrasser, J. Chem. Phys. **98**, 5499 (1993).

[162] M. Nakahara, K. Emi, J. Chem. Phys. **98**, 5418 (1993).

[163] H. Hosoi, Y. Masuda, J. Phys. Chem. B **102**, 2995 (2001).

[164] Y. Masuda, J. Phys. Chem. A **105**, 2989 (2004).

[165] Y. Masuda, A. Muramoto, J. Solution Chem. **33**, 811 (2004).

[166] J. F. Coetzee, W. R. Sharpe, J. Phys. Chem. **75**, 3141 (1971).

[167] C. L Henry, V. S. J. Craig, Langmuir **24**, 7979 (2008).

[168] C. L Henry, S. I. Karakashev, P. T. Nguyen, A. V. Nguen, V. S. J. Craig, Langmuir **25**, 9931 (2009).

6

IONS IN MIXED SOLVENTS

Ions are solvated to different extents and with different strengths in diverse solvent, so that it stands to reason that the immediate surroundings of an ion in a mixture of solvents should differ from the bulk composition of the mixture. This phenomenon is called *preferential solvation*, and if proceeding to an extreme limit, at which one solvent of a binary solvent mixture is excluded from the environment of an ion, this ion is said to be *selectively solvated* in such a mixture. The same may be said for an ion in a multicomponent solvent mixture, and if all but one of the solvents is excluded from the solvation shell of the ion, it is then selectively solvated by the remaining solvent. The properties of an ion in solvent mixtures are generally not the mean of its properties in the individual solvents weighted according to the bulk composition of the mixture, but depend on its preferential solvation. The cations and anions of an electrolyte are generally preferentially solvated to different extents, so that the overall effect depends on the natures of the ions making up the electrolyte. If in a binary mixture the cation and anion of an electrolyte are preferentially solvated by the same solvent component, then the electrolyte is said to be *homo-solvated*, otherwise it is *hetero-solvated*, which is more generally the case.

The physical and chemical properties of (binary) solvent mixtures, necessary for understanding the solvation of ions in the mixtures, are dealt with in Sections 3.4.1 and 3.4.2, respectively. The preferences of ions for certain solvents over others are described in terms of their standard molar Gibbs energies of transfer from a source solvent (water has generally been arbitrarily selected as this source) to neat target solvents in Section 4.3.2.1. These sections should be consulted to complement the present discussions regarding ions in mixed solvents.

Ions in Solution and Their Solvation, First Edition. Yizhak Marcus.
© 2015 John Wiley & Sons, Inc. Published 2015 by John Wiley & Sons, Inc.

6.1   ION TRANSFER INTO SOLVENT MIXTURES

The transfer of ions from a source solvent into a binary solvent mixture has been studied mainly for the cases where water is the source solvent and also one of the components of the solvent mixtures, that is, aqueous solvent mixtures. Few studies are available in which ions are transferred from nonaqueous solvents and into completely nonaqueous mixtures. The source solvent in some of these studies, according to reports by Kondo et al., Cox and Waghorne, Marcus, and Piekarski et al., was methanol [1–6], in other studies acetonitrile [7, 8], and some further systems were also studied [2, 5, 9].

Table 6.1 shows values of the standard molar Gibbs energy for the transfer of ions from water into equimolar aqueous mixtures with cosolvents at 25°C, $\Delta_{tr}G^{\infty}(I^{\pm}, W \rightarrow 0.5W + 0.5S)$, taken from the compilations by Kalidas et al. [10] and by Marcus [11]. Further values for the solvents shown there are available in these compilations at 0.1 mole fraction steps over the entire composition range as far as they have been published. The ionic data were selected from data on electrolytes listed in these references on the basis of their conformation to the TATB or TPTB assumptions (Section 4.3.2). Data are available in these references also for many other aqueous solvents (i-PrOH, t-BuOH, propylene glycol, tetrahydrofuran, 1,4-dioxane, propylene carbonate, pyridine, and N-methylpyrrodin-2-one) for at least a part (the water-rich part) of the composition range, as well as for some other ions that were not measured at $x_s = 0.5$ in the solvents shown in Table 6.1 or that could not be traced to the TATB or TPTB assumptions. On the whole, the $\Delta_{tr}G^{\infty}(I^{\pm}, W \rightarrow W + S) = f(x_S)$ in the water-rich region do not deviate much from linearity, but in the cosolvent-rich region, such deviations could be rather profound.

Table 6.2 shows values of the standard molar enthalpy and entropy for the transfer of ions from water into equimolar aqueous mixtures with cosolvents at 25°C, $\Delta_{tr}H^{\infty}$ or $\Delta_{tr}S^{\infty}(I^{\pm}, W \rightarrow 0.5W + 0.5S)$, taken from the compilation by Hefter et al. [13]. Further values for the solvents shown there are available at 0.1 mole fraction steps over the entire composition range. The ionic data were selected from data on electrolytes listed in these references on the basis of their conformation to the TATB or TPTB assumptions (Section 4.3.1). Data are available in this reference also for many other aqueous solvents (n- and i-PrOH, t-BuOH, glycerol, tetrahydrofuran, 1,4-dioxane, acetone, N,N-dimethylacetamide, and sulfolane) for at least a part (the water-rich part) of the composition range, as well as for some other ions that were not measured at $x_S = 0.5$ in the solvents shown in Table 6.2 or that could not be traced to the TATB or TPTB assumptions. This table also includes data for these functions for the transfer of ions into aqueous urea, which, though urea is not a solvent, it behaves in aqueous solutions as if it were a liquid amide.

In many cases, there exists enthalpy–entropy compensation and relatively small values of $\Delta_{tr}G^{\infty}(I^{\pm}, W \rightarrow W + S)$ may result from much larger values of $\Delta_{tr}H^{\infty}(I^{\pm}, W \rightarrow W + S)$ and $T\Delta_{tr}S^{\infty}(I^{\pm}, W \rightarrow W + S)$. Therefore, it is not advisable to use this path for the estimation of Gibbs energies of transfer for such ions and aqueous solvent compositions that are not available in in Refs. 10 and 11 from data in Ref. 13. On the contrary, if both $\Delta_{tr}G^{\infty}(I^{\pm}, W \rightarrow W + S)$ and

TABLE 6.1 Standard Molar Gibbs Energy of Transfer of Ions from Water into Equimolar Aqueous Mixtures with Cosolvents, $\Delta_t G^\infty (I^\pm, W \rightarrow 0.5W + 0.5S)$/kJ·mol⁻¹ from Refs. 10 and 11

Ion/cosolvent	MeOH	EtOH	EG	Me$_2$CO	MeCN	FA	DMF	HMPT	DMSO
H⁺	−1.6	−4.2	−0.9	−11.4	−2.4	(−27.0)	−16.2		−21.6
Li⁺	3.1	2.5	3.0	−6.4	1.2		−10.9		−16.0
Na⁺	6.1	5.2	1.7	−3.3	0.3	(−6.4)	−5.9		−10.2
K⁺	6.3	5.0	1.8	−4.6	−0.5	(−6.4)	−3.4	−3.7	−8.1
Rb⁺	7.0	5.9		−4.5	0.0	(−6.0)	−3.2	−7.2	−7.2
Cs⁺	6.1	5.5		−5.3	−1.5	(−4.9)	−5.4	−11.8	−10.1
Cu⁺					−53				−35.7
Ag⁺	4.8	−1.4		−6.1	−21.0		−14.7		−21.9
Au⁺		−2.0			−39.9				
Tl⁺	5.5				(3.7)				−13.8
Me$_4$N⁺	2.9	3.3		−7.2		(−1.9)			
Et$_4$N⁺		−1.5				(−5.5)			
Pr$_4$N⁺	−6.8	−6.4							
Bu$_4$N⁺	−12.5	−7.7							−17.1
Ph$_4$As⁺	−17.8[a]	−21.2	−15.9	−30.1	−28.9		−32.1	−35.3	−29.0
Mg²⁺	4.8								
Ca²⁺	10.4								
Sr²⁺	9.3								
Ba²⁺	11.6								
Mn²⁺					1.2				−42.0
Fe²⁺					−2.5				
Cu²⁺	7.9	3.8	2.4	7.0	10.4	(−5.8)	−16.3		−26.9
Zn²⁺	(4.7)	3.6	(−14.8)	(−3.3)	−5.6		(−50.1)	(−64.9)	−26.2
Cd²⁺	(6.5)		(−13.3)					(−56.4)	−39
Pb²⁺	7.3	(4.0)			−5.8			(−70)	−41.5

(continued)

TABLE 6.1 (Continued)

Ion/cosolvent	MeOH	EtOH	EG	Me₂CO	MeCN	FA	DMF	HMPT	DMSO
UO_2^{2+}		(13.3)							
OH^-	(3.4)	(5.3)							
F^-	8.7	16.9	9.0	34	(20.2)				
Cl^-	5.5	12.2	4.3	25.8	12.7				
Br^-	4.1	8.1	3.5	22.6	(22.5)				
I^-	1.7	3.4	0.9	14.0	(7.9)	3.8[b]			
CN^-	−0.2	(7.8)		(18.3)					
SCN^-	0.0			(13.2)					
N_3^-	(10.5)								
ClO_3^-	1.0			(16.)		6.0[b]			
ClO_4^-		(3.9)		(2.7)	(0.7)	−1.4[b]			
MnO_4^-				(−0.9)					
$CH_3CO_2^-$		(−5.1)			(−29.0)				
Pic^-	−3.9	−2.8	(−4.2)			−6.6[b]			
BPh_4^-	−17.0	−19.8	(−15.9)	(−25.9)	28.7				
$C_2O_4^{2-}$	(20.0)	(31.1)							
CrO_4^{2-}		(27.1)							
$Fe(CN)_6^{3-}$	(4.9)								

Values in parentheses are neither recommended nor tentatively recommended as specified in these sources.

[a] Ph_4P^+ rather than Ph_4As^+.

[b] From Ref. 12.

TABLE 6.2 Standard Molar Enthalpies $\Delta_t H^\infty$(I±, W→0.5W + 0.5S)/kJ·mol⁻¹ (roman font) and Entropies $\Delta_t S^\infty$(I±, W→0.5W + 0.5S)/J·K⁻¹·mol⁻¹ (italics font) of Transfer of Ions from Water into Equimolar Aqueous Mixtures with Cosolvents[a]

Ion/cosolvent	MeOH	EtOH	EG	MeCN	FA	DMF	HMPT	DMSO
H⁺	3.1, *16*	−13.2, *−30*	−11.8, *−36*	−22.0, *−66*				−25.1, *−30*
Li⁺	1.6, *−5*	−5.8		−20.5, *−73*	−6.1		−48.6	−21.5, *−38*
Na⁺	3.6, *−8*	−2.1, *−24*	−8.8, *−34*	−20.1, *−67*	−11.0, *−15*	−23.0, *−57*	−42.8	−25.1, *−57*
K⁺	1.8, *−15*	−3.2, *−28*		−23.9, *−78*	−11.6, *−17*	−25.1, *−73*	−41.4, *−126*	−25.1, *−57*
Rb⁺	2.9, *−14*	−6.0, *−40*		−25.3, *−85*	−11.1, *−17*	−24.7, *−72*		−25.1, *−60*
Cs⁺	0.8, *−18*	−6.5, *−40*		−25.9, *−82*	−10.5, *−19*	−24.4, *−64*	−44.4, *−109*	−23.7, *−46*
Ag⁺		−3.2, *−3*		−54.3, *−112*		−6.1		−42.2, *−67*
NH₄⁺	−2.5, *−22*	−9.0			−9.2	−30.6	−63.5	−26.6, *84*
Me₄N⁺				−14.1, *−36*	−0.1, *6*	−11.6	−29.8	
Et₄N⁺				−3.6, *−5*		0.7		
Pr₄N⁺	8.2	16.9		10.0, *46*		14.9		
Bu₄N⁺	34.3, *157*	34.8, *143*		17.6, *60*		26.7		30.4, *159*
Ph₄P⁺	13.7, *106*	11.1, *108*	7.9, *79*	−8.9[b], *67*	0.7	−7.4, *83*	−23.8, *39*	−4.8, *81*
Mg²⁺	2.6, *−26*			−45.3				
Ca²⁺	−3.3, *−11*	−19.2						
Sr²⁺		−19.3						
Fe²⁺		6.0						
Cu²⁺	0.9	5.1, *5*		−39.3, *−162*				−49.5
Zn²⁺	−2.5, *−12.5*							
Cd²⁺	3.1							
OH⁻		−4.1						51.6

(*continued*)

TABLE 6.2 (Continued)

Ion/cosolvent	MeOH	EtOH	EG	MeCN	FA	DMF	HMPT	DMSO
F⁻	1.2	10.0, −22[c]			12.1, −50[c]	9.4	66.9	16.7, −11
Cl⁻	−2.6, −32	4.3, −22	6.4, 39	7.1, −18	0.6	20.1, −12		
Br⁻	−5.0, −33	0.2, −16		5.7, −20	−2.8	9.5, −23	29.8, −17[c]	7.6, −29
I⁻	−6.2, −31	−3.6, −34	−4.2	2.3, −21	−5.4, −29[c]	−4.1	1.6, −83[c]	−2.9, −40
SCN⁻	−8.1, −21[c]			−1.6				
NO₃⁻	−4.1	−1.2		8.1	−5.0		21.1	1.0
ClO₄⁻		−4.3, −24[c]		−5.3				−13.6
Pic⁻		26		1.8				
BPh₄⁻	13.7, 106	11.1, 108	11.2, 79	−8.9	0.7	−7.4, 83	−23.8	−8.8, 81

[a] From Ref. 13.
[b] Ph₄As⁺ rather than Ph₄P⁺.
[c] From Ref. 11.

$\Delta_{tr}H^{\infty}(I^{\pm}, W \rightarrow W + S)$ are available in these sources, but not $\Delta_{tr}S^{\infty}(I^{\pm}, W \rightarrow W + S)$, the latter may be calculated from the appropriate thermodynamic relationship.

Few direct reports on the standard molar heat capacities and volumes of transfer of electrolytes and ions from a source solvent to a solvent mixture are available. Examples are found for transfer from water to aqueous t-butanol [14] and aqueous acetonitrile [15] in the studies by Hefter et al. Since, however, the standard molar heat capacities and volumes of ions in water are known (Sections 2.3.1.1 and 2.3.1.5) and the corresponding values in aqueous cosolvent mixtures are also known (Section 6.2.1 below), the data for the transfer, if needed, can be readily calculated from these sources.

6.2 PROPERTIES OF IONS IN SOLVENT MIXTURES

As said above, the properties of ions in solvent mixtures generally do not depend linearly on the bulk solvent composition (whether in mole or volume fractions). The deviations of the properties from the linear dependence throw light on the interactions that take place between the ions and the components of the solvent mixture and between the latter. In the following, thermodynamic and transport properties are dealt with. Only binary solvent mixtures are handled within the scope of this book, but extension toward multicomponent mixtures should be straight-forward. The information concerning the properties of ions in mixed solvent is very extensive, and hence only representative examples are dealt with here.

6.2.1 Thermodynamic Properties of Ions in Mixed Solvents

The standard molar enthalpies of dissolution of electrolytes in aqueous cosolvent mixtures are monotonic with the composition only if the cosolvent does not involve a sizable alkyl moiety in its molecules. This is the case for formamide, DMF, urea (treated as if it were a solvent), and DMSO. On the other hand, cosolvents that have a hydrophobic group and enhance the structure of the water in the water-rich region according to Marcus [16, 17] show a maximum in the $\Delta_{diss}H^{\infty} = f(x_S)$ curve in this region as reported by Piekarski [18]. Such solvents include alkanols, 2-alkoxyalkanols, THF, 1,2-dimethoxyethane, 2-butanone, DMA, and HMPT.

A useful measure for the interactions in the system is the pair interaction enthalpy h_{ES} defined as:

$$h_{E,S} = \lim\left(m_S \rightarrow 0\right)\left[\Delta_{diss}H^{\infty}(E, W + S) - \Delta_{diss}H^{\infty}(E, W)\right] / 2m_S \qquad (6.1)$$

Although water (W) is specified here as the reference solvent, it might be any other one, and E and m_S denote the electrolyte (or ion, if ionic values have been derived) and the molality of the cosolvent S. Extensive data are available for E = NaCl and NaI, reproduced in Table 6.3 from the work of Piekarski and Tkaczyk [22], but data are available for a few other salts too. Data of $h_{E,S}$ for E = NaI are available for fewer cosolvent when the reference solvents are MeOH, MeCN, and DMF [22]. When

TABLE 6.3 **Pair interaction Enthalpic Parameters, $h_{E,S(R)}$/J·kg·mol^{-2}, and Heat Capacity Parameters $c_{E,S(R)}$/J·mol^{-1}·kg·mol^{-2} for Electrolytes in Mixtures of a Reference Solvent R (W or as Specified) with Cosolvents at 25°C**

Cosolvent	$h_{NaCl,S(W)}$	$h_{NaI,S(W)}$	$h_{Bu4NBr,S(W)}$[a]	$h_{NaI,S(MeCN)}$[b]	$C_{NaI,S(W)}$	$C_{NaI,S(DMF)}$
MeOH	300	314		−2370	−3.4	
EtOH	580	596				
1-PrOH	740	780			−7.0	
2-PrOH	900	1018				
t-BuOH	980	1440	6150[c]		−14.0	−1.3
EG		178				−1.0
2-Methoxyethanol	190	194		−5493	−4.4	
THF	404	344				−3.7
1,4-Dioxane	−290		3564[c]		5[d]	
1,2-Dimethoxyethane	260	210				
Acetone	20	−92	3470		−5[d]	−0.2
2-Butanone	211	140				
PC					3	
MeCN	−286	−494				
Urea	−490	−524			22[d]	
FA	−81	−696	1176		6.5	−1.8
Acetamide	−81	−290				
NMF			1771			
DMF	−79	−350	2414	−2925	2.5	
DMA	99	−124	2927	−3337		
HMPT	842	564	8150			
DMSO	−202	−628	1913	−3050	5[d]	
Water				−2704		<0

[a] Data from Ref. 19.
[b] Data from Ref. 8.
[c] Data from Ref. 20.
[d] Data From Ref. 21, but with E = NaCl, not NaI.

water is the reference solvent, its interactions with the cosolvent predominate over the ion–cosolvent interactions. However, when MeOH and MeCN are the reference solvents, the direct interactions between the ions and the cosolvent play the major role, and when DMF is the reference solvent, its own interactions with the ions are the predominant ones.

It turned out that the $h_{E,S}$ values for aqueous NaCl and NaI correlate linearly well with the heat capacity of interaction of water with the cosolvent, C_{Pint}. This quantity is defined as:

$$C_{Pint} = C_{P(S/W)}^{\infty} - C_{P(S)}^{\circ} - \left[a\sigma_S^2 + b\sigma_S + c\right] - R\left[2\alpha_{PW}T + \left(\partial\alpha_{PW}/\partial T\right)_P - 1\right] \quad (6.2)$$

Here $C_{P(S/W)}^{\infty}$ is the limiting partial molar heat capacity of the cosolvent S in water, $C_{P(S)}^{\circ}$ is the molar heat capacity of the gaseous S, the first term in square brackets is the cavity formation term, depending on the hard-sphere diameter σ_S of S with known coefficients according to French and Criss [23], and α_{pw} is the isobaric expansibility of water.

Other electrolytes featured in studies of cosolvents in aqueous solutions, for example Bu_4NBr with a number of cosolvents as reported by Korolev, Kustov, and coworkers [19, 20] are also shown in Table 6.3. Enthalpic pair interaction parameters for $CaCl_2$ with MeOH, EtOH, and n-PrOh were reported in Ref. 24 and for $NaNO_3$, KNO_3, $Ca(NO_3)_2$, and $La(NO_3)_3$ with MeOH in Ref. 25 by Taniewska-Osinska and coworkers. Many other studies concerned electrolyte interactions with cosolvents in water without reporting pair interaction parameters. Examples are the enthalpy of solvation of $NaNO_2$, KSCN, and NH_4BF_4 in aqueous methanol reported by Manin and Korolev [26] and of NH_4Br and NH_4BF_4 in aqueous HMPT reported by Kustov [27], but such studies are too numerous to be presented here.

Several studies concerned the pair interaction parameters of systems where the reference solvent was not water, but deal with dilute solutions of water and other cosolvents in solutions of electrolytes in these reference solvents. An example is the report by Piekarski and Kubalczyk [8], in which the reference solvent was acetonitrile and the electrolyte again was NaI; see Table 6.3. Pair interaction coefficients of NaI with water in several lower alkanols as reference solvents were reported in Ref. 28 and those of NaI with MeCN in several solvents were reported in Ref. 7 by Piekarski et al.

The thermodynamic properties of electrolytes in mixed solvents include not only the enthalpic pair interaction parameters at one temperature (25°C) but also at several temperatures according to Korolev et al. and Piekarski et al. [20, 28] from which the heat capacity pair interaction parameters, $c_{E,S(ref)}$, (where "(ref)" denotes the reference solvent) may be calculated as may also derived quantities, such as the entropic ones [20]. Several pair interaction parameters for heat capacities of electrolyte interacting with cosolvents in two reference solvents $c_{E,S(W)}$ and $c_{E,S(DMF)}$ reported by Piekarski and Somsen [29] are shown in Table 6.3. At low concentrations of the electrolyte and the cosolvent, it was found by Desnoyers et al. [21] (for E = NaCl) that the values of $c_{E,S(W)} = c_{S,E(W)}$, the latter being the heat capacity pair interaction parameter defined in analogy with Equation 6.1 as $c_{E,S} = \lim(m_E \rightarrow 0)\left[C_P^\infty(S, W+E) - C_P^\infty(S,W)\right]/2m_E$. Data for the partial molar heat capacities of electrolytes in mixed solvents, which were not expressed in terms of the pair interaction parameters, have also been reported by Hefter et al. [14, 15]. As the concentration of S in the reference solvent increases, the curves $C_P^\infty(S, W+E) - C_P^\infty(S,W) = f(m_S)$ lose their (near) linearity and very specific interaction patterns emerge, in particular where the reference solvent is water and hydrophobic interactions with the cosolvent (and hydrophobic ions) are involved.

Similar cases of specific interactions have been found also for the partial molar volumes of electrolytes (ions) in mixed solvents. The reports concerning the partial molar volumes of electrolytes in mixed solvents generally do no pertain to dilute solutions of the cosolvent, but on the contrary, most such studies cover the entire composition range. The curves tend to be very asymmetric in the solvent composition and changes in direction are often encountered, as shown, for example, in Figure 6.1 for some electrolytes in aqueous DMSO according to Letellier et al. [30]. In a few cases were very dilute solutions were studied, so that pair-interaction parameters could be

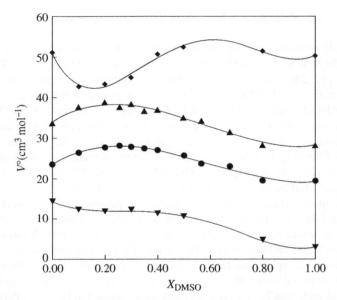

FIGURE 6.1 Representative partial molar volumes of electrolytes in mixed aqueous DMSO solvents: NaBr ($\bullet$), KBr ($\blacktriangle$), Me$_4$NBr ($\blacktriangledown$) with 100 cm^3·mol^{-1} deducted, and Bu$_4$NBr ($\blacklozenge$) with 250 cm^3·mol^{-1} deducted for the clarity of presentation.

determined. Perron et al. [31] reported the values of $V_{E,S(W)}$/cm^3·mol^{-2}·kg^{-1} for E = NaCl that were −0.125 for S = MeOH, −0.05 for S = EtOH, 0 for S = n-PrOH, and −0.10 for S = t-BuOH. Values were reported in [31] for the latter cosolvent also for many other electrolytes. Values of $V_{E,S(DMF)}$/cm^3·mol^{-2}·kg^{-1} for E = NaI were reported for several cosolvent S by Piekarski and Somsen [29].

The standard partial molar volumes of electrolytes in mixed solvents can be modeled, as can those in neat solvents, in terms of the sum of the intrinsic volumes of the ions and their electrostriction. It is assumed that the intrinsic volumes, that is, the volumes of the ions proper and including the voids between ions and solvent molecules, are solvent independent, so that they do not depend on the natures of the solvents near the ions. Then, if no preferential solvation of the ions by the components of the solvent mixture takes place, the electrostriction can be calculated according to Marcus [32] as for neat solvents (Section 4.3.2.5), with the relevant properties of the solvents prorated according to the composition of the mixture. This appeared to be the case for the ions Li$^+$, Na$^+$, K$^+$, ClO$_4^-$, AsF$_6^-$, and CF$_3$SO$_3^-$ in mixtures of PC with MeCN, in which V^∞(I$^\pm$,PC+MeCN) is linear with the composition over nearly the entire composition range. This is the case also for Me$_4$NBr in W+DMSO, as shown in Figure 6.1. Similarly, in aqueous methanol mixtures, smooth curves result for the ions Li$^+$, Na$^+$, K$^+$, Cs$^+$, Cl$^-$, Br$^-$, and I$^-$ like those shown in Figure 6.1 for NaBr and KBr. However, when preferential solvation occurs, the electrostriction could not be calculated in the same manner, because the relevant properties of the local solvent mixture, where the electrostriction takes place, cannot be estimated with confidence. Only if selective solvation of the ions occurs, and only one component of the solvent mixture resides in the vicinity of the ions can the electrostriction be calculated as reported by Marcus [33]. This is the case in W+EtOH (both K^+ and Cl^- are selectively hydrated), W+DMSO (alkali metal and Ag^+ are selectively solvated by DMSO but halide anions by water), and W+MeCN (Cu^+ and Ag^+ are selectively solvated by MeCN, Na^+, Cl^-, and I^- by water).

6.2.2 Transport Properties of Ions in Mixed Solvents

The conductivity of electrolytes and ionic limiting conductivities in mixed solvents are intimately related to the viscosities of these solutions according to the concept of the Walden product: $\Lambda^\infty \eta = $ const for a given electrolyte (ion) irrespective of the temperature and the solvent or mixed solvent composition.

A quasi-thermodynamic treatment of the viscous flow of electrolyte solutions was presented by Feakins et al. [34], including mixed solvents. The treatment involves the breaking of solvent–solvent bonds by the ions and the transition from the behavior in aqueous solutions, where alkali metal ions (except Li^+) are sterically unsaturated and readily exchange solvent molecules, to nonaqueous behavior where such ions are saturated. Specifically, aqueous methanol mixtures were discussed regarding solutions of CsCl and Li_2SO_4. The former salt involves two large ions resulting in negative B_η values in water-rich mixtures, turning to positive values above $x_{MeOH} = 0.36$. The latter salt has a shallow minimum near $x_{MeOH} = 0.06$, explained by increasing importance of the secondary solvation of the ions.

The conductivity of electrolytes in mixed solvents was recently reviewed by Apelblat [35] and only the main features are presented here. References to the data, for example those reported by Barthel and by Kay [36, 37], provide the limiting conductivities Λ_E^∞ mainly, but not exclusively, for aqueous mixtures with cosolvents. It was shown by Apelblat [35] that the ratios of the Λ_E^∞ values (also the ionic values λ_η^∞) for different electrolytes (or ions) tend to universal values in the composition ranges rich in the one or the other solvent component of an A+B solvent mixture. At a given temperature for any two electrolytes E1 and E2:

$$\text{at low } x_B \left[\Lambda_{E2}^\infty (x_B) / \Lambda_{E1}^\infty (x_B) \right] = \left[\Lambda_{E2}^\infty (x_B = 0) / \Lambda_{E1}^\infty (x_B = 0) \right] = \text{const}(0)$$
$$\text{at high } x_B \left[\Lambda_{E2}^\infty (x_B) / \Lambda_{E1}^\infty (x_B) \right] = \left[\Lambda_{E2}^\infty (x_B = 1) / \Lambda_{E1}^\infty (x_B = 1) \right] = \text{const}(1) \quad (6.3)$$

with two different constants for each composition range. KCl was generally taken as the standard electrolyte E1 for which Equation 6.3 was tested. At $0 \le x_B \le x_{B\,max}$ universal curves $\Lambda_E^\infty = f(x_B)$ result that can be represented up to the maximal composition $x_{B\,max}$ by the general expression:

$$\Lambda_{E1}^\infty (x_B) = \Lambda_{E1}^\infty (x_B = 0) - b x_B^a x_A^a \quad (6.4)$$

TABLE 6.4   Coefficients for Equation 6.4 for Aqueous Mixtures with Cosolvents at 25°C from Ref. 35, with the Presumed Standard Electrolyte E1 Noted

Solvent A	Solvent B	$x_{B\,max}$	$\Lambda_{E1}^{\infty}(x_B=0)$/S·cm²·mol⁻¹	a	b/S·cm²·mol⁻¹
Water	Methanol	0.40	149.74 E1 = KCl	0.7570	294.18
Water	Ethanol	0.35	149.74 E1 = KCl	0.6592	413.88
Water	1-Propanol	0.30	126.52 E1 = NaCl(?)^a	0.6145	365.98
Water	t-Butanol	0.22	149.74 E1 = KCl	0.5499	489.84
Water	Tetrahydrofuran	0.40	150.48 E1 = KI(?)^a	0.6149	407.31
Water	1,4-Dioxane	0.35	149.74 E1 = KCl	0.6798	415.00
Water	Acetone	0.25	97.46 E1 = LiPF$_6$(?)^a	0.6421	191.52
Water	Ethylene carbonate	0.35	149.74 E1 = KCl	0.8553	331.15
Water	DMF	0.25	121.63	0.6952	346.04
Acetone	Water	0.55	185.53	0.6297	514.07
DMF	Water	0.50	87.10	0.4365	210.04

^{a} The reported $\Lambda_{E1}^{\infty}(x_B=0)$ fits the suggested E1, but other salts could have been used to produce this value.

The limiting conductivity of the standard electrolyte E1 in pure solvent A, $\Lambda_{E1}^{\infty} = (x_B = 0)$, and the coefficients a and b for some representative examples of water-rich and cosolvent-rich mixtures are shown in Table 6.4. The values for other electrolytes E2 may then be calculated from Equations 6.3. The standard deviations in the limiting conductivities calculated according to Equation 6.4 do not exceed $3\,S\cdot cm^2\cdot mol^{-1}$. As an illustration, the $\Lambda_{E2}^{\infty} = (x_B)$ values of numerous electrolytes E2 referred to E1 = KCl in aqueous ethanol mixtures according to Equation 6.3 are shown in Figure 6.2. Up to $x_{B\,max} = 0.35$, a single curve is obtained.

For the corresponding coefficients a and b for some completely nonaqueous mixtures, the data by Kay [37] should be consulted. The concept of the universal conductivity curve according to Equation 6.3 was subsequently extended from the 1:1 electrolytes dealt with by Apelblat in Ref. 39 to multivalent electrolytes of the 1:2, 2:1, 2:2, 1:3, and 3:3 classes (examples being Na_2SO_4, $MgCl_2$, $MnSO_4$, and $LaFe(CN)_6$), tested in aqueous ethanol and 1,4-dioxane as well as in mixtures of methanol and ethylene glycol [39].

The enhanced conductivities of hydrogen and hydroxide ions in water, due to the Grotthuss mechanism of "hopping" of the charge from one water molecule to another bonded to the former by hydrogen bonds, is sustained in aqueous–cosolvent mixtures under certain conditions. In aqueous acetonitrile, for instance, this mechanism takes place only if sufficient water is present for clusters of a minimal size to be present as shown by Gileadi and Kirowa-Eisner [40], namely above 20 volume% water. When small amounts of water are added to methanol, the conductivity of acids diminishes, because the proton affinity of the water is larger than that of the methanol, so the water molecules trap the hydrogen ions preventing it from "hopping." The existence of this mechanism can be judged by a distinct decline of the Walden product with increasing temperatures (being relatively insensitive to the temperature when movements of the ions themselves through the (mixed) solvent occurs) [40].

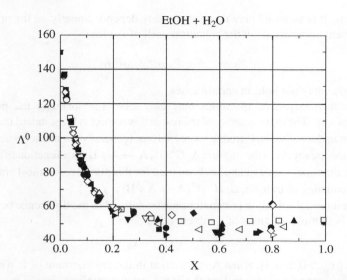

EtOH + H₂O

FIGURE 6.2 Limiting conductivities Λ_{E2}^{∞} of electrolytes in aqueous ethanol mixtures at 25°C adjusted to the value for E1 = KCl according to Equation 6.3. KCl (■), LiCl (□), LiBr (●), LiNO₃ (o), NaCl (▲), NaNO₃ (△), KBr (▼), KClO₄ (▽), and also KNO₃, CsCl, Me₄NBr, Et₄NBr, Pr₄NBr, Bu₄NBr, Bu₄NI (►, ◄, ◊, ◁, and ▷). From Ref. 38 by permission of the publisher, ACS.

6.3 PREFERENTIAL SOLVATION OF IONS

In the following, mainly binary solvent systems are considered, because much less systematic information is available regarding multicomponent mixtures. An ion I± in a binary mixture of solvents A + B has in its solvation shell *local* mole fractions (superscript ᴸ) of the solvent components that generally differ from the bulk composition. Necessarily,

$$x_{A(I)}^{L} + x_{B(I)}^{L} = 1 \qquad (6.5)$$

In certain cases, it is possible to obtain the local mole fractions of the solvent components around the ion from measurements of a spectroscopic property p of the ion that is sensitive to its surroundings. The property p may be the wave number ν of light absorption of a transition metal ion or of the vibrational mode of a polyatomic ion or the chemical shift δ of an NMR signal, etc. Then the local mole fraction of solvent component A around the ion I± is measured (in fact, defined) as:

$$x_{A(I)}^{L} = \frac{p_{AB(I)} - p_{B(I)}}{p_{A(I)} - p_{B(I)}} \qquad (6.6)$$

Here $p_{AB(I)}$ refers to the property p measured in the mixture of bulk mole fraction x_A, and $p_{A(I)}$ and $p_{B(I)}$ refer to the property measured in the neat solvents A and B, respectively. It is assumed here that the property depends linearly on the proportions of the solvent components in the solvation shell of the ion:

$$P_{AB(I)} = x^L_{A(I)} P_{A(I)} + x^L_{B(I)} P_{B(I)} \tag{6.7}$$

This premise may not hold in certain cases.

Certain thermodynamic quantities may also serve for obtaining the preferential solvation of ions. The main quantity of interest in this respect is the standard molar Gibbs energy of transfer of the ion $I^\pm$ from a source solvent (generally but not necessarily one of the components, say A) to the mixture $\Delta_{tr}G^\infty(I^\pm, A \rightarrow A + B)$ as a function of x_B. Rarely have other thermodynamic quantities been used for this purpose, foremost among these being the enthalpy of transfer, $\Delta_{tr}H^\infty(I^\pm, A \rightarrow A + B) = f(x_B)$.

The preferential solvation parameter may be defined as the difference between the local and bulk mole fractions of a solvent component:

$$\delta x_{A(I)} = x^L_{A(I)} - x_A \tag{6.8}$$

Values of $\delta x_{A(I)} > 0$, that is, when A is enriched in the environment of $I^\pm$ with respect to its bulk composition, mean that the ion is preferentially solvated by A and its negative values denote the preferential solvation of $I^\pm$ by B. A further way to express the preferential solvation is by means of the preferential solvation constant:

$$K_{AB(I)} = \frac{\left[x^L_{A(I)} / \left(1 - x^L_{A(I)}\right) \right]}{\left[x_A / \left(1 - x_A\right) \right]} \tag{6.9}$$

Values of $K_{AB(I)} > 1$ denote the preferential solvation of $I^\pm$ by A.

Another way to describe the preferential solvation of an ion in a binary solvent mixture is in terms of the equilibrium constants of the stepwise replacement of one of the solvents by the other:

$$IA_N + B \rightleftarrows IA_{N-1}B + A \quad K_{A/B1} \tag{6.10a}$$

$$IAB_{N-1} + B \rightleftarrows IB_N + A \quad K_{A/B_N} \tag{6.10b}$$

where the $K_{A/Bj}$ are the equilibrium quotients, it being assumed that the coordination number N in the solvation shell is independent of the natures of the solvents. An overall solvent replacement equilibrium quotient $K_{I(A,B)N}$ for the total replacement of A by B in the solvation shell of $I^\pm$ is $IA_N + NB \rightleftarrows IB_N + NA$, which is the product of the stepwise quotients $K_{I(A,B)_N} = \Pi K_{ABj}$. Preferential solvation by A is indicated by the equilibrium quotients being <1.

On purely electrostatic grounds, on the basis of the strict primitive model of the solution (spherical ions in a featureless compressible fluid dielectric), the ratio of the mole fractions of the two solvents at a distance r from the center of the ion $I^\pm$ should be given according to Scatchard [41] by:

$$\frac{\left[x(r)_{A(I)} / \left(1 - x(r)_{A(I)}\right) \right]}{\left[x_A / \left(1 - x_A\right) \right]} = \exp\left[\frac{z_I^2 e^2 \delta}{8\pi RT r^4} \right] \tag{6.11}$$

where the parameter δ is defined in terms of the permittivities of the solvent mixture $A+B$ and the pure solvent A by $\delta = \left(\varepsilon_{AB}^{-1} - \varepsilon_A^{-1} \right)/c_B$. If the distance r is taken as the sum of the ionic radius and the mean diameter of the solvent molecules, representing the solvation shell of the ion, the left-hand side of Equation 6.11 becomes equal to the preferential solvation constant. This so-called solvent sorting by the ion does not take into account the chemical interactions of the solvents with the ion nor does it the existence of solvation shells or the mutual interactions of the solvents (although this is implicit in the quantity δ). In general, Equation 6.11 with this choice of r does not lead to results in conformity with the experimental values of $K_{AB(I)}$ obtained from the property p.

Generally, the preferential solvation measures $x_{A(I)}^L$, $\delta x_{A(I)}$, $K_{AB(I)}$, and $K_{I(A,B)N}$ depend on both the solvation abilities of the two solvent components A and B regarding the ion $I^{\pm}$ and also on the mutual interactions of the two solvent components among themselves.

6.3.1 Spectroscopic Studies

Earlier reports on the preferential solvation of ions by diverse experimental methods were summarized by Schneider [42]. Only a few examples of subsequent studies can be discussed here. NMR has been a major technique used for the study of the preferential solvation of ions, using both chemical shifts $\delta_{AB(I)}$ and spin-lattice relaxation times $\tau_{AB(I)}$ as the properties p investigated.

In a few exceptional cases, it was possible to separate the NMR signals from the two solvated states of an ion and infer from the areas under the separate signals the amount of preferential solvation that occurred. Such a case is the preferential solvation of Mg^{2+} in methanol + water mixtures cooled down to $-75°C$ as reported by Swinehart and Taube [43]. At $x_W = 0.076$, the local water mole fraction in the first solvation shell of Mg^{2+} is 0.128, that is, the ion is preferentially solvated by the aqueous component of the mixture. Other examples of separated signals from the two solvated forms of ions, applicable to room temperature, are confined to ions of a high charge density—small multivalent ions. This was demonstrated by Strehlow et al. [44] for Al^{3+} in mixtures of $A =$ ethylene carbonate and $B =$ acetonitrile, in which at $x_B = 0.6$, the ^1H NMR spectrum shows the cation to be much more solvated by A than by B.

Generally, however, rapid solvent exchange occurs at room temperature so that the NMR signals from the two solvated forms cannot be separated. Still, the average chemical shifts $\delta_{AB(I)} = f(x_B)$ can yield information on the preferential solvation of ions $I^{\pm}$ in the solvent mixture $A+B$. The midpoint of the $\delta_{AB(I)} = f(x_B)$ function between the values for the neat components is the iso-solvation point and the relative solvating abilities of the solvent components are inversely proportional to their mole fractions at this point according to Frankel et al. [45, 46]. This is valid if no ion–ion association occurs and if the total solvation number remains constant throughout the composition range.

For the system $I^{\pm} = Al^{3+}$, $A = W$, and $B = DMSO$, the aluminum is preferentially hydrated at $x_B \leq 0.20$ but preferentially solvated by DMSO at $x_B \geq 0.80$ as found by Thomas and Reyolds [47]. For the system $I^{\pm} = Ag^+$, $A = W$, and $B = DMSO$ at

$0.01 \leq x_B \leq 0.15$, the $\delta_{AB(I)}$ of the DMSO protons shows preferential hydration of the silver ions. At $x_B \geq 0.5$, on the other hand, preferential solvation by DMSO is indicated by the NMR chemical shifts. However, if B=MeCN, then selective solvation by MeCN takes place according to Clausen et al. [48]. The $\delta_{AB(I)}$ of $I^{\pm} = {}^{23}Na^+$ in mixtures of A=NMF and B being one of a series of solvents, yielded the iso-solvation points $x_{A(is)}$ equaling the local mole fractions x_B^L shown in Table 6.5. The iso-solvation points and the local mole fractions x_B^L of some other binary solvent mixtures obtained by Popov and coworkers [50, 51] are also shown there. Some correlation of the preference for B with regard its donor numbers (Section 3.3.2.2) was noted—the larger $DN(B)$, the more is B preferred relative to A.

The $\delta_{AB(I)}$ of $I^{\pm} = {}^{23}Na^+$ in mixtures of A=EG and B=MeCN was interpreted in terms of the stepwise replacement of B in IB_4^+ by A, taking into account the non-ideality of the A+B system. Quantitative evidence for the preferential solvation of sodium ions by ethylene glycol was obtained by Chuang et al. [52]. The $\delta_{AB(I)}$ of ^{13}C of the carbonyl group in 1 M solutions of $LiPF_6$ in binary solvent mixtures involving esters and amides was interpreted in terms of preferential coordination of the solvents to Li^+ with a constant total coordination number of 4. In isomolar mixtures of A=methyl propionate and B=ethyl methyl carbonate $x_B^L = 0.39$, of A=N,N-dimethylacetamide and B=propylene carbonate $x_B^L = 0.40$, and of A=methyl propionate and B=ethyl propionate $x_B^L = 0.43$ according to Matsubara et al. [53]. In all these cases $x_B^L < 0.50$, and hence it is solvent A that preferentially solvates the ion.

NMR relaxation rates of $^{23}Na^+$, $^{87}Rb^+$, ^2H, and ^{14}N in A=water and B=MeCN showed according to Baum and Holz [54] that both the sodium and rubidium cations are strongly preferentially hydrated in the mixtures. The relaxation rates of ^1H and ^2H in solutions of NaI in mixtures of A=water and B=NMF, in which ^1H atoms resided

TABLE 6.5 The Iso-solvation Points $x_{A(is)}$ Equaling the Local Mole Fractions x_B^L for Sodium Ions in Binary Mixtures of Solvents A and B^a

Solvent A	Solvent B	x_B^L at $x_{A(is)}$	Solvent A	Solvent B	x_B^L at $x_{A(is)}$
NMF [49]	HMPT	0.74	DMSO [50]	MeCN	0.11
	Water	0.59		$MeNO_2$	0.06
	DMSO	0.54	[51]	Py	0.10
	DMF	0.50		TU	0.39
	FA	0.50		HMPT	0.15
	MeOH	0.48	Py [51]	MeCN	0.29
	EtOH	0.40		$MeNO_2$	0.12
	Py	0.33		TMU	0.84
	THF	0.19		HMPA	0.90
	Me_2CO	0.12	MeCN [51]	$MeNO_2$	0.15
	MeCN	0.12		TMU	0.89
	$MeNO_2$	0.02		HMPA	0.94
HMPA [51]	$MeNO_2$	0.05	TMU y	$MeNO_2$	0.06
				HMPA	0.77

^{a}Adapted from Covington and Dunn [49].

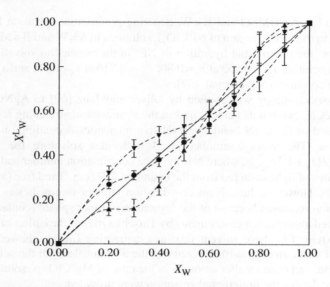

FIGURE 6.3 Local mole fractions of water around Na^+ ions in aqueous mixtures with formamide ($\bullet$), N-methlformamide ($\blacktriangledown$), and N,N-dimethylformamide ($\blacktriangle$), obtained from NMR relaxation rates [56].

in the three possible sites (amide, formyl, and methyl) of deuterated NMF yielded approximate degrees of preferential hydration of the Na^+ cations and of solvation by NMF of the I^- anions according to Finter and Hertz [55]. In a further study of the relaxation rates in solutions of Na^+ in aqueous mixtures with formamide, NMF, and DMF, only small deviations from non-preferential solvation (the straight line in Fig. 6.3) were found by Holz and Rau [56] for formamide and preferential hydration was found for W+NMF, but a preferential solvation of Na^+ by DMF in the non-aqueous-solvent-rich range and by water in the water-rich range can be discerned.

A special NMR technique, ^1H NOESY, using the nuclear Overhauser effect, was applied by Bagno et al. [57] to Me_4NCl and Bu_4NI in $A=W$ and $B=MeCN$ mixtures, the ions of the former salt being preferentially hydrated and those of the latter one preferentially solvated by the organic component. Both ions of $NaCH_3CO_2$ are preferentially solvated by DMSO in its aqueous mixtures according to this technique.

Spectroscopic methods other than NMR have also been employed for studying preferential solvation of ions in mixed solvents. The charge-transfer-to-solvent spectrum of iodide ions in mixtures of EG and MeCN showed preferential solvation of the I^- by the glycol component according to Covington and Dunn [49]. Spectroscopy in the visible region was applied by Kamienska-Piotrowicz and Stangret [58] to Co^{2+} in binary solvent mixtures. Equimolar *local* mole fractions, x_B^L, in the octahedral solvation shells were obtained at *bulk* mole fractions $x_B = 0.018$ in $A = MeCN$ and $B = DMSO$, $x_B = 0.15$ in $A = MeCN$ and $B = W$, $x_B = 0.30$ in $A = MeCN$ and $B = MeOH$, and $x_B = 0.45$ in $A = MeOH$ and $B = W$, denoting preferential solvation by B, but $x_B = 0.67$ in A = DMSO and B = W, denoting preferential solvation by A. Light absorption in the visible region of $Ni(ClO_4)_2$ solutions in A = W and B = MeCN served also to show the preferential hydration of Ni^{2+} in the octahedral solvation shell. At 0.05 M salt solution $x_A^L = 0.833$ at $x_A = 0.80$, $x_A^L = 0.816$ at $x_A = 0.40$, and $x_A^L = 0.533$ at $x_A = 0.10$ according to Miyaji et al. [59].

Raman spectroscopy was applied by Oliver and Janz [60] to $AgNO_3$ in A = W and B = MeCN, two bands being found in the symmetrical stretching regions of the NO_3^- ion and of the C≡N bond, their relative intensities depending on the solvent composition. The average number of B molecules solvating the Ag^+ ion is $Nx_B^L = I_b c_B / (I_b + I_f) c_{AgNO3}$, where N is the total coordination number and I_b and I_f are the intensities of light scattered from the bound (at 2272 cm^{-1}) and free (at 2253 cm^{-1}) C≡N bonds. However, the solvent coordination number diminishes as the $AgNO_3$ concentration increases because of the formation of inner sphere (contact) ion pairs [60]. Infrared absorption measurements by Tutschka [61] of the cation-affected C≡N band (2260 cm^{-1}) in these solvent mixtures containing $LiClO_4$ showed that small amounts of water already suffice to remove the acetonitrile from the solvation shell of the cation, and even smaller amounts in the case of $Mg(ClO4_4)_2$ solutions, but no quantitative data of the preferential solvation were provided.

A titration Raman spectroscopic method, in which multivalent metal perchlorates are gradually added to a binary mixture of amide solvents was developed by Ishiguro and coworkers [62, 63], in which the bending mode $\delta(O–C–N)$ and stretching mode $\nu(N–CH_3)$ of the amides were monitored. In mixtures of A = DMF and B = DMA, small figures show that the cations were slightly preferentially solvated by A; the more so, the smaller the cation (Mg^{2+}, Ca^{2+}, Sr^{2+}, Ba^{2+}, Ni^{2+}, Cu^{2+}, Zn^{2+}, Mn^{2+}, Nd^{3+}, Gd^{3+}, and Tm^{3+}), due to steric crowding by the bulkier DMA. Similar results were obtained in mixtures of DMF and TMU with Mn^{2+} [62]. In mixtures of A = DMF and B = NMF containing perchlorates of Mn^{2+}, Ni^{2+}, and Zn^{2+}, no preferential solvation was observed at $x_A < 0.5$, but at larger DMF contents, the cations were preferentially solvated by NMF [63].

FT-Raman spectroscopy was applied to solutions of $LiCF_3SO_3$ by Alia et al. [64, 65], of $NaBF_4$ by Xuan et al. [66] in A = DMF or DMSO and B = MeCN, and of $LiBF_4$ in binary mixtures of A = 4-methoxyethyl ethylene carbonate with B = dimethyl carbonate or acetonitrile by Qiao et al. [67]. In all these cases, the cations were preferentially solvated by A.

Whether numerical values of, say, $x_B^L = f(x_A)$ were provided in these experimental studies or not, the general qualitative conclusion is that the stronger the bonding of one solvent component to an ion relative to the other component is, the more it is preferred, unless it is more sterically crowded.

6.3.2 Results from Thermodynamic Data

The standard molar Gibbs energies of transfer of an ion from a source solvent (arbitrarily selected as water, W) to a target solvent, $\Delta_{tr} G^\infty (I^\pm, W \rightarrow S)$, described in Section 4.3.2.1, is the basis for a rough estimation of its preferential solvation in mixtures of these two components. However, the mutual interactions of the two components of the mixture affect the thermodynamics and the preferential solvation of the ions in the mixtures. Data for the standard molar Gibbs energies of transfer of an ion from a source solvent to a target solvent mixture, $\Delta_{tr} G^\infty(I^\pm, W \to A + B)$, are required for a firm evaluation of the preferential solvation of the ion in the mixture. A critical compilation by Kalidas et al. [10] of such data for the transfer of cations into mixtures in which water is one of the components of the binary solvent mixture, that is, $\Delta_{tr} G^\infty(I^\pm, W \to W + S)$, is available and a similar compilation by Marcus [11] for anions is also available. The values of the transfer Gibbs energies of ions from water into equimolar mixtures of water with cosolvents are shown in Table 6.1. Fewer data for the transfer of ions between completely nonaqueous solvents are available, and for the sake of convenience, the following discussions are in terms of aqueous mixtures, but may be applied to any solvent mixture on changing the symbols for the solvents.

Such standard transfer data pertain to infinite dilution of the ions in the solvent mixtures so that the ions dealt with are surrounded by the components of the mixture and are remote from other ions. Several methods have been applied to $\Delta_{tr} G^\infty(I^\pm, W \to W + S) = f(x_S)$ data (or the corresponding enthalpies of transfer) to obtain from them the preferential solvation of the ion by the components of the binary solvent mixture.

6.3.2.1 The QLQC Method

The quasi-lattice quasi-chemical method (QLQC) proposed by Marcus [68] employs the following model for the ion surrounded by the molecules of the two solvents. The quasi-lattice part of the model assumes that the ion $I^\pm$ and the molecules of W and S are distributed on sites of a quasi-lattice characterized by a lattice parameter Z. This parameter specifies the number of neighbors that each particle has. The sum of the pair-wise interaction energies, e_{IW}, e_{IS}, e_{WW}, e_{WS}, and e_{SS}, weighted according to the numbers of the corresponding neighbors each particle has, determines the configurational energy of the system. The pairwise energies e_{pq} are assumed to be independent of the natures of the other neighbors of the partners p and q may have. Zero excess volume $V^E = 0$ and excess entropy $S^E = 0$ are assumed, meaning that the Helmholz, Gibbs, and configurational energies are the same.

The quasi-chemical part of the model considers the numbers of neighboring particles: N_{IW}, N_{IS}, N_{WW}, N_{WS}, and N_{SS} according to Guggenheim [69]:

$$\left(\frac{N_{IW}}{N_{IS}}\right)^2 = \left(\frac{N_{WW}}{N_{SS}}\right) \exp\left[\frac{\Delta e_{IW} - \Delta e_{IS}}{k_B T}\right] \tag{6.12}$$

where $\Delta e_{IW} = e_{WW} + e_{II} - 2e_{IW}$ and similarly for Δe_{IS}. The unknown self-interactions of the ions, e_{II}, are irrelevant at infinite dilution, and are eliminated from the difference in the exponent. The local mole fraction of W around the ion is $x_{W(I)}^L = N_{IW}/(N_{IW} + N_{IS}) = 1/(1 + N_{IS}/N_{IW})$, and following Equation 6.12, it can be written as:

$$x_{W(I)}^L = 1/\left[1 + \left(N_{SS}/N_{WW}\right)^{1/2}\right] \exp\left[\left(\Delta e_{IW} - \Delta e_{IS}\right)/2k_B T\right] \tag{6.13}$$

At infinite dilution of the ion, the ratio N_{SS}/N_{WW} depends on the properties of the binary solvent mixture only. The sum of the number of nearest neighbors in the system that has one mole of solvents is:

$$L = N_{WW} + N_{WS} + N_{SS} = \left(\frac{Z}{2}\right)N_A \qquad (6.14)$$

hence $N_{SS}/L = x_S - N_{WS}/ZN_A$ and similarly for N_{WW}/L. The ratio N_{SS}/N_{WW} then becomes:

$$\frac{N_{SS}}{N_{WW}} = \frac{\left(x_S - N_{WS}/ZN_A\right)}{\left(x_W - N_{WS}/ZN_A\right)} \qquad (6.15)$$

The QLQC expression for the binary solvent mixture is similar to Equation 6.12 and results in the value of N_{WS} as a function of the solvent composition:

$$\frac{N_{WS}}{ZN_A} = \frac{\left[1 - \left\{1 - 4x_W x_S\left(1 - \exp\left(e_{WS}/k_B T\right)\right)\right\}^{1/2}\right]}{\left[2\left(1 - \exp\left(e_{WS}/k_B T\right)\right)\right]} \qquad (6.16)$$

The quantity $\exp(e_{WS}/k_B T)$ is obtained from the molar excess Gibbs energy of mixing in the equimolar solvent mixture [68]:

$$\exp\left(\frac{e_{WS}}{k_B T}\right) = \left[\left\{2\exp\left(\frac{-2G_{WS(x=0.5)}^E}{ZRT}\right)\right\}^{1/2} - 1\right]^2 \qquad (6.17)$$

It remains to specify the interaction energy difference $\Delta e_{IW} - \Delta e_{IS}$ appearing in the expression for the local mole fractions around the ion, Equation 6.9:

$$\Delta e_{IW} - \Delta e_{IS} = \frac{\left[\Delta_{solv}G^\infty\left(I,W\right) - \Delta_{solv}G^\infty\left(I,S\right)\right]}{ZN_A} = \frac{\Delta_{tr}G^\infty\left(I,W \to S\right)}{ZN_A} \qquad (6.18)$$

The molar excess Gibbs energy of mixing of the two solvents at the equimolar composition, $G_{WS(x=0.5)}^E$, the standard molar Gibbs energy of transfer of the ion $I^\pm$ from W to S, and the lattice parameter Z are all that are required for obtaining the full dependence of the preferential solvation of $I^\pm$ on the solvent composition: $x_{W(I)}^L = f(x_S)$. The lattice parameter Z, in turn, is obtained from fitting the $G_{WS}^E = f(x_S)$ curve. Setting the variables $P = 1 - \exp(-e_{WS}/k_B T)$ and $Q = (1 - 4x_W x_S P)^{1/2}$ yields the value of Z that fits best the excess Gibbs energy of mixing curve [68]:

$$G_{WS}^E = \left(\frac{ZRT}{2}\right)\left[x_W \ln\left\{\left(x_W - (1-Q)/2P\right)/x_W^2\right\} + x_S \ln\left\{\left(x_S - (1-Q)/2P\right)/x_S^2\right\}\right] \qquad (6.19)$$

There are several weak points in the QLQC approach. One of them is the assumption of a constant value for the lattice parameter Z for a given binary solvent mixture irrespective of its composition. This assumption should hold for the transfer of any ion into such a mixture, but this is not necessarily the case. A further weak point is the assumption of ideal mixing of the particles on the quasi-lattice sites, $S^E = 0$, which is incompatible with the notion of preferential solvation. On the other hand, the QLQC method considers only the nearest environment of an ion, its first solvation shell, and hence is applicable to individual ions at infinite dilution, provided the individual ionic $\Delta_{tr} G^\infty (I, W \rightarrow S)$ values are available, based on a reliable extra-thermodynamic assumption (Section 4.3.2.1).

The replacement of one solvent of a binary solvent mixture A + B in the solvation shell of an ion $I^\pm$ by the other, yielding the set of reactions (6.6a–6.6b), can be described explicitly by the QLQC method. The coordination number N is set equal to the lattice parameter Z and equilibrium quotients depending on the preferential solvation parameter $\delta x_{B(I)}$ result:

$$K_{A/B_1} = \lim (x_B \rightarrow 0) \frac{\left[x_B + \delta x_{B(I)} \right]}{x_B \left[1 - x_B - \delta x_{B(I)} \right]} \qquad (6.20a)$$

$$K_{A/B_N} = \lim (x_B \rightarrow 1) \frac{\left[(1 - x_B)(x_B + \delta x_{B(I)}) \right]}{\left[1 - x_B - \delta x_{B(I)} \right]} \qquad (6.20b)$$

The preferential solvation parameter $\delta x_{B(I)} = x^L_{B(I)} - x_B$ is obtained from the local mole fraction of solvent B around the ion $I^\pm$, $x^L_{B(I)}$, obtained according to Equation 6.13. As an example of the application of the QLQC method, the resulting local mole fractions of water (A) in equimolar mixtures with DMSO (B) are 0.38 around Na^+, 0.23 around Ag^+, and 0.80 around Cl^-, but when B = MeCN, these fractions become 0.62, 0.32, and 0.80, respectively as noted by Marcus [70, 71] and the corresponding replacement equilibrium constants can be calculated from Equations 6.20.

6.3.2.2 The IKBI Method

The inverse Kirkwood-Buff integral (IKBI) method is based on rigorous statistical thermodynamics and does not rely on a model but has drawbacks too. Consider an ion $I^\pm$ at a given position in the mixed solvent W + S and an infinitesimal volume element at a distance r from the center of the ion. The probability of a molecule of S to be in this volume element is the pair correlation function $g_{S(I)}(r)$, averaged over all the mutual orientations. The Kirkwood-Buff space integral

$$G_{S(I)} = \int_0^\infty \left[g_{S(I)}(r) - 1 \right] 4\pi r^2 dr \qquad (6.21)$$

expresses the *affinity* of the solvent S to the ion $I^\pm$ and may have positive or negative values. There is a *correlation volume* around the ion, extending to a distance R_{cor}, only in which $g_{S(I)}(r)$ differs appreciably from unity, hence contributes significantly to $G_{S(I)}$. Only within this correlation volume, $V_{cor} = (4\pi / 3) R^3_{cor}$, does the ion affect the probability of finding molecules of S (or of W). The average number of S molecules in the correlation volume depends on its bulk number density, $\rho_S = N_S/V$, as shown by Ben-Naim [72, 73]:

$$N_{S(I)} = \rho_S V_{cor} + \rho_S G_{S(I)} \qquad (6.22)$$

A similar expression pertains to the solvent component W and the local mole fraction (i.e., within the correlation volume) of S is:

$$X_{S(I)}^L = \frac{N_{S(I)}}{\left(N_{S(I)} + N_{w(I)}\right)} = \frac{\left[x_S V_{cor} + x_S G_{S(I)}\right]}{\left[x_S G_{S(I)} + x_w G_{w(I)} + V_{cor}\right]} \qquad (6.23)$$

noting that $x_S = \rho_S/(\rho_S + \rho_w)$. The preferential solvation parameter is:

$$\delta x_{S(I)} = \frac{x_S x_w \left(G_{S(I)} - G_{w(I)}\right)}{\left[x_S G_{S(I)} + x_w G_{w(I)} + V_{cor}\right]} \qquad (6.24)$$

The Kirkwood-Buff integrals can be obtained from thermodynamic data of the infinitely dilute solution (with respect to $I^\pm$) as follows:

$$G_{S(I)} = RT\kappa_T + V_I^\infty + x_w V_w D/Q \qquad (6.25a)$$

$$G_{w(I)} = RT\kappa_T + V_I^\infty + x_S V_S D/Q \qquad (6.25b)$$

where $D = \left(\partial \Delta_{tr} G^\infty (I, W \rightarrow W + S)/\partial x_S\right)_{T,P}$ and the function Q pertains to the binary solvent mixture

$$Q = 1 + x_S \left(\frac{\partial f_w}{\partial x_S}\right)_{T,P} = 1 + x_S x_S \left(\frac{\partial^2 G_{WS}^E}{\partial x_S^2}\right)_{T,P} \qquad (6.26)$$

where f is the rational activity coefficient. The term with the isothermal compressibility κ_T of the solvent mixture is generally very minor and may be approximated by a linear interpolation between the κ_T values of the neat solvent components. An important feature of the IKBI method is its dependence on *derivative* functions: D being the derivative of the standard molar Gibbs energy of the ion from the source solvent to the solvent mixtures and Q being the derivative of the activity of one component or the second derivative of the molar excess Gibbs energy of the mixture, all with respect to the mole fraction composition of the solvent mixture. Therefore, great accuracy of the quantities to be differentiated is required, this being a weakness of the method, because this requirement is difficult to meet.

The consideration of an individual ion in the function D has the usual problems of thermodynamic quantities pertaining to individual ions connected with it. When the method is applied to the transfer of a symmetrical electrolyte, the result would indicate the same preferential solvation of the cation and of the anion (counter to experience, according to which heterosolvation prevails). This is, because according to Equation 6.21, the Kirkwood-Buff integrals extend to infinity and so comprise both the cation and the anion even at infinite dilution. The application of the TATB assumption for obtaining standard molar Gibbs energies of transfer of individual ions (Section 4.3.1) has been shown by Marcus [74] to deal satisfactorily with this problem.

6.3.2.3 Treatments Based on Stepwise Solvent Replacements

The stepwise replacement of one solvent in a binary mixture of A + B around and ion $I^{\pm}$, Equations 6.6a and 6.6b, has been treated along the lines used in solution coordination chemistry. The equilibrium quotient for the replacement of j molecules of A by j molecule of B can be represented in terms of the volume fractions ϕ of the solvent components according to Cox et al. [75] as:

$$\beta_j = \left[\frac{c(IA_N)}{c(IA_{N-j}B_j)} \right] \left(\frac{\varphi_B}{\varphi_A} \right)^j \tag{6.27}$$

The values of these quotients are obtained from spectroscopic measurements (e.g., NMR) or electrochemically by methods employed in solution coordination chemistry. They lead to the standard molar Gibbs energy of transfer of $I^{\pm}$ from the source solvent A to the mixture:

$$\Delta_{tr}G^{\infty}\left(I^{\pm}, A \rightarrow A+B\right) = -RT\ln\varphi_A - RT\ln\left[1 + \sum \beta_j \left(\frac{\varphi_B}{\varphi_A} \right)^j \right] \tag{6.28}$$

The local mole fraction of B is then:

$$x_{B(I)}^{L} = \sum j\beta_j \left(\varphi_B/\varphi_A \right)^j / N \left[1 + \sum \beta_j \left(\varphi_B/\varphi_A \right)^j \right] \tag{6.29}$$

An alternative treatment is specifically related by Covington and Newman [76] to NMR chemical shift data for specific ions $I^{\pm}$, extrapolated to infinite dilution, in binary solvent mixtures A + B. It is assumed that the chemical shift δ is linear with the fractional solvent composition in the solvation shell of the ion:

$$\delta = \sum x_{Bi(I)}^{L}\delta_i \tag{6.30}$$

where $\delta_i = (i/N)\delta_A$ is proportional to the number of B molecules that have replaced A ones and δ_A is the chemical shift for the ion in pure solvent A. The final expression is:

$$\beta_N^{(1/N)}\left(\frac{x_A}{x_B} \right) = \frac{(\delta/\delta_A)}{(1-\delta/\delta_A)} \tag{6.31}$$

The expression for the standard molar Gibbs energy of transfer is similar to Equation 6.24, but in terms of the mole fractions rather than the volume fractions used by Cox et al. [75]. However, $\Delta_{tr}G^{\infty}(I^{\pm}, A \rightarrow A+B)$ has an additional term that is related to the electrostatic effect of the transfer from the source solvent with permittivity ε_A to the solvent mixture with its different permittivity (unless transfer is between isodielectric solvents).

In a subsequent development by Covington and Newman [77], a relationship with the Kirkwood-Buff approach was established, valid for infinite dilution of the ion $I^\pm$. The preferential solvation parameter $\delta x_{B(I)} = x_{B(I)}^L - x_B$ pertaining to the solvation shell of the ion is:

$$\delta x_{B(I)} = \sum i x_{Bi(I)}^L - \left(\frac{4\pi}{3}\right)\rho_B V_{cor} \tag{6.32}$$

where ρ_B is the number density of component B. The NMR chemical shift is used to obtain the first term on the right-hand side of Equation 6.32 according to Equation 6.30. The values of β_N obtained in this manner for some univalent ions in A = MeOH and B = W were as follows: 1.64 for Na^+, 1.42 for Rb^+, 1.32 for Cs^+, 1.03 for F^-, and 1.43 for Cl^-; for A = DMSO and B = W, the values were less than unity: 0.76 for Li^+ and 0.87 for Cs^+, signifying reluctance of water to replace DMSO near these cations.

REFERENCES

[1]   Y. Kondo, M. Ittoh, S. Kusabayashi, J. Chem. Soc. Faraday Trans. 1 **78**, 2793 (1982).

[2]   B. G. Cox, W. E. Waghorne, J. Chem. Soc. Faraday Trans. 1 **80**, 1267 (1984).

[3]   A. Piekarska, H. Piekarski, S. Taniewska-Osinska, J. Chem. Soc. Faraday Trans. 1 **82**, 513 (1986).

[4]   A. Piekarska, S. Taniewska-Osinska, Thermochim. Acta **194**, 109 (1992).

[5]   Y. Marcus, Z. Naturforsch. A **50**, 51 (1995).

[6]   H. Piekarski, A. Pietrzak, J. Therm. Anal. Calorim. **110**, 917 (2012).

[7]   H. Piekarski, K. Kubalczyk, J. Mol. Liq. **121**, 35 (2005).

[8]   H. Piekarski, K. Kubalczyk, J. Chem. Eng. Data **55**, 1945 (2010).

[9]   G. Gritzner, A. Lewandowski, J. Chem. Soc. Faraday Trans. 1 **89**, 2007 (1993).

[10] C. Kalidas, G. T. Hefter, Y. Marcus, Chem. Rev. **100**, 819 (2000).

[11] Y. Marcus, Chem. Rev., **107**, 3880 (2007).

[12] M. Suzuki, J. Electroanal. Chem. **384**, 77 (1995).

[13] G. Hefter, Y. Marcus, W. E. Waghorne, Chem. Rev. **102**, 2773 (2002).

[14] G. T. Hefter, J.-P. E. Grolier, A. H. Roux, J. Solution Chem. **18**, 229 (1989).

[15] G. T. Hefter, J.-P. E. Grolier, A. H. Roux, G. Roux-Desgranges, J. Solution Chem. **19**, 207 (1990).

[16] Y. Marcus, J. Mol. Liq. **158**, 23 (2011).

[17] Y. Marcus, J. Mol. Liq. **166**, 62 (2012).

[18] H. Piekarski, J. Therm. Anal. Calorim. **108**, 537 (2012).

[19] A. V. Kustov, A. V. Bekeneva, V. I. Saveliev, V. P. Korolev, J. Solution Chem. **31**, 71 (2002).

[20] V. P. Korolev, N. L. Smirnova, A. V. Kustov, Thermochim. Acra **427**, 43 (2005).

[21] J. E. Desnoyers, O. Kiyohara, G. Perron, L. Avedikian, Adv. Chem. Ser. **155**, 274 (1976).

[22] H. Piekarski, M. Tkaczyk, J. Chem. Soc. Faraday Trans. 1 **87**, 3661 (1991).

[23] R. N. French, C. M. Criss, J. Solution Chem. **10**, 713 (1981).

[24] S. Taniewska-Osinska, J. Barczynska, J. Chem. Soc. Faraday Trans. 1 **80**, 1409 (1984).

[25] B. Palecz, J. Barczynska, S. Taniewska-Osinska, Thermochim. Acta **150**, 121 (1989).

[26] N. G. Manin, V. P. Korolev, Russ. J. Phys. Chem. **72**, 1256 (1998).

[27] A. V. Kustov, Russ. J. Phys. Chem. **75**, 221 (2001).

[28] H. Piekarski, A. Piekarska, S. Taniewska-Osinska, Can. J. Chem. **62**, 856 (1984).

[29] H. Piekarski, G. Somsen, J. Solution Chem. **19**, 923 (1990).

[30] P. Letellier, R. Gaboriaud, R. Schaal, J. Chim. Phys. **77**, 1051 (1980).

[31] G. Perron, D. Joly, J. E. Desnoyers, L. Avedikian, J.-P. Morel, Can. J. Chem. **56**, 552 (1978).

[32] Y. Marcus, J. Solution Chem. **33**, 549 (2004).

[33] Y. Marcus, J. Solution Chem. **34**, 317 (2005).

[34] D. Feakins, F. M. Bates, W. E. Waghorne, J. Solution Chem. **37**, 727 (2008).

[35] A. Apelblat, Acta Chim. Slov. **56**, 1 (2009).

[36] J. Barthel, Angew. Chem. Int. Ed. **7**, 260 (1968).

[37] R. L. Kay, in F. Franks, ed., Water: A Comprehensive Treatise, Plenum Press, New York, Vol. **3**, 173 (1973).

[38] A. Apelblat, J. Phys. Chem. B **112**, 7032 (2008).

[39] A. Apelblat, J. Solution Chem. **40**, 1544 (2011).

[40] E. Gileadi, E. Kirowa-Eisner, Electrochim. Acta **51**, 6003 (2006).

[41] G. Scatchard, J. Chem. Phys. **9**, 34 (1941).

[42] H. Schneider, in J. F. Coetzee, C. D. Ritchie, eds., Solute-Solvent Interactions, Dekker, New York, Chapter 5 (1969).

[43] J. H. Swinehart, H. Taube, J. Chem. Phys. **37**, 1579 (1962).

[44] H. Strelow, W. Knoche, H. Schneider, Ber. Bunsenges. Phys. Chem. **77**, 760 (1973).

[45] L. S. Frankel, T. R. Stengle, C. H. Langford, J. Chem. Soc. Chem. Commun. 393 (1965).

[46] L. S. Frankel, T. R. Stengle, C. H. Langford, J. Phys. Chem. **74**, 1376 (1970).

[47] S. Thomas, W. L. Reyolds, Inorg. Chem. **9**, 78 (1970).

[48] A. Clausen, A. A. El-Harakany, H. Schneider, Ber. Bunsen. Ges. Phys. Chem. **77**, 994 (1973).

[49] A. K. Covington, M. Dunn, J. Chem. Soc. Faraday Trans. 1 **85**, 2835 (1989).

[50] R. H. Erlich, M. S. Greenberg, A. I. Popov, Spectrochim. Acta A **29**, 543 (1973).

[51] M. S. Greenberg, A. I. Popov, Spectrochim. Acta A **31**, 697 (1975).

[52] H.-J. Chuang, L.-L. Soong, G. E. Leroi, A. I. Popov, J. Solution Chem. **18**, 759 (1989).

[53] K. Matsubara, R. Kaneuchi, N. Maekita, J. Chem. Soc. Faraday Trans. **94**, 3601 (1998).

[54] B. M. Baum, M. Holz, J. Solution Chem. **12**, 685 (1983).

[55] C. K. Finter, H. G. Hertz, J. Chem. Soc. Faraday Trans. 1 **84**, 2735 (1988).

[56] M. Holz, C. K. Rau, J. Chem. Soc. Faraday Trans. 1 **78**, 1899 (1982).

[57] A. Bagno, M. Campulla, M. Pirana, G. Scorrano, S. Stiz, Chem. Eur. J. **5**, 1291 (1999).

[58] E. Kamienska-Piotrowicz, J. Stangret, J. Mol. Struct. **440**, 131 (1998).

[59] K. Miyaji, K. Nozawa, K. Morinaga, Bull. Chem. Soc. Jpn. **62**, 1472 (1989).

[60] B. G. Oliver, G. J. Janz, J. Phys. Chem. **74**, 3819 (1970).

[61] J. Tutschka, Z. Phys. Chem. (Leipzig) **261**, 512 (1980).

[62] S.-I. Ishiguro, Y. Umebayashi, R. Kanzaki, Anal Sci. **20**, 415 (2004).

[63] K. Fujii, T. Kumai, T. Takamuku, Y. Umebayashi, S.-I. Ishiguro, J. Phys. Chem. A **110**, 1798 (2006).

[64] J. M. Alia, H. G. M. Edwards, F. J. Garcia, J. Mol. Struct. **43**, 565 (2001).

[65] J. M. Alia, H. G. M. Edwards, E. E. Lawson, Vibrat. Spectrosc. **34**, 187 (2004).

[66] X. P. Xuan, J. J. Wang, Y. Zhao, H. C. Zhang, Spectrochim. Acta A **62**, 2097 (2005).

[67] H. Qiao, H. Luan, X. Fang, Z. Zhou, W. Yao, X. Wang, J. Li, C. Chen, Y. Tian, J. Mol. Struct. **878**, 185 (2008).

[68] Y. Marcus, Austr. J. Chem. **36**, 1719 (1983).

[69] E. A. Guggeheim, Proc Roy Soc. (London) A **169**, 134 (1938).

[70] Y. Marcus, J. Chem. Soc. Faraday Trans. 1 **84**, 1465 (1988).

[71] Y. Marcus, Pure Appl. Chem. **62**, 2069 (1990).

[72] A. Ben-Naim, Cell Biophys. **12**, 255 (1988).

[73] A. Ben-Naim, J. Phys. Chem. **93**, 3809 (1989).

[74] Y. Marcus, Solvent Mixtures, Dekker, New York, 213 (2002).

[75] B. G. Cox, A. J. Parker W. E. Waghorne, J. Phys. Chem. **78**, 1731 (1974).

[76] A. K. Covington, K. E. Newman, Pure Appl. Chem. **51**, 2041 (1979).

[77] A. K. Covington, K. E. Newman, J. Chem. Soc., Faraday Trans. 1 **84**, 973 (1988).

7

INTERACTIONS OF IONS WITH OTHER SOLUTES

Solvated ions interact in solution with other solutes, whether the latter are ions of the same charge sign (they repulse each other electrostatically), ions of the opposite sign (they attract each other), or non-ionic solutes. If the other solute is non-ionic, it may be non-polar (ions salt such solutes out) or polar (where several kinds of interaction may take place). In some cases, such interactions may cause changes in the solvation structure and dynamics of the ions in question, and such issues are dealt with in this chapter.

7.1 ION–ION INTERACTIONS

At extreme dilution of an electrolyte in any solvent, its ions are remote from each other so that they interact only with the surrounding solvent molecules and have their individual properties as described in Chapters 4–6. As the concentration of the electrolyte is increased and the distances between the ions diminish, each ion is surrounded by an ionic atmosphere, that is, other ions of either sign (mainly of opposite sign). The change from ideal dilute solutions is expressed by the osmotic coefficient of the solvent and the activity coefficient of the electrolyte departing from unity, their value at infinite dilution as the reference state. The well-known Debye–Hückel limiting law describes the resulting changes in the properties of the solution only up to quite low concentrations (say, 0.01 m). Various extensions of the Debye–Hückel limiting law expression are available for practically encountered higher concentrations.

The osmotic coefficient of the solvent in an m_E molal electrolyte solution is related to the activity of the solvent, a_S (roughly, its vapor pressure in the solution divided by the vapor pressure of the neat solvent), as:

$$\varphi = -(M_S \nu m_E)^{-1} \ln a_S \tag{7.1}$$

Ions in Solution and Their Solvation, First Edition. Yizhak Marcus.
© 2015 John Wiley & Sons, Inc. Published 2015 by John Wiley & Sons, Inc.

where M_S is the molar mass of the solvent (in $kg \cdot mol^{-1}$) and ν is the number of ions per formula unit of the electrolyte. The mean ionic activity coefficient of the electrolyte, on the molal scale, is designated by $\gamma_\pm$. On the mole fraction scale, the mean ionic activity coefficient is designated by $f_\pm$ and is related to $\gamma_\pm$ as:

$$f_\pm = \gamma_\pm (1 + M_S \nu m_E) \tag{7.2}$$

On the molar scale, it is designated by $y_\pm$ and is related to $\gamma_\pm$ as:

$$y_\pm = \gamma_\pm \left(\frac{m_E \rho_S}{c_E} \right) \tag{7.3}$$

where ρ_S is the density of the solvent. At a given temperature, the Gibbs–Duhem rule relates the osmotic and activity coefficients:

$$\ln \gamma_\pm = (\varphi - 1) + \int_0^{m_E} (\varphi - 1) d \ln m_E \tag{7.4}$$

$$\varphi = 1 + m_E^{-1} \int_0^{m_E} m_E \, d \ln \gamma_\pm \tag{7.5}$$

7.1.1 Activity Coefficients of Electrolyte Solutions

Values of φ and $\gamma_\pm$ for many aqueous electrolytes at 25°C are reported in the books by Harned and Owen [1], by Robinson and Stokes [2], in subsequent reports from the US National Bureau of Standards (now NIST) [3–8], and have been reevaluated more recently by Partanen and coworkers [9, 10]. Values of the mean ionic activity coefficients of representative aqueous electrolytes at 25°C at several molalities m are shown in Table 7.1, in order for the trends with the natures of the ions making up these electrolytes to be seen, and some of these data are shown in Figure 7.1. In all the cases there is a decrease in $\gamma_\pm$ at low concentrations but the values tend to increase again at higher molalities. For the univalent cation chlorides, for instance, at 1 m the trend of the $\gamma_\pm$ values is HCl > LiCl > NaCl > KCl ~ NH$_4$Cl > RbCl > CsCl. For the acids and lithium, sodium, and potassium salts, the trend at this molality among the anions is $NO_3^- < Cl^- < Br^- < ClO_4^- < I^-$ (KClO$_4$ is insoluble), but for rubidium and cesium the trend is $I^- < Br^-$ (not shown) $< Cl^-$. These trends are commented on in Section 7.2. The $\gamma_\pm$ values of 1 : 2 and 2 : 1 electrolytes are smaller than those for 1 : 1 electrolytes and the diminution becomes larger as the charge numbers of the ions increase.

The values of the mean ionic activity coefficients can be expressed by the following expression:

$$\ln \gamma_\pm = -A I^{1/2} (1 + B I^{1/2})^{-1} + C m_E + D m_E^2 + E m_E^3 + \cdots \tag{7.6}$$

where $I = 1/2 \sum_I \nu_I m_I z_I^2$ is the ionic strength of the electrolyte solution, where the summation extends over all the ionic species I. The coefficient $A = 1.8246 \times 10^6 (\varepsilon T)^{-3/2} = 1.1765$ at 25°C for aqueous 1 : 1 electrolytes is the Debye–Hückel

TABLE 7.1   The Mean Ionic Activity Coefficients of Representative Aqueous Electrolytes at 25°C at Selected Molalities [2]

m_E/mol·kg	0.1	0.3	1.0	3.0	6.0
Electrolyte					
HCl	0.796	0.756	0.809	1.316	3.22
HBr	0.805	0.777	0.871	1.674	
HI	0.818	0.811	0.963	2.015	
HNO_3	0.791	0.735	0.724	0.909	
$HClO_4$	0.803	0.768	0.823	1.448	4.76
LiOH	0.718	0.628	0.523	0.467	
LiCl	0.790	0.744	0.774	1.156	2.72
LiBr	0.796	0.756	0.803	1.341	3.92
LiI	0.815	0.804	0.910	1.715	
$LiNO_3$	0.788	0.736	0.743	0.966	1.506
$LiClO_4$	0.812	0.792	0.887	1.582	
NaOH	0.764	0.706	0.677	0.782	1.296
NaCl	0.778	0.710	0.657	0.714	0.968
NaBr	0.782	0.719	0.687	0.812	
NaI	0.787	0.735	0.736	0.963	
$NaNO_3$	0.762	0.666	0.548	0.437	0.371
$NaClO_4$	0.775	0.701	0.629	0.611	0.677
KOH	0.776	0.721	0.735	1.051	2.14
KF	0.775	0.700	0.645	0.705	
KCl	0.770	0.688	0.604	0.569	
KBr	0.772	0.693	0.617	0.595	
KI	0.778	0.707	0.645	0.652	
KNO_3	0.739	0.614	0.443	0.249	
NH_4Cl	0.770	0.687	0.603	0.561	0.564
NH_4NO_3	0.740	0.636	0.504	0.368	0.279
RbCl	0.764	0.675	0.583	0.536	
RbI	0.762	0.671	0.575	0.516	
CsCl	0.756	0.656	0.544	0.479	0.480
CsI	0.754	0.651	0.533	0.434	
$MgCl_2$	0.528	0.476	0.569	2.32	
$Mg(NO_3)_2$	0.522	0.467	0.536	1.449	
$Mg(ClO_4)_2$	0.577	0.576	0.925	8.99	
$CaCl_2$	0.518	0.455	0.500	1.483	11.11
$Ca(NO_3)_2$	0.488	0.397	0.338	0.382	0.596
$Ca(ClO_4)_2$	0.557	0.532	0.743	4.21	63.7
H_2SO_4	0.266	0.183	0.132	0.142	0.257
Li_2SO_4	0.478	0.369	0.283	0.294	
Na_2SO_4	0.452	0.325	0.204	0.139	
K_2SO_4	0.436	0.313			
$(NH_4)_2SO_4$	0.423	0.300	0.189	0.125	
$LaCl_3$	0.314	0.263	0.342		
$Th(NO_3)_4$	0.279	0.203	0.207	0.486	

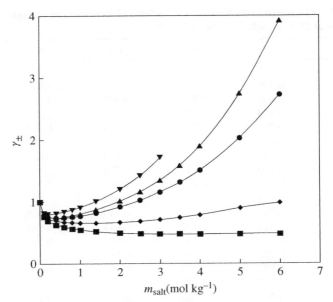

FIGURE 7.1 Molal scale mean ionic activity coefficients of aqueous LiCl($\bullet$), LiBr ($\blacktriangle$), LiI ($\blacktriangledown$), NaCl ($\blacklozenge$), and CsCl ($\blacksquare$) at 25°C (Data from Ref. 2).

coefficient of the limiting law, and B, C, D, and E, etc. are empirical fitting coefficients. Another expression for the concentration dependence of the activity coefficients is due to Pitzer and coworkers, summarized in Ref. 11:

$$\ln \gamma_\pm = |z_+ z_-| f_\gamma + \left(\frac{2v_+ v_-}{v} \right) B_\gamma m_E + ((2v_+ v_-)^{3/2} v) C_\gamma m_E^2 \qquad (7.7)$$

The Debye–Hückel coefficient f_γ is:

$$f_\gamma = -A_\varphi [I^{1/2}/(1+1.2I^{1/2}) + (2/1.2)\ln(1+1.2I^{1/2})] \qquad (7.8)$$

where the coefficient A_φ is:

$$A_\varphi = \left(\frac{2\pi N_A \rho_W}{9000} \right)^{1/2} \left(\frac{e^2}{\varepsilon k_B T} \right)^{3/2} \qquad (7.9)$$

For aqueous solutions at 25°C $A_\varphi = 0.3910$ and the values at other temperatures have been listed, as also values for many electrolytes [11]. Pitzer's expressions with listed B_γ and C_γ values are generally valid up to $m_E \sim 6m$.

For aqueous electrolyte solutions at temperatures other than 25°C, the activity coefficients are obtained from integration of the expression:

$$\frac{d \ln \gamma_\pm}{dT} = \frac{-L_E}{v R T^2} \qquad (7.10)$$

where L_E is the partial molar heat content of the solution. Values of L_E of many electrolytes have been reported by Parker [12].

Various aspects of the ion–ion interactions are concealed beneath the expressions (7.6) and (7.7) for the activity coefficients and related ones for the osmotic coefficients. The curtailed form of Equation 7.6 is:

$$\ln \gamma_{\pm} = -AI^{1/2}(1 + BI^{1/2})^{-1} + Cm_E \qquad (7.11)$$

where parameter A is generally taken as the value from the Debye–Hückel theory, namely:

$$A/\mathrm{mol}^{-1/2} \cdot \mathrm{kg}^{1/2} = |z_+ z_-| (2\pi N_A/1000)^{1/2} \left(e^3 k_B^{-3/2}\right)(\varepsilon T)^{-3/2}$$
$$= 1.8246 \times 10^6 (\varepsilon T)^{-3/2} |z_+ z_-| \qquad (7.12)$$

The last equality pertains to any solvent and any temperature. For aqueous solutions at 25°C $A/\mathrm{mol}^{-1/2} \cdot \mathrm{kg}^{1/2} = 0.510|z_+ z_-|$. The quantity $BI^{1/2}$ in the denominator equals κ, the reciprocal of the "thickness" of the ionic atmosphere in the Debye–Hückel theory [2]. This thickness is to be construed as pertaining to the distance beyond the distance of closest approach of the ions, a, to the periphery of the ionic atmosphere. Therefore, $BI^{1/2}$ may be replaced by $B'aI^{1/2}$, where, according to the theory:

$$B'/\mathrm{nm}^{-1} \cdot \mathrm{mol}^{-1/2} \mathrm{kg}^{1/2} = \left(\frac{8\pi N_A e^2}{1000\ k_B}\right)^{1/2} (\varepsilon T)^{-1/2} = 5.014 \times 10^9 (\varepsilon T)^{-1/2} \qquad (7.13)$$

The numerical value pertains to any solvent and any temperature. For aqueous solutions at 25°C, the numerical constant B' is 3.281. For many aqueous electrolytes and not too large m_E, say 1–2 m, the universal value $B = B'a = 1.5$ may be used in Equation 7.11, and the onus of fitting the activity coefficient data is placed on the linear term Cm_E, that is, a single electrolyte-dependent coefficient.

7.1.2 Ion Hydration Related to Ion–Ion Interactions

The linear term, Cm_E, was dealt with and interpreted in several ways. Empirically, $C = 0.1|z_+ z_-|$ was proposed by Davies [13] for fitting activity coefficients of aqueous electrolytes at 25°C up to 0.1 m. Stokes and Robinson [14] suggested that the amount of solvent bound by the solvated ions should be deducted from the total amount of solvent in order to represent the entropic part of the chemical potential of the solute appropriately. Therefore, the following expression results for the linear term:

$$Cm_E = M_S h\varphi - \ln[1 + M_S(\nu - h)m_E] \qquad (7.14)$$

where M_S is the molar mass of the solvent (in kg·mol^{-1}), h is the solvation number (number of solvent molecules bound per formula unit of electrolyte), $\varphi = -(\nu M_S m_E)^{-1} \ln a_S$ is the osmotic coefficient of the solvent (having an activity a_S), and ν is the number of ions per formula unit of the electrolyte. A two-parameter fitting expression resulted [2] for the activity coefficients, on setting $BI^{1/2} = B'aI^{1/2}$ in

Equation 7.11. It involves the distance of closest approach of the ions, a, and the sum of their hydration numbers, h, yielding the following expression for the mean ionic activity coefficients of aqueous electrolytes:

$$\ln \gamma_{\pm} = -AI^{1/2}\left(1+B'aI^{1/2}\right)^{-1} - \left(\frac{h}{v}\right)\ln a_{S} - \ln[1+M_{S}(v-h)m_{E}] \qquad (7.15)$$

Numerical values of a and h were shown by Robinson and Stokes [2] for fitting the activity coefficients of aqueous alkali metal and alkaline metal halides and perchlorates at 25°C with Equation 7.15. The a values ranged from 0.348 nm for RbBr to 0.618 nm for MgI_2 and had little relationship to the sums of the radii of the cation and anion, whether bare or with their first hydration shells.

As the concentration of the electrolyte is increased beyond ca. 2 m, the curtailed expression (7.11) becomes inadequate for fitting the experimental activity coefficients and either the full expression (7.6), where the parameters $D, E \ldots$ are completely empirical, or Pitzer's expression (7.7) may be employed up to ca. 6 m. In the latter case, the parameters $B\gamma$ and $C\gamma$ do have physical meanings, expressing according to Pitzer [15] the short-range forces between pairs and triple groups of ions of opposite charge sign.

Zavitsas [16, 17] took the concept, according to which the amount of water bound by the hydrated ions should be deducted from the total amount of solvent, a step further, but apparently too far as is demonstrated below. According to Zavitsas' ideas, the mole fraction of the ions of a strong aqueous electrolyte of molality m_E dissociating into v ions per formula is:

$$x'_{E} = \frac{vm_{E}}{55.51-(h-v)m_{E}} \qquad (7.16)$$

and the corresponding mole fraction of the free water (that amount of water which is not in the hydration shells) is then:

$$x'_{wf} = 1-x'_{E} = \frac{55.51-hm_{E}}{55.51-(h-v)m_{E}} \qquad (7.17)$$

where the hydration number h is specified to pertain to water molecules bound to the ions with a binding energy more negative than -56 ± 6 kJ·mol⁻¹. The colligative properties of aqueous solutions of chloride, bromide, iodide, and perchlorate salts should then follow Raoult's law. The freezing point depression, boiling point elevation, and water vapor pressure depression functions of the electrolyte content should present straight lines when plotted against $\ln x'_{wf}$ up to quite large molalities m_E, provided that a correct fixed value of $h = h_{cp}$ (for the colligative properties) is found. These straight lines should pass through the origin and have the correct theoretical slope. The resulting h_{cp} values have an estimated average uncertainty of ± 0.5. They have been assigned to the cations of the salts on the premise that the h_{cp} values of the halide anions are essentially zero (the successive water-binding energies of the isolated anions being less negative than -56 kJ·mol⁻¹, Table 2.4).

The problem with this approach is that it does not recognize the requirement that the hydration numbers of ions diminish with increasing concentrations as the hydration shells of the ions overlap and water molecules are eliminated from these shells, joining the free water. The average distance apart of the ions of a symmetrical electrolyte of concentration c is on purely geometrical grounds: $d^{av}/nm = 0.94(c/M)^{-1/3}$ as shown by Marcus [18]. At a molar concentration of >1.9 M, the average distance apart of the ions is <0.76 nm, not permitting two water molecules of diameter 0.278 nm (their collision diameter) to reside between two ions having radii ~ 0.1 nm, and hence their solvation shells overlap to some extent; see also the following text. Therefore, constant values of h_{cp} that fit the colligative properties of the electrolyte solutions at concentrations >1.9 M cannot have physical meaning.

At very high electrolyte concentrations, where a large fraction of the solvent is expected to be bound to the ions in their solvation shells, a different approach applies, namely one that considers the solvent to be "adsorbed" on the ionic sites of the electrolyte, according to the BET method. This was suggested by Stokes and Robinson [14] and recently taken up by Marcus [19] as applying to molten salt hydrates. The terms "adsorption" and "binding sites," taken over from the BET method for sorption of neutral small molecules on solid surfaces, should not be taken too literally. The operative expression is:

$$\left(\frac{m_E}{55.51}\right)\left(\frac{a_w}{1-a_w}\right) = (cr)^{-1} + \left[\frac{c-1}{cr}\right]a_w \qquad (7.18)$$

employing only water activities a_w at such large salt molalities m_E that the left-hand side of Equation 7.18 is linear with a_w. The numerical constant is the number of moles of water per kilogram and the two parameters of the method, c and r, have the following meanings. The parameter $c = \exp(\Delta_{ads}H/RT)$, where $\Delta_{ads}H$ is the difference between the molar enthalpy of "adsorption" of water on the salt and the molar enthalpy of liquefaction of water ("adsorption" of water onto liquid water). The parameter r represents the number of "binding sites" of water on the ions of the salt, conceptually equivalent to its hydration number. The values of c and r are only moderately dependent on the temperature. In cases where the experimental water activity data do not reach sufficiently low values to make Equation 7.18 linear, a modification can be employed to obtain the c and r values, introducing another parameter $K \neq 1$ suggested by Anderson [20] and endorsed by Stokes and Robinson [14]:

$$\left(\frac{m_E}{55.51}\right)\left(\frac{a_w}{1-Ka_w}\right) = (Kcr)^{-1} + \left(\frac{c-1}{cr}\right)a_w \qquad (7.19)$$

where K should be regarded as an empirical fitting parameter. The data for cobalt bromide are shown in Figure 7.2 and the values of c and r (and $K \neq 1$ where necessary) at 25°C are shown for many salts in Table 7.2 taken from Marcus [19]. It should be noted that only few 1 : 1 electrolytes have sufficient solubilities in water at 25°C to be treatable by the BET method, which at this temperature is more applicable to 2 : 1 electrolytes, as shown in Table 7.2. The BET parameters for many lanthanide salts at

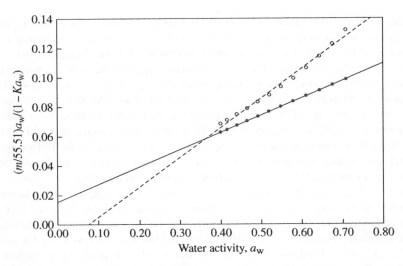

FIGURE 7.2 BET plots for $\geq 3\,mol\cdot kg^{-1}$ aqueous cobalt bromide at 25°C: - - o - - Equation 7.18 yielding a physically inacceptable negative ordinate intercept, —•— Equation 7.19 with $K = 0.86$ (From Ref. 19 with permission from the publisher, Springer).

TABLE 7.2 **Hydration Numbers $h(c)$ of Electrolytes Obtained from Isothermal Compressibilities at 25°C [21] Compared with h_{cp} Values Obtained From Colligative Properties at Several Temperatures [16], Independent of the Concentration**

Electrolyte	$h(c), c = 0\,M$	$h(c), c = 1\,M$	$h(c), c = 3\,M$	$h(c), c = 6\,M$	h_{cp}
HCl	1.9	1.8	1.5		6.7 ± 0.5
LiOH	9.1	8.5	6.8		
LiCl	6.0	5.6	4.8	3.9	6.7 ± 0.5
LiBr	5.2	4.8			
NaOH	10.5	9.5	7.8		
NaF	8.8	8.1	6.8		
NaCl	6.9	6.3	5.4		3.9 ± 0.5
NaBr	6.2	5.8	5.1	4.3	3.6 ± 0.5
NaI	5.4	5.1	4.6	4.0	5.5
Na_2SO_4	14.6	14.5	10.6	4.8	
KCl	6.8	6.1	5.1		1.5 ± 0.5
KBr	5.9	5.5	4.6		1.6 ± 0.5
KI	5.3	4.9	4.2		2.1 ± 0.5
K_2SO_4	16.9	14.7	10.3		
$MgCl_2$	10.7	9.9	8.2		13.7 ± 0.5
$CaCl_2$	17.6	15.4	11.0		12.7 ± 0.5
$Ca(NO_3)_2$	10.0	9.1	7.5	5.8	
$SrCl_2$	21.5	17.8	10.4		13.1 ± 0.5
$BaCl_2$	9.3^a				11.2 ± 0.5
$AlCl_3$	26.4^a				21.8 ± 0.5
$FeCl_3$	22.6^a				17.9 ± 0.5

$^a h_{IWemp}{}^\infty$ values from Table 4.9.

25°C are also available in Ref. 19, as are data for other salts (including some $1:1$ electrolytes) at temperatures higher than 25°C.

The values of r, construed as hydration numbers of electrolytes at high concentrations, are considerably smaller than the sum of the ionic hydration numbers at infinite dilution h^{∞} shown in Tables 4.7, 4.8, and 4.9. Theoretically, increasing concentrations necessarily lead to hydration numbers $h(c)$ that diminish with increasing electrolyte concentrations as shown by Padova [22]:

$$h(c) = h^{\infty}\left[1 - \left\{\frac{S_V}{\left(V_{intr} - V^{\infty}\right)}\right\}c^{1/2}\right] \tag{7.20}$$

Here $S_V > 0$ is the Debye–Hückel limiting slope for electrolyte partial molar volumes, V_{intr} is the intrinsic volume of the electrolyte, and V^{∞} is its standard partial molar volume, and theory requires that $(V_{intr} - V^{\infty}) > 0$ because of electrostriction. In fact, methods such as those involving isothermal compressibility data at increasing electrolyte concentrations, lead to hydration numbers that diminish with increasing concentrations, and $h(c)$ of some two dozen electrolytes at 25°C have been reported by Marcus [21]. These hydration numbers tend to converge toward the BET r values where known. Because the hydration shells of oppositely charged ions of an electrolyte start to overlap at concentrations ≥ 1.9 M, it is inevitable that $h(c) < h^{\infty}$ already at this limit, but ion–ion interactions cause diminishing hydration numbers already at lower concentrations. The values of $h(c)$ at $c = 0, 1, 3,$ and 6 M for the electrolytes obtained from isothermal compressibilities at 25°C [21] are shown in Table 7.3 and the curves for some electrolytes are shown in Figure 7.3.

7.2 ION ASSOCIATION

The eventual upturn of the mean ionic activity coefficients as the electrolyte concentration increases, described in Equation 7.11 by the linear term Cm_E and in Equations 7.6 and 7.7 by higher powers of the concentration, can be interpreted in several ways. A concept that differs from that involving hydration numbers, Equations 7.14 and 7.15, is in terms of the association of ions of opposite charges. In fact, ion association competes with the solvation of the ions and in certain cases an ion of opposite charge may replace some of the solvent in the solvation shell of a given ion.

When the ions of opposite charge sign in an electrolyte C^+A^- (the $1:1$ case is selected for simplicity) approach each other, they attract each other and may associate to ion pairs of sufficiently long lifetimes. The Eigen–Tamm [23] scheme for the successive formation of solvent-separated ion pairs (SSIPs or 2SIPs), solvent-shared ion pairs (SIPs), and contact ion pairs (CIPs), Figure 7.4, with each step being characterized by its equilibrium constant, is:

$$\underset{\text{free ions}}{C^+(\text{solv}) + A^-(\text{solv})} \underset{K_1}{\rightleftarrows} \underset{\text{2SIP}}{C^+(\text{solv})(\text{solv})A^-} \underset{K_2}{\rightleftarrows} \underset{\text{SIP}}{C^+(\text{solv})A^-} \underset{K_3}{\rightleftarrows} \underset{\text{CIP}}{C^+A^-} \tag{7.21}$$

TABLE 7.3 The BET Parameters r and ΔH and K (if $\neq 1$) for Highly Concentrated Aqueous Salts at 25°C [19]

Salt	r	ΔH/kJ·mol^{-1}	K
LiCl	3.64	7.05	
LiBr	3.82	9.32	
LiClO$_4$	3.18	7.50	
NaOH	3.20	7.34	
KOH	3.25	8.26	
MgBr$_2$	7.10	9.19	
Mg(CH$_3$CO$_2$)$_2$	3.49	9.10	0.9735
Mg(ClO$_4$)$_2$	8.34	8.47	0.86
CaCl$_2$	6.73	5.58	
CaBr$_2$	7.06	9.30	
Ca(NO$_3$)$_2$	3.86	5.55	
Ca(ClO$_4$)$_2$	6.83	8.68	
MnCl$_2$	4.21	6.61	
MnBr$_2$	5.37	8.98	
Mn(NO$_3$)$_2$	4.96	7.25	
Mn(ClO$_4$)$_2$	8.58	8.08	0.86
CoCl$_2$	5.38	8.36	0.94
CoBr$_2$	7.68	5.62	0.86
Co(NO$_3$)$_2$	5.57	11.10	0.91
Co(ClO$_4$)$_2$	8.41	11.82	0.86
NiCl$_2$	5.20	10.45	
NiBr$_2$	7.06	9.82	
Ni(NO$_3$)$_2$	6.06	8.34	0.89
CuCl$_2$	3.13	10.12	
Cu(NO$_3$)$_2$	4.70	9.97	
Cu(ClO$_4$)$_2$	8.44	7.63	0.86
ZnCl$_2$	3.69	7.73	
ZnBr$_2$	4.01	7.40	
Zn(NO$_3$)$_2$	5.23	10.25	
Zn(ClO$_4$)$_2$	8.63	9.32	0.85
Pb(ClO$_4$)$_2$	5.33	8.36	
UO$_2$(NO$_3$)$_2$	5.86	1.14	
UO$_2$(ClO$_4$)$_2$	10.13	7.66	0.82
Al(NO$_3$)$_3$	10.1	10.16	0.86

The overall association constant is $K_{ass} = K_1 + K_1 K_2 + K_1 K_2 K_3$. Most methods for studying ion association (such as conductivity or potentiometry, see below) provide values only for K_1 or K_{ass}, but some methods (Section 7.2.2 and volumetric data according to Hemmes [25]) are able to distinguish between the three kinds of ion pairs.

Qualitative views of ion association have been derived from the trends dealt with above of activity coefficients of series of ions with a common counterion. The trends for $\gamma_\pm$ at 1 m, being RbCl > RbI and CsCl > CsI, may be compared with the opposite, increasing trends of the lighter alkali metal halides, MCl < MBr < MI, see Table 7.1.

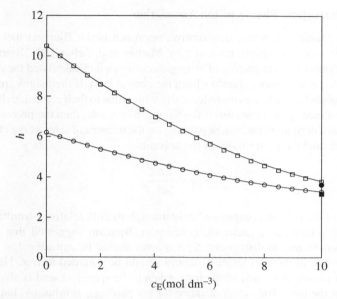

FIGURE 7.3 The hydration numbers $h(c)$ of aqueous salts at 25°C obtained from isothermal compressibility data for LiCl (-o-) and NaOH (-□-). At the right-hand side are shown the BET parameters r (number of water-binding sites per formula unit of the salt) of these salts (large filled symbols) (From Ref. 21 with permission from the publisher, ACS).

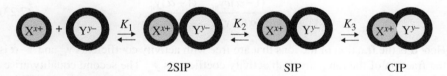

FIGURE 7.4 The Eigen–Tamm scheme for stepwise formation from free solvated cations X^{x+} and solvated anions Y^{y-} of 2SIP ion pairs, then SIP ion pairs, and finally CIP ion pairs, with elimination of solvent molecules from the solvation shells of the ions (From Ref. 24 by permission of the publisher, IUPAC).

This trend extends to the fluorides of the lighter alkali metal cations, but not to the $\gamma_\pm$ of KF and of the heavier fluorides (data from Hamer and Wu [3]). The reversed trend for the poorly hydrated larger ions of the alkali metal halide salts has been ascribed to *water structure enforced* ion association, suggested by Diamond [26, 27], because such ions break the structure of the water (see Section 5.1.1). Less water structure is broken when the cation and anion associate and the gain of the Gibbs energy of the hydrogen bonding is reflected in the decreased chemical potential, and hence the activity coefficients of electrolytes containing the large ions.

7.2.1 Electrostatic Theory of Ion Association

A more sophisticated, hence quantitative, approach is the Bjerrum theory [28] of electrostatic ion association, reviewed by Marcus and Hefter [29]. Bjerrum introduced the concept of ion pairing of strong electrolytes and calculated the probability of an ion i to be at a given distance r from the central ion j. If their signs are the same, then the probability increases monotonically with r, due to their repulsion diminishing with the distance apart. However, if the signs are opposite, then the probability has a minimum at a certain distance q, depending on the numerical sizes of the charges, the temperature, and the permittivity of the solvent:

$$q = \frac{z_i z_j e^2}{2\varepsilon k_B T} \tag{7.22}$$

The effect of the solvent is expressed only through its bulk relative permittivity, ε, the solvent being taken as a dielectric continuum. Bjerrum suggested that oppositely charged pairs of ions at distances $a \leq r \leq q$ apart should be considered as associated ion pairs, whereas those at larger distances should be regarded as free. Here a is the distance of closest approach of the ions, taken to be spherical, and is also the mean diameter of the ions. The cutoff distance for ion pairing q is arbitrary, but is reasonable, the work required to separate such ion pairs being at least twice the thermal energy. Free ions participate in the ionic atmosphere and contribute to the electrostatic effects described by the extended Debye–Hückel theory.

The mass action law is applicable to the equilibrium between the free and associated ions and the association constant K_{ass} is

$$K_{ass} = \frac{(1-\alpha)cy_{ip}}{(\alpha c)^2 y_{f\pm}^2} = \frac{(1-\alpha)y_{ip}}{cy_\pm^2} \tag{7.23}$$

Here α is the fraction of the ions that are free with activity coefficients $y_{f\pm}$ and $1-\alpha$ is the fraction of the ion pairs with activity coefficients y_{ip}. The second equality arises from the fact that the experimentally determinable mean ionic activity coefficient squared is $y_\pm^2 = \alpha^2 y_{f\pm}^2$. The electrostatic considerations in the Bjerrum theory yield the expression:

$$K_{ass} = \left(\frac{4\pi N_A}{1000}\right) b^3 Q(b) \tag{7.24}$$

where $b = q/a$ and $Q(b)$ is an integral of an auxiliary variable x that has to be solved numerically:

$$Q(b) = \int_2^b x^{-4} \exp(x)\, dx = -1.5669 + 1.1431b - 0.2179b^2 + 0.0164b^3 \tag{7.25}$$

the polynomial applying for values of $b \leq 10$, relevant to aqueous $1:1$ electrolytes at 25°C. The fraction of the electrolyte present as ion pairs, $1-\alpha$, is obtained by equating the K_{ass} in Equation 7.23 and 7.24 but requires an expression for the activity coefficient of the ion pair, y_{ip}.

If the ion pair is charged (e.g., $MgCl^+$ or $NaSO_4^-$), y_{ip} may be estimated according to $\ln \gamma_{ip} = -AI^{1/2}\left(1 + B'aI^{1/2}\right)^{-1}$, but if it is neutral (e.g., CsI or $MgSO_4$), it has traditionally been ignored, that is, set equal to unity. This is incorrect, however, and y_{ip} of neutral ion pairs can be estimated according to two approaches: on the assumption that the ion pairs are salted-out by the free ions or of them to be zwitterions as shown by Marcus [30].

7.2.1.1 Activity Coefficients of Neutral Ion Pairs

Neutral ion pairs can be treated as if they were nonelectrolytes being salted-out by the free ions of the electrolyte. For the salting-out of nonelectrolytes (Section 7.3) according to McDevit and Long [31] (see Section 7.3.2), the treatment requires the isothermal compressibility of the solvent κ_T, the intrinsic volume of the electrolyte V_{Eintr}, its standard partial molar volume V_E^∞, and the latter quantity of the nonelectrolyte (ion pair) V_{ip}^∞. The value of V_{Eintr} can be estimated from values for the ions tabulated by Marcus et al. [32] and in Table 2.8 and are independent of the solvent, and V_E^∞ values are known for many salts in many solvents (Table 4.6). The value of V_{ip}^∞ must be estimated from the dimensions of the ions and may be assumed to be equal to V_{Eintr} for the symmetrical ion pair. Then its activity coefficient becomes according to Marcus [30]:

$$\ln y_{ip} = \frac{V_{Eintr}\left(V_{Eintr} - V_E^\infty\right)/1000RT\kappa_T}{\alpha c_E} \tag{7.26}$$

For the analogy of a symmetrical ion pair with a zwitterion, glycine, $^+H_3NCH_2CO_2^-$, may be chosen for the sake of being explicit. Kirkwood's theory [33] considers the activity coefficient of a zwitterion with a distance d between the (unit) charges (centers of the ions). The dipole moment of glycine is $\mu = zed = 11.50$ D ($1D = 3.33564 \times 10^{-30}$ Cm) according to Khoshkbarchi and Vera [34], and hence the length of its dipole is $d = \mu/ze = 0.239$ nm. For β-alanine, $^+H_3NCH_2CH_2CO_2^-$, the dipole moment is 17.55 D [34], so that the length is even larger, $d = 0.399$ nm, of a size similar to the distance apart of the charges of common ion pairs. The activity coefficients of many zwitterionic amino acids have been determined experimentally in aqueous solutions and in some aqueous–organic mixtures containing added electrolytes and have been fitted to the Pitzer expression [11]. For small amounts of the amino acid, this reduces for the analogous neutral ion pair of similar distance d to:

$$\ln y_{ip(1-\alpha \to 0)} = 2\chi_{ip-E}c_E \tag{7.27}$$

where χ_{ip-E} is the ion pair (amino acid)—electrolyte interaction coefficient. (This can be separated into $\chi_{ip-+} + \chi_{ip--}$ for the interactions with the cations and the anions.) The activity coefficients of the amino acids need to be obtained at the isoelectric point, where their zwitterionic form predominates.

Contrary to most studies of ion pairing that ignore deviations of y_{ip} from unity, a recent study by Marcus [35] of the change of the static permittivity of aqueous electrolytes with their increasing concentration c_E did calculate the activity coefficient of the ion pair, y_{ip}, values iteratively, using Equation 7.26. The fraction of

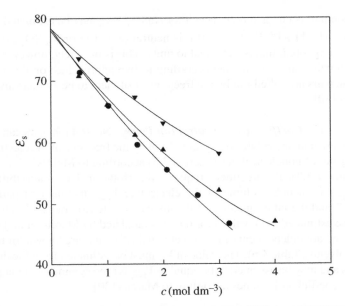

FIGURE 7.5 The static permittivities ε_s of aqueous CsF($\bullet$), KF($\blacktriangle$), and KOH($\blacktriangledown$) at 25°C as functions of their molar concentration c and their fits according to Equation 7.28 (———) (Data from Ref. 35).

ion-paired dipolar species, $1-\alpha$, was calculated from the Bjerrum theory, using Equations 7.23 and 7.24. The permittivity of the solution $\varepsilon = f(c_E)$ was expressed by means of these $1-\alpha$ values as:

$$\varepsilon = \varepsilon_W - \delta' c_E + (\beta' + \delta')(1 - \alpha)c_E \qquad (7.28)$$

see Figure 7.5. A linear dielectric decrement $\delta' c_E$ is produced by the free ions as in dilute solutions, but electrostatically paired ions of opposite charges represent dipolar species, characterized by the β' parameters, that are orientable by an external field, contrary to the spherical hydrated free ions, hence contribute positively to the permittivity of the solution, as observed experimentally.

7.2.2 Methods for Studying Ion Association

The Bjerrum theory of electrostatic ion pairing, as applied to conductivity data, has been well substantiated by Justice and Justice [36]. The interpretation of such data that have traditionally been one of the main methods for studying ion association according to Martell and Motekaitis [37] is according to the following expression by Fernandez-Prini and Justice [38]:

$$\Lambda_E = \alpha \left[\Lambda_E^\infty - S(\alpha c)^{1/2} + E\alpha c \ \ln(\alpha c) + J_1(R_1)\alpha c - J_2(R_2)(\alpha c)^{3/2} \right] \qquad (7.29)$$

Here Λ_E is the molar conductivity of the electrolyte, Λ_E^∞ is that quantity at infinite dilution, αc is, as above, the concentration of free ions, and S, E, J_1, and J_2 are known explicit expressions, containing contributions from relaxation and electrophoretic effects. The latter two functions depend also on the ion distance parameters R_1 and R_2 that are set equal to q according to the Bjerrum theory. If the departure of y_{ip} from unity is ignored, only two parameters, K_{ass} (yielding α from Eq. (7.23)) and Λ_E^∞ have to be modeled.

Thermodynamic methods, such as potentiometric, solubility, and electrolyte and solvent activity coefficient data have also been frequently used for studying ion pairing of strong electrolytes [37]. If a theoretical expression, such as the first term in the extended Debye–Hückel expression (7.11) with $B = B'a$, namely $\ln y_{f\pm} = -AI^{1/2}(1 + B'aI^{1/2})^{-1}$, is assumed for the mean ionic activity coefficients $y_{f\pm}$ of the *free* ions in Equation 7.23, and the stoichiometric (measured) activity coefficients are $y_\pm$, then the fraction of free ions is $\alpha = y_\pm / y_{f\pm}$ as mentioned above. The conductivity and thermodynamic methods yield the standard association constant, K_{ass}°, when appropriate corrections for activity coefficients at the applicable ionic strengths of the solutions studied are applied, noting the difference between the total electrolyte concentration and that pertaining to the ionic strength.

In recent years, dielectric relaxation spectroscopy (DRS) has become prominent among methods for studying ion association. It has the unique advantage of being able to distinguish between solvent-separated ion pairs (SSIPs or 2SIPs), solvent-shared ion pairs (SIPs), and contact ion pairs (CIPs), because of the dependence of the deconvoluted signals from the individual dipolar species on their differing dipole moment distances d and relaxation times as shown in the review by Buchner [24]. When DRS is employed, the complex permittivity ε^* of the electrolyte solution is obtained over an appropriate range of frequencies ν, from megahertz to terahertz, and concentrations c. After correction for the conductivity of the solution, the resulting $\varepsilon^*(\nu, c)$ is the sum of the individual relaxation processes of the dipoles in the solution:

$$\varepsilon^*(\nu, c) = \varepsilon_\infty + \sum_j S_j(c_j)\left[1 - (i2\pi\nu\tau_j)^{1-\alpha j)}\right]^{-\beta j} = \varepsilon_\infty + \sum_j S_j(c_j)\left[1 - (i2\pi\nu\tau_j)\right]^{-1} \quad (7.30)$$

Here $\varepsilon_\infty \approx n_\infty^2$ is the infinite frequency permittivity (the square of the refractive index at infinite optical wavenumber), $S_j(c_j)$ is the amplitude of the signal from the individual dipolar species j having a relaxation (dipole orientation) time τ_j, and $i = (-1)^{1/2}$ is the imaginary constant. The second, simplified, equality is the Debye limit, where $\alpha_j = 0$ and $\beta_j = 1$ for all the processes. This simplification is generally applicable and at most three species j, Equation 7.21, can be discerned with individual relaxation times τ_j. Once the amplitudes $S_j(c_j)$ have been extracted from the $\varepsilon^*(\nu, c)$ data, they are related to the concentrations of the relaxing species as:

$$c_j = S_j(c_j)\left(\frac{N_A}{3k_B T\varepsilon_0}\right)^{-1} \varepsilon^{-1}[1 + (1-\varepsilon)A_j](g_j\mu_j)^{-1}(1 - f_j\alpha_{jpol})^2 \quad (7.31)$$

Here A_j is a geometric factor reflecting the ellipsoid of rotation of the ion pair to be obtained from its dimensions and f_j is the reaction field, expressions for which having been reported by Barthel et al. [39], μ_j and α_{jpol} are the dipole moment and polarizability of the dipolar species, and g_j is the dipole orientation correlation factor, generally set as $g_j = 1$. Eventually the concentrations of the ion pair species, c_j, are used for the calculation of the partial association equilibrium quotients, K_1, K_2, and K_3 and the overall constant K_{ass}, which may be compared with values obtained by other methods.

Ultrasonic relaxation measurements have also been used for the study of stepwise ion association, but in contrast with the DRS method, it detects the ion pairing equilibria rather than the species formed, which must be assumed. NMR and Raman spectroscopy are sensitive only to CIPs and treat SIPs and 2SIPs, with solvent molecules intervening between the partner ions, as if they were free ions. A thorough treatment of the theories relating to ion association of strong electrolytes and the experimental methods used to study it are presented in the review by Marcus and Hefter [29].

7.2.3   Thermodynamic Quantities Pertaining to Ion Association

Results for the standard thermodynamic functions, $-RT \ln K_{ass}^\circ = \Delta_{ip} G^\circ$ and the temperature derivatives thereof leading to $\Delta_{ip} H^\circ$ and $T \Delta_{ip} S^\circ$ are shown in the review by Marcus and Hefter [29] for many electrolytes in several solvents. Some of these data are shown in Table 7.4. For the ion pairing of the thallium halides in the solvents shown in this table, the association is enthalpy driven, pointing to CIP formation with some coordinative bonding. For the other ion pairs shown, the association is entropy driven, for even if there are only 2SIPs being formed, fewer particles result in the solution than when only free solvated ions exist.

The volume change on ion association has led to information regarding the stepwise pairing in Equation 7.21. According to Hemmes [25], when SIPs or CIPs are formed and solvent molecules are liberated from the solvation shell of the ions, the volume changes must reflect this process. The relevant quantity, $\Delta_{ip} V^\circ$, is obtained from the pressure derivative of the association constant:

$$\Delta_{ip} V^\circ = -RT \left[\left(\partial \ln \left(K_{ass}^\circ / dm^3 mol^{-1} \right) / \partial P \right)_T + \kappa_T \right] \tag{7.32}$$

The $-RT\kappa_T$ term is needed for K_{ass}° expressed in M^{-1}. It may also be calculated according to the Bjerrum theory (Section 7.2.1) as:

$$\Delta_{ip} V^\circ = -RT \left[-3 - b (\partial \ln \varepsilon / \partial P)_T + \kappa_T \right] \tag{7.33}$$

provided $b = q/a$ is sufficiently large, as occurs in solvents of relatively low permittivity. Positive values of $\Delta_{ip} V^\circ$ denote that some solvent has been liberated by the ion pairing process from being electrostricted in the solvation shells of the ions, that is, that a SIP or a CIP has been formed.

TABLE 7.4 Thermodynamic Functions of Ion Pairing of Some Electrolytes in Several Solvents at 25°C[a]

Solvent	Ion pair	$\Delta_{ip}G°$/kJ mol^{-1}	$\Delta_{ip}H°$/kJ mol^{-1}	$T\Delta_{ip}S°$/kJ mol^{-1}
Water	TlCl	−21.3	−26.3	−5.0
	TlBr	−31.4	−34.6	−3.2
	TlI	−41.4	−71.1	−29.7
	NaSO$_4^-$	0.5		
	MgSO$_4$	−12.6	5.8	18.4
	CaSO$_4$	−13.0	6.7	19.7
	MnSO$_4$	−13.0	7.8	20.8
	CoSO$_4$	−13.2	5.7	18.9
	NiSO$_4$	−13.3	5.4	18.7
	ZnSO$_4$	−13.2	6.2	19.4
	CdSO$_4$	−13.6	8.4	22.0
Methanol	CaClO$_4^+$	−11.9	16.9	28.8
	SrClO$_4^+$	−13.0	17.2	30.2
	BaClO$_4^+$	−13.9	16.3	30.2
	ZnClO$_4^+$	−10.3	15.5	25.8
2-Propanol	NaI	−13.2	18.9	32.1
	KI	−14.7	19.1	33.8
	RbI	−15.5	17.5	33.0
	Et$_4$NI	−15.6	6.6	22.2
	Pr$_4$NI	−15.5	6.0	21.5
	Bu$_4$NI	−15.6	5.7	21.3
Tetrahydrofuran	Bu$_4$NI	−37.2		
	Pe$_4$NI	−38.9		
	Hx$_4$NI	−38.2		
Propylene carbonate	TlCl	−64.5	−77.3	−12.8
	TlBr	−61.9	−70.4	−8.5
	TlI	−57.0	−46.6	10.4
	NaI	−1.0		
	KI	1.3		
	Et$_4$NBr	−3.3		
	Pr$_4$NBr	4.0		
	Bu$_4$NBr	−2.9		
Acetonitrile	Et$_4$NBr	−7.4		
	Et$_4$NI	−7.0		
	Pr$_4$NBr	−6.4		
	Bu$_4$NBr	−6.7		
Dimethylformamide	TlCl	−49.0	−30.5	18.5
	TlBr	−45.3	−29.4	15.9
	TlI	−39.1	−26.9	12.2
	LiSCN	−1.3	1.8	3.1
Dimethyl sulfoxide	LiSCN	0.9	0.3	−0.6

[a] Adapted from Ref. 29.

Marcus [40, 41] has used the $\Delta_{ip}V°$ data, some of which are shown in Table 7.5, to calculate the number of solvent molecules released when the ion pairing proceeds beyond the 2SIP stage to form a SIP and/or a CIP:

$$\Delta_{ip}h(V) = \frac{\Delta_{ip}V°}{V_S^* - \Delta_{els}V_S} \tag{7.34}$$

Here V_S^* is the molar volume of the pure solvent and $\Delta_{els}V_S$ is its mean molar volume change on the solvent being electrostricted, so that $V_S^* - \Delta_{els}V_S$ is the mean molar volume of the solvent in the solvation shells of the ions that is released on ion pairing.

Similar information can also be obtained from the standard molar entropy change on ion pairing, $\Delta_{ip}S°$ [41], Table 7.5. This number is obtained from the entropy of desolvation, $\Delta_{des}S°$, divided by the entropy of fusion of the solvent extrapolated to the temperature employed, $\Delta_{fus}S$. The former of these two quantities is obtained according to Marcus [41] when the translational, $\Delta_{tr}S°$, rotational, $\Delta_{rot}S°$, and electrostatic, $\Delta_{el}S°$, contributions of the ion pairing process are subtracted from the overall entropy change, $\Delta_{ip}S°$:

$$\Delta_{des}S° = \Delta_{ip}S° - [\Delta_{tr}S° + \Delta_{rot}S° + \Delta_{el}S°] \tag{7.35}$$

In the following expressions, $a = r_+ + r_-, d$ is the diameter of a solvent molecule and M denotes the molar masses of the species:

$$\Delta_{tr}S° = 1.5R\ln\left(\frac{M_{ip}}{M_+ M_-}\right) - 82.2 \text{ J K}^{-1}\text{mol}^{-1}$$

$$\Delta_{rot}S° = R\ln\left(\frac{M_{ip}}{M_+ M_-}\right) + 2R\ln a + 67.5 \text{ J K}^{-1}\text{mol}^{-1}, \text{ and} \tag{7.36}$$

$$\Delta_{el}S° = -\left[\left(\frac{N_A e^2}{8\pi\varepsilon_0}\right)\left(\frac{d\varepsilon}{dT}\right)\varepsilon^2\right]\left[\frac{z_+^2/(r_+ + d) + z_-^2/(r_- + d) - (z_+ - z_-)^2}{(r_+ r_- a)^{1/3} - d}\right]$$

The number of solvent molecules released on ion pairing $\Delta_{ip}h(S) = \Delta_{des}S°/\Delta_{fus}S°$ is shown for a few ion pairs in Table 7.5 and for some more in Ref. 41. Unfortunately, such numbers have been obtained for only a few ion pair/solvent systems from both the volume and entropy changes on ion pairing for comparisons to be made. The agreement between the $\Delta_{ip}h$ values is not perfect, but the disagreements are not excessive, in view of the different methods used for their calculation.

The ion–solvent interactions leading to ion solvation and the ion–ion interactions leading to ion pairing are seen to compete with each other. At infinite dilution of an electrolyte in any solvent, only ion solvation takes place, of course, but at finite concentrations ion association becomes more and more prevalent, the solvation numbers diminish (see Table 7.3 for hydration numbers), and the concentrations of ion pairs, following the sequence from 2SIPs to SIPs to CIPs, increases. It must be remembered that the ion pairs in the solution, even the CIPs, are themselves solvated. If the ion pair is charged (e.g., $NaSO_4^-$ or $MgCl^+$), it can be treated as any ionic species, taking into account its nonspherical shape. If it is neutral (e.g., CsI or $MgSO_4$), it still is dipolar and is solvated

TABLE 7.5 Volume Change on Ion Pairing, $\Delta_{ip}V$, and the Number of Solvent Molecules Released from the Ions on Pairing, Obtained from the Volume, $\Delta_{ip}h(V)$, and Entropy, $\Delta_{ip}h(S)$, Changes at 25°C [19, 29, 40, 41]

Solvent	Ion pair	$\Delta_{ip}V°/cm^3mol^{-1}$	$\Delta_{ip}h(V)$	$\Delta_{ip}h(S)$
Water	LiF	7.9	2.7	
	NaF	4.6	1.6	
	KF	3.8	1.2	
	RbF	4.0	1.4	
	CsF	4.2	1.4	
	LiSO$_4$	5.8	1.9	
	NaSO$_4^-$	7.3	2.5	
	KSO$_4^-$	5.9	2.1	
	RbSO$_4^-$	3.3	1.1	
	CsSO$_4^-$	6.2	2.2	
	MgSO$_4$	7.4	2.5	4.9
	CaSO$_4$	11.7	4.0	5.1
	MnSO$_4$	7.4	2.5	5.4
	CuSO$_4$	11.3	3.9	5.3
	ZnSO$_4$	10.0	3.5	5.1
	CdSO$_4$	9.3	3.3	5.5
	LaSO$_4^+$	21	6.8	6.3
Methanol	LiCl	18	3.0	
	LiBr	17	2.8	
	KCl	29	4.8	
Ethanol	LiCl	17		
2-Propanol	LiCl	17.4	0.7	
	NaI	15	0.4	
	Bu$_4$NCl	11.7	0.3	
	Bu$_4$NBr	8.8	0.3	
	Bu$_4$NI	8.7	0.3	
	Bu$_4$NClO$_4$	7.8	0.3	
Acetone	LiI	21	0.8	
	NaI	25	1.0	1.9
	KI	23	0.9	
	CsI	24	0.9	
Diethyl ether	Bu$_4$NPic	115	3.4	
Benzene	Bu$_4$NPic	59	2.8	

as any dipolar nonelectrolyte would be and would interact with the free ions accordingly. It is, therefore, appropriate to deal in the subsequent sections with the effects of ions in solution on nonelectrolytic species that are also present in the solution.

7.2.4 Aggregation of Ions in Solutions

In solvents of low permittivity, ion aggregation may proceed beyond ion pairing to form triple ions or larger aggregates. With 1 : 1 electrolytes composed of small ions—cations C^+ and anions A^-—triple ions such as $C^+A^-C^+$ and $A^-C^+A^-$ can be found.

Dipolar ion pairs may also aggregate to form quadrupoles $C_2^+A_2^-$ as pointed out by Petrucci et al. [42]. The formation of triple ions was suggested first by Fuoss and Kraus [43], noting a minimum in the conductivity curve $\Lambda = f(c_E)$, which they interpreted as the formation of conducting species, that is, the charged $C^+A^-C^+$ and/or $A^-C^+A^-$ species, at concentrations c_E beyond this minimum. This idea has been taken up by many other researchers who studied the conductivity of electrolytes in nonaqueous solvents, examples being $Et_4N^+CF_3CO_2^-$ formed in acetonitrile studied by Forcier and Oliver [44], and formation of triple ions in solutions of $iPe_4N^+NO_3^-$ in chlorobenzene studied by Grunwald and Effio [45], of $Li^+ClO_4^-$ in tetrahydrofuran studied by Jagodzinski and Petrucci [46], of several lithium and tetraalkylammonium salts in solvents having relative permittivities below 10 studied by Salomon and Uchiyama [47], and of tetraalkylammonium halides and tetrafluoroborate in 1, 3-dioxolane studied by Roy and coworkers [48, 49].

However, doubts were cast upon this interpretation of the minimum in the conductivity curve, because effects other than the formation of new conducting species could account for the upturn of the $\Lambda = f(c_E)$ curve. The use of modern conductivity equations, such as (7.29), involves also the consideration of a change of the viscosity of the solution and in particular an increase in the permittivity, due to the presence of polar species (the ions and neutral ion pairs) suggested by Petrucci, by Grigo, and by Salomon and their respective coworkers [42, 50, 51]. Nevertheless, there are theoretical arguments in favor of the formation of triple ions and eventually also of the quadrupoles, not dependent on the interpretation of $\Lambda = f(c_E)$ conductivity curves. These species may be formed not only in the media of low relative permittivity but also in solvents of moderate permittivity if the solvation abilities of the solvents are small (Section 3.3.2), as in the case of acetonitrile as proposed by Hojo and coworkers [52, 53]. The Pitzer treatment of activity coefficients, Equation 7.7 does provide for triple ion formation with the $C\gamma$ term, and was applied to aqueous solutions of large ions: Pr_4N^+ and Br^- by Roy et al. [54]. The hypernetted chain equation does, according to Friedman and Larsen [55], yield nonzero pair correlation functions g_{++} for 1 : 1 electrolytes in ethanol, interpreted as the formation of triple ions.

Spectroscopic observations of triple ion formation are independent of the interpretation of conductivity curves and provide more direct evidence for them. The stretching vibrations ν_{CN} and ν_{CS} in the infrared and Raman spectra of alkali metal thiocyanates in nitromethane and tetrahydrofuran measured by Bacelon et al. [56] provided such evidence. Similarly, the internal vibrations of the trifluoroacetate anion in its lithium salt dissolved in acetonitrile yield evidence for triple ion formation, whereas in the more strongly cation solvating solvent, dimethyl sulfoxide– only ion pairs were detected according to Regis and Corset [57]. More recently, Raman spectra of aqueous magnesium sulfate solutions provided evidence for the $Mg_2SO_4^{2+}$ triple ion according to Rudolph et al. [58], consistent with dielectric relaxation spectroscopic (DRS) results obtained by Buchner et al. [59]. Direct observation of $Ni_2SO_4^{2-}$ obtained from aqueous solutions by electrospray ionization experiments by Schröder et al. [60] is consistent with their detection (as well as that of $Co_2SO_4^{2+}$) in such solutions by DRS according to Chen et al. [61].

Ions having large hydrophobic groups (alkylammonium cations, carboxylate anions, etc.) tend to be driven out from aqueous solutions, and when the hydrophobic group becomes large enough, the solute becomes only poorly soluble in water. The exclusion of the hydrophobic parts of ions is intensified when long-chain nonpolar groups (tails) are attached to polar groups (ionic heads). The resultant solutes are amphiphiles and surfactants, and their sorption at the interface conveys onto the solutions special properties and structures. Typical ionic surfactants are sodium dodecyl sulfonate and cetyltrimethylammonium bromide. This field is outside the scope of this book (but see its discussion by Marcus [62] for a broad outline of the subject and the main concepts involved).

Polyelectrolytes are soluble polymers that carry ionically dissociable groups, ionizing partly or completely to polyions. Polystyrene sulfonic acid $-[CH_2CH$ $(C_6H_5SO_3^-H^+)]_n-$ is a typical example of a strong polyelectrolyte, practically completely ionized. The permittivity of the water near the surface of biological polyelectrolytes (proteins) and also synthetic polyelectrolytes is drastically diminished relative to that of pure water due to the effects of the ionic electric fields. Consequently, the fixed ions ($-SO_3^-$ in the above example) associate partly with the mobile counter ions in the solution. This subject is also outside the scope of this book (but see Ref. 62 for review of it).

7.3 SALTING-IN AND SALTING-OUT

Nonelectrolytes in electrolyte solutions interact with both the free solvent molecules and with the solvated ions and ion associates. A general approach to the study of such interactions is in terms of the solubility of the nonelectrolyte, subscript N, in the solvent in the presence of an electrolyte, $s_{N(E)}$, compared with the solubility in its absence, $s_{N(0)}$. The logarithm of the ratios of these solubilities is generally proportional to the electrolyte concentration, c_E, up to at least 1 M of the latter, a relationship called the Setschenow expression:

$$\log\left[\frac{s_{N(0)}}{s_{N(E)}}\right] = k_{NE}c_E \qquad (7.37)$$

Decadic logarithms are usually employed and the proportionality constant, k_{NE}, depends on both the nonelectrolyte and on the electrolyte. It may be positive (in the more usual cases), describing what is called *salting-out*, or negative, describing *salting-in*. It is noteworthy that there hardly exist studies on the salting-in and salting-out of nonelectrolytes, henceforth called solutes, by electrolytes dissolved in nonaqueous solvents, so the rest of this section deals with aqueous solutions. It is of interest to explore the dependence of the sign and magnitude of the Setschenow constants k_{NE} on the properties of the solutes and of the ions constituting the electrolyte as well as on the properties of the solutes.

7.3.1   Empirical Setschenow Constant Data

Empirical expressions have been reported for presenting the Setschenow constants for a large variety of solutes and ions. Weisenberger and Schumpe [63] presented an expression for gaseous solutes that do not interact chemically with the solvents (excluding, e.g., NH_3, HCl, and CO_2 in aqueous solutions):

$$k_{NE} = \sum_I \nu_I \left[k_{NI} + k_{N25} + h_{N25}((t/°C) - 25) \right] \tag{7.38}$$

The summation extends over all the ionic species present in the (aqueous) solution, including cases of electrolyte mixtures, k_{NI} is an ion-specific parameter, relatively independent of the temperature, and k_{N25} and h_{N25} are gas-specific parameters. Because of the proportionality in Equation 7.37, the k_{NE} values are valid down to infinite dilution, hence should be additive with respect to the individual ions. Conventional ionic values of k_{NI}, based on the convention that $k_{NI}(H^+, aq) = 0$ and on the basis that also $k_{N25}(O_2) = 0$, have been reported [63], as well as values of k_{N25} and h_{N25} for several gases. The values of k_{NI}, pertaining to the salting of O_2 at 25°C, are reproduced in Table 7.6 and are approximately proportional to the charge numbers of the ions—for 19 cations $k_{NI}/z_+ = 0.080 \pm 0.002$ (Na$^+$ and Ba^{2+} are outliers) and for 25 anions $k_{NI}/z_- = -0.072 \pm 0.007$. Within this approximate proportionality, the k_{NI} diminish with increasing sizes of the ions. The values of k_{N25} generally increase with the sizes of the gaseous solute molecules, for example, from -0.0353 for He to $+0.0133$ for Xe and from 0.0022 for CH_4 to 0.0240 for C_3H_8.

Xie et al. [66] reported the salting constants for benzene in aqueous electrolyte solutions, $k_{PhH,E}$, and on the basis of the convention that $k_{PhH}(H^+, aq) = 0$ ionic values could be obtained from the data, as shown in Table 7.6. It is noted that several ions have a salting-in effect: iodide and perchlorate mildly and tetraalkylammonium ions strongly. Similar constants have also been reported by Xie et al. [64, 66] for other organic compounds, and several theories have been explored in order to systematize these constants.

Ni and Yalkowski [67] presented extensive data for the salting-out of various liquid and solid organic solutes in aqueous sodium chloride solutions. The solutes included not only small organic molecules with various functional groups but also solutes with large molecules, such as drugs.

An unsophisticated view of the change of solubility of the solute in the presence of an electrolyte is in terms of the diminution of the amount of free solvent available for the solute to dissolve in, because some of the solvent is bound by the ions, solvating them. This view, implicit in the papers by Zavitsas [16, 17] but not stated there, invariably leads to salting-out of the solute and cannot explain instances of salting-in and ignores any interactions of the solute with the ions.

7.3.2   Interpretation of Salting Phenomena

Debye and McAulay [68] presented an electrostatic theory according to which the salting constant is proportional to the product of several factors. These are the differences between the permittivity of water and that of the solute, $\varepsilon_W - \varepsilon_N$, the molar

TABLE 7.6 Empirical Setschenow Constants and the Ionic Electrostriction Parameter

Ion	k_{NI} [63]	$k_{PhH,I}$ [64]	ΔV_I [65]
H^+	0	0	
Li^+	0.0754	0.063	2.2
Na^+	0.1145	0.106	5.5
K^+	0.0922	0.077	1.6
Rb^+	0.0839	0.061	−0.8
Cs^+	0.0759	0.009	−8.9
NH_4^+	0.0556	0.024	−4.9
Me_4N^+		−0.183	−34.2
Et_4N^+		−0.246	−52.6
Pr_4N^+		−0.373	−77.1
Bu_4N^+		−0.568	−117.1
Mg^{2+}	0.1694	0.143	22.7
Ca^{2+}	0.1762		21.9
Sr^{2+}	0.1881	0.145	24.0
Ba^{2+}	0.2168	0.207	22.6
Mn^{2+}	0.1463		20.0
Fe^{2+}	0.1523		24.8
Co^{2+}	0.1680		27.7
Ni^{2+}	0.1654		31.0
Cu^{2+}	0.1675		29.5
Zn^{2+}	0.1537		28.5
Cd^{2+}	0.1869		16.9
Al^{3+}	0.2174		45.7
Cr^{3+}	0.0648		37.8
Fe^{3+}	0.1161		37.7
La^{3+}	0.2297		44.3
Ce^{3+}	0.2406		44.4
Th^{4+}	0.2709		58.0
F^-	0.0920	0.151	10.7
Cl^-	0.0318	0.084	6.0
Br^-	0.0269	0.054	5.6
I^-	0.0039	−0.006	0.7
OH^-	0.0839		10.0
HS^-	0.0851		−1.2
CN^-	0.0679		7.5
SCN^-	0.0627	0.004	11.3
NO_2^-	0.0795		8.3
NO_3^-	0.0128	0.018	6.0
ClO_3^-	0.1348	0.020	4.5
BrO_3^-	0.1116		−17.5
IO_3^-	0.0913		−10.4
ClO_4^-	0.0492	−0.005	−9.2
HCO_3^-	0.0967		−13.8
HSO_3^-	0.0549		−20.9

(continued)

TABLE 7.6   (Continued)

Ion	k_{NI} [63]	$k_{PhH,I}$ [64]	ΔV_I [65]
HSO_4^-			−18.4
$H_2PO_4^-$	0.0906		−11.5
HCO_2^-		0.065	7.9
$CH_3CO_2^-$		0.064	10.3
CO_3^{2-}	0.1423		18.5
SO_3^{2-}	0.1270		11.3
SO_4^{2-}	0.1117	0.310	16.7
$S_2O_3^{2-}$	0.1149		5.4
HPO_4^{2-}	0.1499		−11.5
PO_4^{3-}	0.2119		48.1
$Fe(CN)_6^{4-}$	0.3574		133.8

volume of the solute, and the sum of the squares of the charge numbers of the ions divided by their radii. A modification of this theory by Givon et al. [69] showed that instead of the ionic radii, the molar dielectric decrement of the electrolyte, δ_E, divided by its hydrated volume $\left(V_E^\infty + h^\infty V_W^*\right)$, should be employed in addition to $\varepsilon_W - \varepsilon_N$. Such approaches predict salting-in only for such solutes that have permittivities larger than that of water, such as HCN or formamide, for which $\varepsilon_W - \varepsilon_N < 0$. For most solutes, that is, those with $\varepsilon_W - \varepsilon_N > 0$, salting-out should occur for any electrolyte, even salts of tetraalkylammonium that are known to salt-in benzene and other solutes. Some other theories have been considered by Xie et al. [66], but it appears that the most successful one is that indirectly dependent on the internal pressure/electrostriction caused by the electrolyte.

This theory is traceable in principle to that of McDevit and Long [31], the operative quantity of which is the difference between the intrinsic molar volume, V_{Eintr}, and the standard partial molar volume, V_E^∞, of the electrolyte. According to Ref. 31:

$$k_{NE} = \frac{V_N(V_{Eintr} - V_E^\infty)}{\ln(10)RT\kappa_T} \tag{7.39}$$

Here, $V_{Eintr} - V_E^\infty$ may be replaced by $\sum_I \nu_I \Delta V_I$, the weighted sum of the ionic electrostriction volumes. Conventional ionic electrostriction volumes, $\Delta V_I = V_{Iintr} - V_I^{\infty conv}$, for the convention that $V_I^{\infty conv}(H^+) = 0$, are shown in Table 7.6, the molar intrinsic volumes being calculated according to Mukerjee's expression for small ions, $V_{Iint}/cm^3 mol^{-1} = 2522(1.213 r_I/nm)^3$, but with the bare ionic radii for large, poorly hydrated ions, $V_{Iint}/cm^3 mol^{-1} = 2522(r_I/nm)^3$ as suggested by Marcus [65]. It turns out, however, that the salting constants calculated according to Equation 7.39 for benzene and similar solutes overestimate the experimental values by a factor of ~3 as shown by Deno and Spink [70], because they ignore direct interactions of the solutes with the ions that do occur beyond the effects of the molar volumes of the solutes, V_N, as related to the change in the internal pressure of the solvent caused by the ions. Properties of the solutes additional to V_N, used in Equation 7.39, are required. One approach to this problem, that of Ni and Yalkowski [67], employed the partition coefficients of the solutes between water and 1-octanol, P_N, quantities known for a large number of solutes or that can be calculated from group contributions. Their operative expression for the salting of 101 solutes from aqueous sodium chloride solutions is:

$$k_{N,NaCl} = 0.114 + 0.040 \log P_N \tag{7.40}$$

However, the quantities P_N already include the interactions of the solutes with water, but properties foreign to such interactions are preferable for predictive purposes. Two such properties have been proposed and tested by Marcus [65]: the Kamlet–Taft polarity/polarizability index π^* (Section 3.3.2.1) and the Hildebrand solubility parameter δ_H (Section 3.2.2). These as well as the intrinsic volumes of the solutes, expressed as their McGowan volumes V_X, are independent of interactions with water, hence true predictive parameters that are known or calculable from group contributions for a large number of solutes. The predictive expression (for ambient temperatures) is:

$$k_{NE} = -0.10 + 10^{-4}(2.0 + 0.729V_X - 6.6\delta_H)\sum_I \gamma_I \Delta V_I \tag{7.41}$$

and a similar one can be written with π^* replacing δ_H (but with a different coefficient) or with some other measure of the intrinsic volume of the solute, for

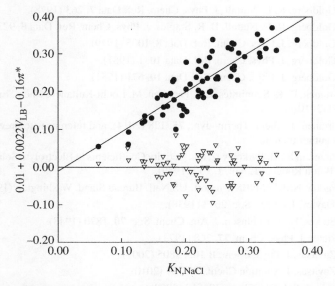

FIGURE 7.6 Predicted salting coefficients of various solutes by aqueous sodium chloride according to Equation 7.42 (——) plotted against the experimental values [67] (●) and the differences (∇) (From Ref. 65 by permission from the publisher, Elsevier).

example, its LeBas molar volume, V_{LB}, used by Ni and Yalkowski [67], as shown by Marcus [65]:

$$k_{NE} = -0.10 + 10^{-4}(0.94V_{LB} - 68\pi^*)\sum_I v_I \Delta V_I \tag{7.42}$$

The direct solute–ion interactions expressed by the terms with δ_H or π^* counteract the effect of the volume of the solute that disrupts the hydrogen bonding of the water solvent. They may eventually cause salting-in where without them salting-out would have been predicted for a given electrolyte for which $\sum_I v_I \Delta V_I$ is positive but rather small. Comparison with experimental data shows fits with an uncertainty in k_{NE} of $\pm 0.05\,M^{-1}$, commensurate with the uncertainty of the experimental data, see Figure 7.6. The temperature dependence of the parameters is mild, hence Equations 7.41 and 7.42 should be valid also for other temperatures than 25°C, for which the parameters are generally listed. This was demonstrated [65] for the salting behavior of chloroform, benzene, chlorobenzene, and anisole studied by Görgenyi et al. [71] in terms of their Henry's law constants at 40°C in two dozen aqueous electrolytes.

REFERENCES

[1] H. S. Harned, B. B. Owen, The Physical Chemistry of Electrolyte Solutions, McGraw-Hill, New York, 3rd ed. (1958).

[2] R. A. Robinson, R. H. Stokes, Electrolyte Solutions, Butterworths, London, 2nd ed. (1959).

[3] W. J. Hamer, Y.-C. Wu, J. Phys. Chem. Ref. Data **1**, 1047 (1972).

[4] R. N. Goldberg, R. L. Nuttall, J. Phys. Chem. Ref. Data **7**, 263 (1978).

[5] R. N. Goldberg, R. L. Nuttall, B. R. Staples, J. Phys. Chem. Ref. Data **8**, 923 (1979).

[6] R. N. Goldberg, J. Phys. Chem. Ref. Data **8**, 1005 (1979).

[7] R. N. Goldberg, J. Phys. Chem. Ref. Data **10**, 1 (1981).

[8] R. N. Goldberg, J. Phys. Chem. Ref. Data **10**, 671 (1981).

[9] J. I. Partanen, R. K. Salmimies, L. J. Partanen, M. Louhi-Kultanen, Chem. Eng. Technol. **33**, 730 (2010).

[10] J. I. Partanen, J. Chem. Thermodyn., **75**, 128 (2014), and references to previous reports by this author therein.

[11] K. S. Pitzer, in R. M. Pytkowycz, ed., Activity Coefficients in Electrolyte Solutions, CRC Press, Baton Rouge, FL, Vol. **1**, 157 (1979).

[12] V. B. Parker, NSRDS-NBS-2 report, US Natl. Bureau Stand, Washington (1965).

[13] C. W. Davies, J. Chem. Soc. 2093 (1938).

[14] R. H. Stokes, R. A. Robinson, J. Am. Chem. Soc. **70**, 1870 (1948).

[15] K. S. Pitzer, J. Phys. Chem. **77**, 268 (1973).

[16] A. A. Zavitsas, J. Phys. Chem. B **105**, 7805 (2001).

[17] A. A. Zavitsas, J. Solution Chem. **39**, 301 (2010).

[18] Y. Marcus, J. Solution Chem. **38**, 513 (2009).

[19] Y. Marcus, J. Solution Chem. **34**, 297 (2005).

[20] R. A. Anderson, J. Am. Chem. Soc. **68**, 686 (1946).

[21] Y. Marcus, J. Phys. Chem. B **118**, 10471 (2014).

[22] J. Padova, J. Chem. Phys. **40**, 691 (1964).

[23] M. Eigen, K. Tamm, Z. Elektrochem. **66**, 93, 107 (1992).

[24] R. Buchner, Pure Appl. Chem. **80**, 1239 (2008).

[25] P. Hemmes, J. Phys. Chem. **76**, 895 (1972).

[26] R. M. Diamond, J. Am. Chem. Soc. **80**, 4808 (1958).

[27] R. M. Diamond, J. Phys. Chem. **67**, 2513 (1963).

[28] N. Bjerrum, K. Dan, Vidensk. Selsk. **7**, 9 (1926).

[29] Y. Marcus, G. Hefter, Chem. Rev. **106**, 4585 (2006).

[30] Y. Marcus, J. Mol. Liq. **123**, 8 (2006).

[31] W. F. McDevit, F. A. Long, J. Am. Chem. Soc. **74**, 1773 (1952).

[32] Y. Marcus, H. B. D. Jenkins, L. Glasser, J. Chem. Soc. Dalton Trans. 3795 (2002).

[33] J. G. Kirkwood, Chem. Rev. **24**, 233 (1939).

[34] M. Khoshkbarchi, J. H. Vera, Ind. Eng. Chem. Res. **35**, 4319 (1996).

[35] Y. Marcus, unpublished results (2014).

[36] M. C. Justice, J.-C. Justice, J. Solution Chem. **6**, 819 (1977).

[37] A. E. Martell, R. J. Motekaitis, The Determination and Use of Stability Constants, VCH, New York (1992).

[38] F. Fernandez-Prini, J.-C. Justice, Pure Appl. Chem. **56**, 541 (1984).

[39] J. Barthel, H. Hetzenauer, R. Buchner, Ber. Bunsenges. Phys. Chem. **96**, 1424 (1992).

[40] Y. Marcus, Z. Naturforsch. A **38**, 247 (1983).

[41] Y. Marcus, J. Solution Chem. **16**, 735 (1987).

[42] S. Petrucci, M. C. Masiker, E. M. Eyring, J. Solution Chem. **37**, 1031 (2008).

[43] R. M. Fuoss, C. A. Kraus, J. Am. Chem. Soc. **55**, 2387 (1933).

[44] G. A. Forcier, J. W. Oliver, Electrochim. Acta **15**, 1609 (1970).

[45] E. Grunwald, A. Effio, J. Solution Chem. **2**, 393 (1973).

[46] P. Jagodzinski, S. Petrucci, J. Phys. Chem. **76**, 917 (1974).

[47] M. Salomon, M. C. Uchiyama, J. Solution Chem. **16**, 21 (1987).

[48] A. Sinha, M. N. Roy, J. Mol. Liq. **140**, 39 (2008).

[49] I. Banik, M. N. Roy, J. Chem. Thermodyn. **63**, 52 (2011).

[50] M. Grigo, J. Solution Chem. **11**, 529 (1982).

[51] M. Salomon, S. Slane, E. Plichta, M. C. Uchiyama, J. Solution Chem. **18**, 977 (1980).

[52] M. Hojo, T. Takiguchi, M. Hagiwara, H. Nagai, Y. Imai, J. Phys. Chem. **93**, 955 (1989).

[53] M. Hojo, Pure Appl. Chem. **80**, 1539 (2008).

[54] R. N. Roy, J. J. Gibbons, R. Snelling, J. Moeller, T. White, J. Phys. Chem. **81**, 391 (1977).

[55] H. L. Friedman, B. Larsen, Pure Appl. Chem. **51**, 2147 (1979).

[56] P. Bacelon, J. Corset, C. de Loze, J. Solution Chem. **9**, 129 (1980).

[57] R. Regis, J. Corset, Chem. Phys. Lett. **32**, 462 (1975).

[58] W. W. Rudolph, G. Irmer, G. T. Hefter, Phys. Chem. Chem. Phys. **5**, 5253 (2003).

[59] R. Buchner, T. Chen, G. Hefter, J. Phys. Chem. B **108**, 2365 (2004).

[60] D. Schröder, L. Duchackova, J. Tarabek, M. Karwowska, K. J. Fijalkowski, M. Ončak, P. Slavček, Am. Chem. Soc. **133**, 2444 (2011).

[61] T. Chen. G. Hefter, R. Buchner, J. Solution Chem. **34**, 1045 (2005).

[62] Y. Marcus, Ions in Water and Biophysical Implications, Springer, Dordrecht (2012).

[63] S. Weisenberger, A. Schumpe, AIChE J. **42**, 298 (1996).

[64] W.-H. Xie, J.-Z. Su, X.-M. Xie, Thermochim. Acta **169**, 271 (1990).

[65] Y. Marcus, J. Mol. Liq. **177**, 7 (2013).

[66] W.-H. Xie, W.-Y. Shiu, D. Mackay, Mar. Environ. Res. **44**, 429 (1997).

[67] N. Ni, S. H. Yalkowski, J. Pharm. Sci. **254**, 167 (2003).

[68] P. Debye, J. McAulay, Phys. Z. **26**, 22 (1925).

[69] M. Givon, Y. Marcus, M. Shiloh, J. Phys. Chem. **67**, 2495 (1963).

[70] N. C. Deno, C. H. Spink, J. Phys. Chem. **67**, 1347 (1963).

[71] M. Görgenyi, J. Dewulf, H. Van Langenhove, K. Heberger, Chemosphere **65**, 802 (2006).

8

APPLICATIONS OF SOLUTIONS OF IONS

Ions in solution play important roles in many fields of science and the interaction of the ions with the solvent (their solvation) or with solvent mixtures (leading to preferential solvation) affects the use of such solution. So does also the interaction of ions with other solutes, be they other ions, leading to ion association, or nonionic solutes, causing them to be salted-out (or -in, in rare cases).

When solutions of ions are needed either for studying them or for using them, the primary requirement is that their source, generally solid electrolytes but in few cases liquids (e.g., sulfuric acid, yielding hydrogen, hydrogen sulfate, and sulfate ions) or gases (e.g., ammonia, yielding ammonium, and hydroxide (in water) ions), be soluble to sufficient extent in the solvent (or solvent mixture) to be employed. The solvation of the ions produced on ionic dissociation of these sources or their direct reaction with the solvent is the key process that leads to the needed sufficient solubility. If the ion of interest is only poorly solvated, then its counter ion in the electrically neutral electrolyte needs to be well solvated, since otherwise the solubility will be too small for practical purposes. For instance, the reference electrolyte that is often used in order to obtain individual ionic properties, namely tetraphenylarsonium or tetraphenylphosphonium tetraphenylborate (TATB or TPTB) (Sections 2.3.1.1, 4.2.3, and 4.3.1), is only sparingly soluble in water and similar solvents, because both ions are very poorly solvated by such solvents due to their bulk, shielded electric charges, and hydrophobic surfaces. On the contrary, salts such as tetraphenylarsonium chloride and sodium tetraphenylborate are well soluble in water, because the chloride and sodium ions in these salts are sufficiently well solvated to overcome the electrostatic forces holding the ions together in the crystalline solid, expressed by its lattice energy, and so permit the ions to dissolve and move away from each other.

Ions in Solution and Their Solvation, First Edition. Yizhak Marcus.
© 2015 John Wiley & Sons, Inc. Published 2015 by John Wiley & Sons, Inc.

Once the ions are in solution at an appreciable concentration, they can manifest usefully their properties for the required applications. In the following, some applications of ions, being dependent on their solvation, in a number of fields of chemistry and technology are dealt with. They are illustrated by representative examples from the extensive literature, but by no means is any exhaustive or even systematic discussion of such applications within the scope of this book.

8.1 APPLICATIONS IN ELECTROCHEMISTRY

8.1.1 Batteries and Supercapacitors

Electrochemical devices have come in recent years to the forefront in many applications. An example is the provision of electrical energy for electrical vehicles, where high energy storage density is provided by (rechargeable) batteries. Another example is pulsed lasers, where high power, as delivered by supercapacitors, is needed.

Batteries employ electrodes, an electrolyte solution, and some auxiliary materials, and the proper choice of the materials is essential for achievement of economic use. It is interesting to note that the first rechargeable battery was the lead-acid accumulator, invented in 1859 by Gaston Plantè. Nowadays it is in extensive use where its bulk and weight are of little consequence, although it has a rather low energy storage density, $30–50\,Wh\cdot kg^{-1}$, a nominal voltage of $2\,V$, but on the other hand a cycling ability of 200–300 cycles. It employs Pb and PbO_2 as the electrodes and aqueous sulfuric acid ($\sim 4.2\,M$) as the electrolyte. The total reaction during discharge can be written as $Pb(s) + PbO_2(s) + 2H_2SO_4(aq) \rightarrow 2PbSO_4(s) + 2H_2O(aq)$ that is reversed during charging, but the ions involved are $H^+(aq)$ and $HSO_4^-(aq)$. Another type of aqueous battery in extensive use (as a primary, not rechargeable battery) is the alkaline battery, where the reaction during discharge is $Zn(s) + 2MnO_2(s) \rightarrow ZnO(s) + Mn_2O_3(s)$ with aqueous KOH as the electrolyte, providing $K^+(aq)$ and $OH^-(aq)$ ions. Its initial energy storage density is $80\,Wh\cdot kg^{-1}$ and it has a nominal voltage of $1.5\,V$. The same electrolyte, aqueous KOH, is used in rechargeable nickel metal hydride (NiMH) batteries, where the overall reaction is $M + Ni(OH)_2 \rightarrow MH + NiO(OH)$, where M is a suitable metal. The energy storage density is $60–120\,Wh\cdot kg^{-1}$, the cycling ability is 300–500 cycles, and the nominal voltage is $1.2\,V$. However such batteries are prone to self-discharge of up to 30% per month (compared with $\sim 5\%$ for lead acid and $\sim 0.5\%$ for alkaline batteries).

For applications where high energy storage density is required, lighter materials are preferably used, as in lithium ion batteries widely employed in portable electronic devices. The electrodes are $LiCoO_2$ (or $LiFePO_4$ or several other oxides containing lithium) and carbon (graphite) that intercalate lithium metal. The electrolyte is a salt of lithium, such as $LiBF_4$, $LiPF_6$, or $LiClO_4$, and large anions are used in order to minimize the lattice energy of the electrolyte, and hence increase its solubility in the solvent employed. Lithium metal reacts with water, and therefore aprotic non-aqueous solvents are used in such batteries. From the point of view of this book on the solvation of ions, following are criteria for the choice of the solvent to be used in the electrochemical devices.

a) A sufficiently high electric permittivity that ensures essentially complete ionic dissociation of the dissolved electrolyte and avoidance of its aggregation.

b) A low viscosity over the temperature range of the intended application of the device, in order to ensure adequate mobility of the ions.

c) A sufficiently high solvating power for both the cation and anion of the electrolyte for achievement of its adequate solubility. For small cations this requires good electron pair donicity and for large anions this requires large ability to interact via dispersion forces.

d) Stability against attack by reactive elements in the electrode materials and possibly present depolarizing materials. If not stability in the thermodynamic sense, then at least in the kinetic sense, involving passivation.

e) A suitable liquid range for applications involving extreme temperatures, such as for energy storage in remote cold places.

f) A suitable electrochemical voltage window for stability against oxidation or reduction.

g) A low vapor pressure to avoid loss by evaporation and possible explosion hazards in unvented systems.

h) Ready availability, low cost, ease of purification, low flammability, and non-toxicity.

Chapter 3 should be consulted regarding the properties of solvents that may meet the criteria (a) to (c) and (e) to (g). Adequate solubility of the electrolyte is at least 0.3 M. Solvents that meet one criterion may not meet another: a solvent with good solvating power may have a too large viscosity. In such cases, mixtures of solvents provide the overall best properties. Ethers, such as tetrahydrofuran, 1,3-dioxolane, or dimethoxyethane may be used for provision of the low viscosity of the mixture, in spite of their low permittivities, and certain esters, such as ethylene carbonate, dimethyl carbonate, or γ-butyrolactone, provide the needed permittivity and solvating ability. The energy storage density achievable for lithium ion batteries reaches up to $260\,Wh\cdot kg^{-1}$ and the cycling ability reaches 1000 cycles at 3.2 V and the self-discharge is ~2% per month, increasing with the temperature.

Some recent developments are directed toward improvement of some features of such batteries. The amount of storable energy is proportional to the operative voltage of the battery and to the number of electrons involved in the cell discharge/charge reaction. Therefore, solvents that are stable to oxidation at higher potentials vs. the Li/ Li+ electrode than ethers and esters have been sought. Methyl ethyl (or methoxyethyl) sulfone is a good candidate proposed by Angell and coworkers [1, 2] and adiponitrile proposed by Abu-Lebdeh and Davidson [3] is another one. This sulfone withstands a voltage of $\geq 5.5\,V$ vs. the Li/Li+ electrode with $LiClO_4$ or $Li((CF_3)_2SO_2)_2N$ (bis(trifluoromethylsulfinyl)imide) as the electrolytes, whereas adiponitrile mixtures with ethylene carbonate with the latter salt showed good performance up to 4.4 V.

It is remarkable that lithium ion batteries can even operate with an aqueous electrolyte, when lithium metal is avoided. A cell with a $LiTi_2(PO_4)_3$ electrode against the well-known $LiFePO_4$ one with an aqueous Li_2SO_4 electrolyte is usable for 1000 cycles, provided that air (oxygen) is strictly excluded according to Luo et al. [4]. Another way to increase the energy storage density is to use multi-electron reactions but light-weight materials, possibly magnesium or aluminum, but so far such efficient batteries have not been commercialized as reported by Gao and Yang [5].

Table 8.1 compares the features of three types of widely used batteries: lead acid, nickel metal hydride, and lithium ion, and the advantages and disadvantages become more apparent.

Supercapacitors (electrical double layer capacitors, EDLCs) are distinguished from batteries in that they are able to store electrical energy that can be delivered very fast, that is, they have a large power generating ability, some 10- to 100-fold of that of batteries, but not necessarily a large energy storage capacity, mostly only ca. 10% of that of batteries. They consist of two electrodes separated at a small distance (0.3–0.8 nm) by an electrolyte that provides a Helmholz double layer. The electrodes have very large surface areas, made either of graphite, other (activated) carbon forms, or of metal oxides. The energy density of EDLCs is proportional to the square of the operative voltage, and hence the electrochemical stability window of the electrolyte/solvent system is of prime importance.

The choice of the electrolyte devolves around its adequate solubility on the one hand and avoidance of ion pairing in the resultant solutions on the other in order to provide adequate conductivity of the solution. Electrolytes, such as H_2SO_4, KOH, $NaClO_4$, $LiClO_4$, $LiAsF_6$, or quaternary phosphonium salts may be dissolved in water as the solvent to produce useful EDLCs. However, the use of aqueous solutions limit the voltage of the supercapacitor to 2.3 V. Organic solvents provide a somewhat larger applicable voltage, 2.7 V, and a wider temperature range than obtainable with aqueous solutions. The criteria listed above for the selection of the solvent apply to supercapacitors as well.

Acetonitrile and propylene carbonate (PC) are currently widely used in commercial devices. Acetonitrile-based electrolytes have high electrochemical stability and high conductivities (due to low viscosity) even at low temperatures (down to −40°C). On the other hand, this solvent has a low flashpoint (6°C) and is toxic and hence

TABLE 8.1   Comparison of Features of Widely Used Batteries

Feature	Lead acid	NiMH	Li ion
Energy storage density, $Wh\,kg^{-1}$	30–40	60–120	110–260
Cyclability (to 80% of initial capacity)	200–300	300–500	500–1000
Charging time, h	8–16	2–4	2–4
Overcharge tolerance	High	Low	Very low
Self-discharge per month, %	5	30	2
Nominal cell voltage, V	2.0	1.2	3.6
Operating temperature,°C	−20 to 60	0 to 40	−20 to 60
Cost	Medium	Low	High

presents safety limitations. PC has an appreciably higher flashpoint (135°C) but shows poor conductivities at low temperatures, while altogether it is considered to be a safer solvent.

Acetonitrile and PC were tested with a quaternary ammonium tetrafluoroborate as the electrolyte in devices described by Beguin et al. [6]. The ions of the electrolyte need to be able to enter the pores of the electrodes and hence should not be too large, or else must be (partly) desolvated in order to fit the pores of the activated carbon electrodes. As for batteries, asymmetric linear sulfones and dinitriles may permit operation at higher voltages than acetonitrile and PC permit.

Solvent mixtures may be advantageous in these respects, and mixtures of ethylene carbonate (EC) and 2-methoxypropionitrile have been proposed by Perricome et al. [7], for example, as superior solvents for the commonly used Et_4NBF_4 electrolyte. The use of solvent mixtures was taken a step further, leading to the quaternary mixture of one acetate and three carbonate esters, permitting with $1\,M\;NaPF_6$ electrolyte operation at -30 to $60°C$ at $3.2\,V$ with a capacity of $120\,Fg^{-1}$ according to Vali et al. [8]. Many other combinations of solvents (and their mixtures) and electrolytes can be found in recent publications, but the main point is that appropriate choice of electrolyte/solvent/electrode material is essential for improving the performance of present day commercial applications of EDLCs as supercapacitors for high-power devices.

8.1.2 Solvent-Independent pH and Electrode Potential Scales

The activity of hydrogen ions in a solution, $a\left(H^+\right)$, such a widely used concept, cannot be measured directly because it pertains to the thermodynamic quantity of an individual ion. Therefore, $pH = -\log a\left(H^+\right)$ of the solution must be determined operationally in terms of the method employed for its measurement. It is related to internationally agreed-on and defined standards. The pH is generally measured electrochemically in a cell such as:

$$\text{Reference electrode} \parallel \text{solution} \mid \text{measuring electrode} \qquad (8.1)$$

where $\parallel$ denotes a salt bridge producing a liquid junction and $\mid$ denotes a phase boundary. If the electromotive force (EMF) of the cell with the standard solution (standard buffer) is E_s and that of the solution, the pH of which is to be measured in the same cell, is E, then:

$$pH = pH_s + \left(E - E_s\right)\left(\frac{F}{RT\ln(10)}\right) \qquad (8.2)$$

Several operational standard buffers have been defined and their pH values assigned by the use of cells without liquid junctions according to Covington et al. and Kristensen et al. [9, 10]. The primary standard, $0.05\,m$ aqueous potassium hydrogen phthalate (KHPh) was established. One implementation of the cell used for assignment of the reference standard pH is:

$$Pt(s), H_2\left(g, P°\right) \mid KHPh(S)\left(m_{KHPh}\right) + KCl(S)\left(m_{KCl}\right) \mid AgCl(s), Ag(s) \qquad (8.3)$$

where S is the solvent and for aqueous solutions the solvent S=W. For this cell

$$pH = \lim \left(m_{KCl} \to 0\right)\left[\left(E - E°(Ag, AgCl)\right)\left(F/RT \ln(10) + \log m_{KCl}\right)\right] + \log \gamma_{Cl}$$

(8.4)

This expression, however, involves the activity coefficient of an individual ion, γ_{Cl}, that cannot be measured but is assigned *conventionally* (the Bates–Guggenheim convention) as:

$$\log \gamma_{Cl} = \frac{-A(T)I^{1/2}}{1 + 1.5I^{1/2}}$$

(8.5)

where A is the Debye Hückel theoretical coefficient (Section 7.1.1, the first term on the left-hand side of Equation 7.11 and the numerical constant represents B). The assigned reference value standard is $pH_s = 4.005$ at 25°C, and the theoretical slope for E_s of 59.159 mV·K^{-1} is used for other temperatures, endorsed by IUPAC (International Union of Pure and Applied Chemistry). The conventional value 1.5 in the denominator of Equation 8.5 is temperature-independent, so it causes the closest approach distance of the ions a (Equation 7.13) to depend on the temperature, and covers up for the lack of a linear term in the ionic strength that involves the hydration of the ions (see Section 7.1). The thus-defined pH scale [9, 10], endorsed by IUPAC, has an inherent uncertainty of ±0.02 units, of which ±0.01 units are due to the Bates-Guggenheim convention regarding γ_{Cl}, applicable to aqueous solutions having $2 \leq pH \leq 12$ and ionic strengths $I = \frac{1}{2}\sum z_i^2 m_I$ below a few tenths of 1 m. The practical use of glass (membrane) electrodes and reference electrodes with salt bridges and liquid junctions and the uncertainties to be expected in pH measurements in such aqueous solutions are most recently described by Buck et al. [11].

These considerations, however, do not apply to aqueous solutions beyond the above-stated limits of pH and concentrations and to nonaqueous and mixed aqueous–nonaqueous solutions, for which the nominal hydrogen ion activity, that is, the pH, is required. Sea water is one example of aqueous solutions for which special treatment is needed and highly saline brines is another one. Standard sea water of salinity 0.35% has an ionic strength of about 0.7 m and at 25°C has, because of this, pH=pH(dilute aqueous solutions)+0.076 as shown by Bates [12], provided that the cell is standardized with the same electrodes against the standard buffer used in dilute aqueous solutions described above.

As commonly practiced, cells with liquid junctions to the reference electrode, for example, a saturated calomel electrode (SCE), are employed in pH measurements. Therefore, the liquid junction potential occurring for the standardizing buffer measurement, E_{js}, and that in a highly saline water or brine, subscript x, the pH of which is to be measured, E_{jx}, must be taken into account. A term $(E_{jx} - E_{js})(F/RT \ln(10))$ must be added on the right-hand side of Equation 8.2, both E_j terms of which are not well known but are not expected to cancel out entirely in the difference. On rearranging, we get:

$$pH_x = \left[\left(pH_s - E_{js}\right)\left(\frac{F}{RT \ln(10)}\right)\right] + \left(E_x - E_s\right)\left(\frac{F}{RT \ln(10)}\right) + E_{jx}\left(\frac{F}{RT \ln(10)}\right)$$

(8.6)

The first term (the one in square brackets) is a constant but is unknown and the last term may be negligible for certain well-designed liquid junctions. This is, because the liquid junction potential between the generally 3.5 m aqueous KCl in the salt bridge (used also in the SCE reference electrode) and the saline solution should be quite small. The avoidance of a liquid junction by dipping the reference electrode, Ag(s), AgCl(s), directly in the standard and test solutions cannot solve the problem. This is because of the appreciable solubility of AgCl in concentrated chloride solutions and the "poisoning" of this reference electrode by bromide or sulfide ions that may be present in the saline water, as pointed out by Marcus [13]. However, a reproducible liquid junction between the salt bridge solution and the test solution can be achieved by proper design of the junction. The pH of the standard, 0.04 m equimolar tris (tris(hydroxymethyl)methanamine hydrochloride) buffer solution in synthetic sea water (devoid of weak acids and bases) should be used for calibration according to Khoo et al. [14]. This method results in reproducible values, but the meaning of the resulting pH in terms of hydrogen ion concentration or activity is obscure. Else, 0.01 m HCl in the brine can be used to establish its $-\log m_H = 2.00$ for calibration purposes and subsequent measurements of other acidities in this brine are then in terms of the hydrogen ion molality as proposed by Marcus [13].

Measurements of the pH in aqueous–organic solvent mixtures and in nonaqueous mixtures pose similar problems, but the need for such data is extensive, for instance for mobile phases used in chromatography, for electrochemistry, and for corrosion studies. When in cell (8.3) the solvent S is not pure water but its mixture with another solvent or a non-aqueous solvent altogether, Equation 8.4 still holds, but the activity coefficient of the chloride ion must be changed, using the modified Bates-Guggenheim convention:

$$\log \gamma_{Cl} = \frac{-A(S,T)I^{1/2}}{\left[1 + 1.5\left(\varepsilon_w \rho_S / \varepsilon_S \rho_w\right)I^{1/2}\right]} \tag{8.7}$$

where $A(S,T) = 1.8246 \times 10^6 \left(\varepsilon_S T\right)^{-3/2}$ according to Mussini et al. [15]. A problem exists in cases of solvents with sufficiently low permittivities ε_S, where extensive ion pairing takes place, since then the ionic strength I in Equation 8.7, which pertains only to the free ions, cannot be readily calculated from the nominal composition. The difference δ_w^S in the resulting reference pH of cell (8.3) with 0.05 m KHPh buffer at 25°C in several aqueous solvent mixtures S from that in water (4.005) can be described with quadratic expressions in the mole fractions of several cosolvents $\delta_w^S = a x_S + b x_S^2$ up to some maximal x_S. The coefficients of this expression are shown in Table 8.2. The quantity δ_w^S is, in fact, the relationship of pH values measured in the solvent S $\left(_S^S pH\right)$ relative to that measured in water $\left(_w^S pH\right)$, and is the transfer activity coefficient for hydrogen ions from water to S, $_w^S \gamma_{H^\infty}$, or the so-called primary solvent effect according to Rosès [16]:

$$_S^S pH = _w^S pH + \log _w^S \gamma_H^\infty = \frac{_w^S pH + \Delta_{tr} G^\infty\left(H^+, W \to S\right)}{RT \ln(10)} \tag{8.8}$$

TABLE 8.2   The Coefficients of the Expression $\delta^S_{\ W} = ax_S + bx_S^2$ for the Difference Between the Reference pH of (0.05 mol (kg solution)$^{-1}$ KHPh buffer) at 25°C in Several Aqueous Solvent Mixtures S from that in Water (4.005), According to Data in [15]

Cosolvent S	a	b	x_{max}
Methanol	3.03	−0.19	0.75
Ethanol	5.95	−6.09	0.48
2-Propanol	7.69	−9.91	0.41
Ethylene glycol	3.86	−2.00	0.40
Acetonitrile	4.05	−3.78	0.50
1,4-Dioxane	14.39	−23.34	0.17
Dimethylsulfoxide	8.77	−3.99	0.10

The standard molar Gibbs energies of transfer of ions from water to nonaqueous solvents are dealt with in Section 4.3.2.1 and those for transfer into mixed aqueous–organic solvents in Section 6.1. Specifically, the standard molar Gibbs energies of transfer of hydrogen ions from water to solvents S, $\Delta_{tr} G^\infty (H^+, W \rightarrow S)$, are available in Table 4.2 for nonaqueous solvents, in Table 6.1 for equimolar mixtures of water with cosolvents , and in the compilations by Kalidas et al. [17] and by Marcus [18] for other compositions. The $^S_W pH$ scale is a universal one, because it refers to the same standard state, infinite dilution of hydrogen ions in pure water, where its activity coefficient is unity. The acidity in other solvents, $^S_S pH$, is related to this universal one by Equation 8.8.

A cognate issue is the establishment of a universal standard electrode potential scale in nonaqueous and mixed solvents, based on the aqueous standard hydrogen electrode (SHE), for which $E^\infty (H^+, aq/H_2) = 0$ at all temperatures. The standard potentials E° are obtained on extrapolation of the EMF of a suitable cell to zero of the concentration of the electroactive electrolyte, the one that responds to the electrodes irrespective of the eventual presence of a constant inert background electrolyte, in the solutions of the two half-cells. A proper procedure for such an extrapolation that assures accuracy has been described by Mussini et al. [15]. The standard electrode potentials for a cation/metal pair M^{z+}/M in a solvent S vs. the SHE is related to its standard potential in water and the standard molar Gibbs energy of transfer of the cation from water to the solvent (or solvent mixture):

$$^S_W E^\circ \left(\frac{M^{z+}}{M} \right) = {}^W_W E^\circ \left(\frac{M^{z+}}{M} \right) + \Delta_{tr} G^\infty \left(M^{z+}, W \rightarrow S \right) / zF \qquad (8.9)$$

For an anion X^{z-}, using the sparingly soluble silver salt, that is, $X^{z-}/Ag_{z-}X,Ag$ electrode, the corresponding expression is:

$$^S_W E^\circ (X^{z-}/Ag_{z-}X, Ag) = {}^W_W E^\circ (X^{z-}/Ag_{z-}X, Ag) - \Delta_{tr} G^\infty (X^{z-}, W \rightarrow S)/|z| F \qquad (8.10)$$

Between aqueous solutions and solutions that contain another solvent component (mixed aqueous–organic or nonaqueous solvents), there exists a liquid junction potential that cannot be determined within the frame of thermodynamics, because it is determined by transport properties of ions. The same problem pertains to the standard molar Gibbs energies of transfer involved in Equations 8.9 and 8.10. Therefore, an extra-thermodynamic assumption is required in order to relate electrode potentials in such solvents to the SHE. Several approaches have been used for this purpose, including (i) the use of a suitable salt bridge that minimizes the liquid junction potential, (ii) using a reference electrolyte to establish single ion potentials, and (iii) using a solvent-independent reference redox couple.

(i) This approach assumes that the potential difference (E_j) which develops at the phase boundary between solutions in a galvanic cell can be rendered independent of the solvent by separating them with an appropriate salt bridge solution. It is called the negligible liquid junction potential (NLJP) assumption. Parker and coworkers [19] advocated the use of cells such as:

$$Ag, AgClO_4 \left(0.01 \text{ M in } S_1\right) \| 0.1 \text{ M } Et_4 \text{Npicrate in the bridge solvent} \|$$
$$AgClO_4 \left(0.01 \text{ M in } S_2\right), Ag$$
$$(8.11)$$

The large ions of the salt bridge solution have similar electrical mobilities and (low) solvation energies in many solvents. The theory of liquid junction potentials (see the book by McInnes [20]) suggests this should lead to $E_j \rightarrow 0$. The choice of the bridge electrolyte and solvent is to some extent arbitrary, but depends to some extent on the natures of S_1 and S_2. A solution of 0.1 M Et_4Npic (picrate) in acetonitrile has been widely adopted, but satisfactory results were obtained with other solutions too.

(ii) The reference electrolyte assumption, for example, the TATB assumption, has already been discussed in Section 4.3.1 and need not to be elaborated on here.

(iii) The reference redox couple assumption considers that the differences between Gibbs energies of transfer of the oxidized and reduced species of a chemically related couple comprising large ions are negligible. Examples of such couples that have been extensively used are ferrocene/ferricinium (Fc) proposed by Strehlow and Wendt [21] and bisbiphenyl-chromium(0/I) (BBCr) proposed by Gritzner [22]. The oxidized and reduced species have low charge/radius ratios and typically have their charge "buried" inside a large organic "cage," thereby shielded from direct interaction with the solvent in which the species are dissolved. Despite the inevitable charge difference, the two species of the couple are sufficiently chemically similar and thus relatively little affected by transfer from solvent to solvent.

An extensive compilation of standard electrode potentials of metals vs. their univalent cations obtained mainly by polarography and based on the BBCr redox couple assumption has been reported by Gritzner [23]. The values for two such electrodes Na^+/Na and Ag^+/Ag are shown in Table 8.3, and those for Li^+/Li, K^+/K, Rb^+/Rb,

Cs^+/Cs, and Tl^+/Tl are also available [23]. They are compared in this table with values obtained by Inerowicz et al. [24] using the TATB assumption and with those obtained by Cox et al. [25] using the NLJP assumption. The drawback of the polarographic method is that it is not directly applicable to the electrode potentials involving anions, say with an X^-/AgX, Ag electrode obtained potentiometrically. Such potentials for $X = Cl$ and I obtained using the TATB assumption are shown in Table 8.3. Some additional values of standard electrode potentials are in the reports by Parker and coworkers and by Johnsson and Persson [27–29].

TABLE 8.3 Standard Electrode Potentials, $E°/V$, vs. the SHE at 25°C According to Various Extra-thermodynamic Assumptions

Solvent/electrode	Na+/Na			Ag+/Ag			Cl-/AgCl, Ag	I-/AgI, Ag
References	23	24	25	23	24	25	26	26
Water	−2.71	−2.62	−2.62	0.800	0.800	0.800	0.222	−0.152
Methanol	−3.09			0.82	0.88	0.89	0.09	−0.23
Ethanol	−3.02			0.80		0.87	0.01	−0.29
n-Propanol	−2.97			0.83			−0.05	−0.35
Trifluoroethanol				1.28		1.33	0.33	−0.07
Ethylene glycol	−2.70			0.75			0.13	−0.18
Tetrahydrofuran	−2.70			0.83				
Acetone	−2.67			0.85		0.92	−0.37	−0.41
Propylene carbonate	−2.52		−2.55	1.05		1.03	−0.19	−0.29
γ-Butyrolactone	−2.61			0.90				
Formamide				0.73		0.68	0.08	−0.23
N-Methylformamide	−2.77		−2.82	0.60				
DMF	−2.79			0.64		0.66	−0.28	−0.36
DMA	−2.82			0.56		0.59	−0.34	−0.37
Tetramethylurea	−2.84			0.57				
NMPy	−2.81		−2.88			0.59	−0.31	−0.35
Pyridine	−2.65	−2.56		0.14	0.19			
Acetonitrile	−2.56	−2.57	−2.57	0.56	0.57		−0.21	−0.33
Benzonitrile	−2.48			0.64				
Nitromethane							−0.16	−0.33
Nitrobenzene							−0.14	−0.34
1.1-Dichloroethane		−2.41[a]					−0.38	−0.47
1,2-Dichloroethane		−2.46[a]					−0.38	−0.42
Dimethylsulfoxide	−2.81			0.49	0.45	0.46	−0.20	−0.26
TMS	−2.59		−2.74	0.90		0.79	−0.26	−0.37
Tetrahydrothiophene				0.23	0.27			
DMThF	−2.35	−2.85	−2.85	−0.21	−0.22			
Trimethylphosphate	−2.81			0.71				
HMPT	−2.97		−2.71	0.42		0.39	−0.38	−0.46

TATB [24, 26], and negligible liquid junction potential [25].
DMA, N,N-dimeyhylacetamide; DMF, N,N-Dimethylformamide; DMThF, N,N-Dimethylthioformamide; HMPT, Hexamethyl phosphoric triamide; NMPy, N=methylpyrrolidinone; TMS, Tetramethylenesulfone at 30°C.
[a]From Ref. 26.

The electrode potentials in Table 8.3 and its sources are given to two decimals only, because their uncertainty is expected to be ±0.05 V, due to both experimental errors and the inherent uncertainty involved in the assumptions, as is seen in the differences between the values derived from the three assumptions compared there.

8.2 APPLICATIONS IN HYDROMETALLURGY

Reprocessing of spent nuclear fuel is a major hydrometallurgical process that depends on ion solvation. The purpose of the process is the recovery of the (generally isotopically enriched) uranium and any by-product plutonium from the nuclear fuel after the nuclear reactor has run for the specified time and their separation from the fission products that would hamper the further operation of the nuclear reactor. Some of the fission products have large neutron absorption cross-sections and act as neutron "poisons" and others are long-lived radioisotopes that emit intense gamma rays, and all are sent to further processing and eventual storage as nuclear waste. The spent fuel, after a period of "cooling" during which the level of gamma radiation is diminished considerably, is sent to a "head-end" process in which the cladding and the fissionable materials (generally in the form of UO_2), the plutonium and the fission products, are dissolved in nitric acid. The most widely employed reprocessing is the PUREX (**P**lutonium **U**ranium **Ex**traction) process, which is based on the solvation of the U(VI) and Pu(IV) cations by tri-n-butyl phosphate (TBP) in a suitable hydrocarbon diluent extracting them from the nitric acid medium that may contain also nitrate salts from the cladding or added specifically. The reactions are:

$$UO_2^{2+}(aq) + 2NO_3^{-}(aq) + 2TBP(org) \rightarrow UO_2(NO_3)_2 \cdot 2TBP(org) \quad (8.12)$$

$$Pu(IV)(aq) + 4NO_3^{-}(aq) + 2TBP(org) \rightarrow Pu(NO_3)_4 \cdot 2TBP(org) \quad (8.13)$$

The fission products are not extracted by the TBP and the aqueous solution is directed to the "tail-end" processes of nuclear waste management. The organic phase is washed by nitric acid, the plutonium is reduced to Pu(III) that is stripped from the organic phase by moderately concentrated nitric acid leaving the uranium behind in the organic phase. The latter is finally stripped by dilute nitric acid and sent to refabrication of nuclear fuel elements.

The standard molar Gibbs energy of reaction (8.12) on the mole fraction scale when dodecane is the diluent of the TBP is $\Delta G^\circ = -46$ kJ mol^{-1}, the reaction being dominated by its enthalpy change, $\Delta H^\circ = -55$ kJ mol^{-1} as determined by Marcus [30]. The net enthalpy change arises from the amount invested in desolvation (dehydration) of the uranyl ion UO_2^{2+} (aq) and the two nitrate ions NO_3^{-} (aq) but regained on their electrostatic association and the solvation of the neutral species by the TBP. A solvating solvent with a larger electron pair donation ability than TBP (it has a donor number $DN = 23.7$) would extract the uranium more efficiently (at a lower concentration in the inert diluent). A possible such reagent is tri-n-octylphosphine oxide (TOPO, $DN \sim 32$ was reported by Modin and Schill [31]), but its use would interfere with the final recovery of the uranium by stripping it from the organic phase with dilute nitric acid. (Nevertheless, TOPO has been used according to Flett [32] for recovery of uranium from wet-process phosphoric acid, in which it is present at the 100 ppm level.) The role of the nitrate salts, such as aluminum or zirconium nitrate from the fuel rod cladding is to reduce the water activity in the aqueous phase and salt out the neutral uranyl nitrate species into the organic phase where it is solvated by the TBP. All these considerations apply also to Pu(IV) present in the solution, but trivalent actinides (including Pu(III)) and lanthanides, as well as di- and univalent fission products are neither well complexed by nitrate ions nor solvated as neutral species by TBP.

Hydrometallurgical processes on a large commercial scale, using solvent extraction, are being applied for extraction of metals from ores and from scrap metal. The ores are leached by sulfuric acid whereas chloride solutions may result from scrap metal dissolution, but the variations of the solvent extraction processes are manyfold, the literature on this subject being very extensive, and only a few recent examples are dealt with here. It must be stressed that complexation by the organic extractant, rather than solvation by it (as in the case of uranium extraction by TBP), is the rule, although dehydration of the metal ions to be extracted does play a role. Extractants under the commercial name LIX® are hydroxyoximes or hydroxamic acids, and some extractants under the commercial name CYANEX® are dialkylphosphinic acids that exchange their acidic hydrogen atoms for the metal ions as reported by Flett [32]. However, some large-scale processes do employ solvating agents, rather than or in addition to complexing agents, in the latter case being employed as synergists. For example, Cyanex923 is a mixture of trialkylphosphine oxides (alkyl=hexyl and octyl) and has been suggested by Reddy et al. [33] for the extraction and separation of nickel and cadmium in chloride media from spent batteries.

A different application of solvating solvents to hydrometallurgy is at the stage of ore leaching. For example, the dissolution of chalcopyrite ($CuFeS_2$) in sulfuric acid is facilitated by addition to the acid of acetone or ethylene glycol that helps in the prevention of passivation according to Solis-Marcial and Lapidus [34]. A more striking example is in the application of acetonitrile to the leaching of silver and copper from ores or scrap metals and their separation. Cu(I) and Ag(I) are better solvated by MeCN than by water, but the reverse is valid for Cu(II) (see Table 4.2). These observations apply also to mixed aqueous-acetonitrile solutions; see Table 8.4 for the relevant thermodynamic data. When slightly acidified (pH=2) solutions of $CuSO_4$ in 6 M acetonitrile in water ($x_{MeCN} \sim 0.1$) is applied to copper and silver sulfide ores or to copper scrap metal, then dissolution to form Cu_2SO_4 and Ag_2SO_4 occurs according to Parker and coworkers [36, 37]. Base metals, such as iron, do not accompany the copper and silver into the solution. The acetonitrile is then removed from the solution by codistillation with low-grade steam and high-purity metallic silver and copper precipitates out from the remnant aqueous solution. If chloride, rather than sulfate, media are employed, the low solvating power of aprotic dipolar solvents for anions leaves the chloride to complex effectively the group Ia cations, forming $CuCl_2^-$, $AgCl_2^-$, and $AuCl_4^-$, and hence effective dissolution of these metals in dimethyl sulfoxide takes place. Addition of water precipitates AgCl and metallic

TABLE 8.4　Thermodynamics of Transfer (in kJ mol⁻¹) from Pure Water into Aqueous Acetonitrile Pertinent to the Dissolution of Copper and Silver [17, 35]

x_{MeCN}	Cu(I)		Cu(II)		Ag(I)	
	$\Delta_{tr}G^{\infty}$	$\Delta_{tr}H^{\infty}$	$\Delta_{tr}G^{\infty}$	$\Delta_{tr}H^{\infty}$	$\Delta_{tr}G^{\infty}$	$\Delta_{tr}H^{\infty}$
0.057	−27.8[a]	−71.6[a]			−7.3[a]	−24.9[a]
0.100	−36.1		+3.4	−15.2	−11.6	−36.8
0.200	−44		+4.9	−38.0	−17.6	−45
0.500	−53		+10.4	−45.5	−21.0	−54.3
0.700	−54.4		+13.5	−42.8	−22.3	−59
1.000	−55.7		+66.8	+8	−24.1	−53.2

[a]From Ref. 26.

gold from the solution, because the Cl⁻ becomes hydrated and much less available for complexation according to Parker et al. [38].

The (electro-) deposition of adherent coatings on suitable substrates reported by Gores and Barthel [39] as well as the formation of nanoscale metallic particles (as catalysts) is facilitated by the presence of organic cosolvents in the aqueous media in which salts of the metallic elements are dissolved. The presence of 33% by volume of ethanol in the solution helps the precipitation of submicron silver powders by reduction of aqueous Ag(I) with hydrazine hydrate according to Ghosh and Dasgupta [40]. Aluminum that is readily hydrolyzed in aqueous solutions can be electroplated from a solution of $AlCl_3$ in tetrahydrofuran containing some benzene according to Yoshio and Ishibasi [41]. These are just a few examples of the extensive literature relating to the use of solvating solvents in hydrometallurgy.

8.3　APPLICATIONS IN SEPARATION CHEMISTRY

8.3.1　Solvent Extraction of Alkali Metal Cations

The selective extraction of alkali metal ions from aqueous solutions by solvating solvents is a challenging problem and of importance in the case of removal of the relatively long lived fission product ¹³⁷Cs from active nuclear waste. One approach to the problem is the use of crown ethers that solvate alkali metal ions selectively with regard to the size of their cavities. These cyclic ethers, $-(C_2H_4O)_n-$, with $4 \leq n \leq 10$, having possibly substitutions of one or two 1,2-benzo- or 1,2-cyclohexano groups for the carbons of the O–C–C–O skeleton (to make them water-insoluble), are of interest because the univalent alkali metal cations are not extractable by normally employed (chelating) extractants of metal cations. Crown ethers are denoted by the numbers m of atoms in the ring of which n are oxygen atoms, in the manner m-C-n, and prefixed if necessary by DC- (for dicyclohexano), B- (for benzo), DB (for dibenzo), etc. Thus, DB-18-C-6 is the symmetrical dibenzohexaoxocyclooctadecane.

The water molecules hydrating the alkali metal cations are replaced by the ether oxygen atoms with little change of the bonding energy, provided that the mean distances of these oxygen atoms from the metal cation are appropriate. The selectivity of the crown ether rings with regard to the alkali metal cations is according to their sizes: 12-C-4 for Li$^+$, 15-C-5 for Na$^+$, 18-C-6 for K$^+$, 24-C-8 for Rb$^+$, and 30-C-10 for Cs$^+$. However, smaller ions are solvated also by crown ethers with larger rings than required for a perfect fit, though with smaller equilibrium constants and the cations then "rattle" within the rings. Cations that are too large to fit well into the ring are still solvated by the crown ether, being partly outside the ring and thus partly hydrated on the far side, again with a smaller equilibrium constant than for a cation fitting perfectly in the ring. A requisite for the extraction of the crown ether solvated alkali metal cations into a water-immiscible liquid phase is that the anion is also taken care of by a suitable solvent for it to be coextracted with the cation. Large, poorly hydrated anions, such as picrate, have been used for this purpose as reported by Marcus [26], by Eisenman et al. [42], by Takeda [43], and by Tanigawa et al. [44]. If small, hydrophilic anions, such as chloride or nitrate, accompany the alkali metal cation, say K$^+$, in the aqueous solutions from which they are to be selectively extracted, provision for the coextraction of the anions should be made. A water-immiscible protic solvent, SH, having an acceptor number AN or a Kamlet-Taft α parameter (Section 3.3.2) larger than that of water should be provided. The anions prefer such a solvent over water, if such a solvent is capable of solvating them by stronger hydrogen bonds. This protic solvent constitutes also the organic phase in which the crown ether, Cw, is dissolved:

$$K^+(aq) + Cl^-(aq) + Cw(SH) \rightleftarrows K^+Cw(SH) + Cl^-(SH) \tag{8.14}$$

Whether the solvated cation and the solvated anion are dissociated in the organic phase or form an ion pair K$^+$CwCl$^-$(SH) depends primarily on the permittivity of this solvent (see Section 3.2.3). Substituted phenols were proposed by Marcus and coworkers [45–47] for this purpose. 4-Fluorophenol is the best solvent in combination with DB-18-C-6 for the selective extraction of KCl from aqueous solutions according to Marcus et al. [47].

The selective extraction of Cs$^+$ (with particular regard to the removal of fission product ^{137}Cs from active nuclear waste) was effectively achieved by the use of the large dicarbolide anion (cobalt bis(dicarbolide), $[(1,2\text{-}C_2B_9H_{11})_2\text{-}3\text{-}Co]^-$). The history of the use of this anion for the solvent extraction of alkali metal cations from aqueous solutions was reviewed by Kyrš [48] and its subsequent development by Rais and Gruener [49].

A comprehensive paper on the distribution of alkali metal and ammonium cations between mutually saturated liquids (water and an organic solvent) is that by Rais and Okada [50]. Contrary to the data in Table 4.2, which pertain to pure water and neat nonaqueous solvents, the data for quite immiscible but still mutually saturated phases can be obtained from distribution and electrochemical data. The latter provide the standard ionic distribution potential, $\Delta^*\varphi^\infty(I)$, usually obtained from cyclic voltammetry, the asterisk denoting an abbreviation of $\Delta^{o(a)}_{a(o)}\varphi^\infty(I)$, where o(a) and a(o) designate the mutually saturated organic and aqueous phases. This potential is related to the standard molar Gibbs energy of transfer of the ion I between these phases, $\Delta * G^{\infty}(I)$, and to the individual ion extraction constant $*K^{\infty}(I)$ by:

$$\Delta * \varphi^{\infty}(I) = -\frac{\Delta * G^{\infty}(I)}{Fz_I} = \left(\frac{RT\ln(10)}{Fz_I}\right)\log * K^{\infty}(I) \qquad (8.15)$$

The exchange equilibrium constant $*K_{exch}^{\infty}(Cs/M)$ involves a solution of cesium dicarbolide, denoted by CsB, dissolved in the organic phase contacted with an aqueous solution of the cation M^+ and some hydrophilic anion, such as nitrate, for exchange partition equilibrium to be attained, at which stage $*K_{exch}^{\infty}(Cs/M) = *K^{\infty}(Cs^+)/*K^{\infty}(M^+)$. The values of $\Delta * G^{\infty}(I)$ for two solvents, o-nitrophenyl octyl ether and 1,2-dichloroethane, at 25°C are shown in Table 8.5 for the univalent cations M^+ studied by Rais and Okada [50]. Exchange constants, $\log * K_{exch}^{\infty}(Cs/M)$, were also reported for these ions for the solvents 1-nitropropane, 1-octanol, and dioctyl sebacate ($OcOOC)(CH_2)_8(COOOc)$), but no individual ionic values could be shown. The individual ionic $\Delta * G^{\infty}(I)$ and $\log * K_{exch}^{\infty}(Cs/M)$ are linear with the standard molar ionic Gibbs energies of hydration of the cations as expected, but Li^+ and to a lesser extent Na^+ deviate from these relationships because of the presence of water in the organic phase.

The solvent extraction of hydrophilic anions can be effected by the use of crown-ether solvated cations as in Equation 8.14. The order of extractability of potassium salts from aqueous solutions by 0.1 M DB-18-C-6 in m-cresol is according to Marcus et al. [45, 46]:

$$SO_4^{2-} \ll Cl^- < Br^- < I^- < NO_3^- < CH_3CO_2^- < F^- \qquad (8.16)$$

This sequence is *not* that of the molar Gibbs energy of hydration of the anions (Table 4.1) because it depends also on the solvating ability of the protic solvent, m-cresol. The order of the sequence (8.16) results from a balance of the hydration ability of water molecules and of the protic solvent, as well as from the space available around the anion for solvation by the small water molecules and the bulky protic solvent ones. A further complicating factor is possible ion pairing of the crown-solvated

TABLE 8.5 The Standard Molar Gibbs Energy for Transfer, $\Delta * G^{\infty}(I)/\text{kJ mol}^{-1}$, of Univalent Ions Between Mutually Saturated Water and o-nitrophenyloctyl Ether and Water and 1,2-dichloroethane at 25°C, Obtained from Exchange Equilibria [49]

Cation	o-Nitrophenoloctyl ether	1,2-Dichloroethane
H^+	34.4	49.4
Li^+	43.5	54.1
Na^+	40.0	51.7
K^+	29.7	47.7
Rb^+	26.3	41.8
Cs^+	22.9[a]	34.9[a]
NH_4^+	32.7	44.0

[a] Reference value from electrochemical measurements.

cation, axially to the crown ring, with the solvated anion. Protic solvents with increasing hydrogen bond donation abilities (measured, e.g., by their E_T^N values) extract with increasing effectiveness the anion with a given cation–crown ether solvate as shown by Hormodaly and Marcus [51].

8.3.2 Solvation of Ionizable Drug Molecules

The solvation of drug molecules that dissociate into ions (many drugs that are weak acids or bases that ionize on adjustment of the pH) in aqueous or mixed aqueous–organic solutions is used for their separation by high-performance liquid chromatography for preparative or analytical purposes. The stationary phase may be hydrophilic, for example, water adsorbed on silica gel, and then direct extraction chromatography occurs, or hydrophobic, for example, a water-immiscible organic solvent adsorbed on polystyrene, and then reversed-phase extraction chromatography takes place. The affinity of the drug molecules to the aqueous phase is primarily governed by their ionization, the ions being better hydrated than the neutral molecules, and in the second place by the solvation of the ionized and neutral molecules by the respective solvents in the two phases.

The hydrophilicity/hydrophobicity balance of the counter-ion present or added affects the affinity of the compound of interest to the two phases. Examples for such general statements can be provided by partition into a solvent of low polarity and low solvating ability, such as chloroform. When Bu_4N^+ is the counter-ion, the order of the extractability of aromatic anions is phenolate < benzoate < toluene-4-sulfonate < salicylate < 1-phenylpropyl sulfate. When the counter-ion is chloride, the extractability of amines with one long chain R, $RNMe_nH_{3-n}^+$ follows the sequence quaternary ($n = 3$) < primary ($n = 0$) < secondary ($n = 1$) < tertiary ($n = 2$). When a solvent with much better donor properties than chloroform is used, such as methyl isobutyl ketone, tertiary amines are no longer extracted preferentially as reported by Marcus [52].

For the purpose of purification of drugs, their solubilities in mixed aqueous-organic solvents and their preferential solvation in them are of consequence. Various approaches have been suggested for dealing with these issues, but only a few of them are pertinent to ionized or ionizable substances according to Jouyban and Acree [53], Jouyban-Gharamaleki [54], and Marcus [55]. Many drugs are only poorly soluble in water and a cosolvent is employed for increasing their solubility. The log-linear expression is widely used in drug design, which uses the 1-octanol/water partition coefficient of the solute drug D, P_{Dw}°, as a parameter relating the mole fraction solubility of D in water and the solvent (or mixture) being studied and the volume fraction of the latter, φ_S:

$$\log x_{D(W+S)} = \log x_{D(W)} + \left(S_0 + S_1 \log P_{Dw}^\circ \right) \varphi_S \qquad (8.17)$$

where S_0 and S_1 are solute- and cosolvent-independent coefficients according to Yalkowski and Rosennab [56]. The mole fraction solubility of the drug D (or a drug-related substance) in the solvent mixture is also related to those in the neat components according to the volume fractions of the components as shown by Jouyban and Acree [53]:

$$\log x_{D(W+S)} = \varphi_W \log x_{D(W)} + \varphi_S \log x_{D(S)} + \varphi_W \varphi_S \sum A_j \left(\varphi_S - \varphi_W \right)^j \qquad (8.18)$$

The summation in the last term is from $j = 0$ to $j = 2$ and pertains to specific interactions of the solute with itself and with the solvent components and those among these components themselves in the absence of the solute. This term increases the predictive accuracy on the average threefold over a similar expression without this term, the log-linear expression (8.17).

These expressions are applicable to drugs and drug-related ionizable substances, such as asparagine, aspartic acid, caffeine, leucine, nalidixic acid, paracetamol, salicylic acid, and sulfanilamide. However, the log-linear expression (8.17) or the first two terms on the right-hand side of Equation 8.18 and even its modification in terms of the complete Equation 8.18 cannot model cases where the solubility curve exhibits a maximum as is observed in many cases, a problem solved when fluctuation theory is employed in the calculations as pointed out by Ruckenstein and Shulgin, [57]. If the solute is rather poorly soluble in the solvents and their mixtures, then solute–solute interactions can be ignored and the mutual interactions of the solvent components can then be treated either as ideal or as non-ideal. The molar volume of the binary solvent mixture is expressed as $V_{(W+S)} = x_W V_W^* + x_S V_S^* + e\ x_W x_W$, where e is an empirical parameter that in the ideal case is zero (Section 3.4.1.1). The solubility of the drug is given by:

$$\log x_{D(W+S)} = \frac{\left[\left(\log V_{(W+S)} - \log V_W^*\right)\log x_{D(S)} - \left(\log V_{(W+S)} - \log V_S^*\right)\log x_{D(W)}\right]}{\left[\log V_S^* - \log V_W^*\right]} \quad (8.19)$$

When the aqueous cosolvent mixture is nonideal, which is the usual case, the nonideality is best dealt with in terms of the Wilson expression [58]. This approach was applied to ionizable drugs that carry phenolic or carboxylic groups that ionize at high pH or amino groups or nitrogen atoms in rings that can be protonated at low pH. Such compounds include caffeine, paracetamol, phenacetine, sulfanilamide, oxalinic acid, and theophylline in aqueous mixtures with DMF, dioxane, ethanol, and ethylene glycol. Caffeine, niflumic acid, diazepam, benzocaine, phenacetin, paracetamol, and nalidixic acid were also studied by both the quasi-lattice quasi-chemical (QLQC) method and the inverse Kirkwood–Buff integral (IKBI) method (Sections 6.1.2.1 and 6.1.2.2) by Marcus [55] to yield the local mole fractions (i.e., those around the drug molecule) of the components of the binary solvent mixtures: ethanol+ethyl acetate, water+ethanol, and water+1,4-dioxane. Both methods were also applied by Marcus [59] to the preferential solvation of ibuprofen and naproxen in water+1,2-propanediol and by Martinez and coworkers [60] to the preferential solvation of several drugs, such as meloxicam, in water+ethanol. The QLQC method was applied to indomethacin in water+1,4-dioxane [61], and the IKBI method to meloxicam in water+1,2-propanediol [62]. In the cases of all these drugs, the abilities of their molecules (having acidic and/or basic groups) to donate hydrogen bonds to and/or accept them from the components of the binary solvent mixture affect their preferential solvation.

The role of ion solvation in nonaqueous and mixed aqueous–organic solvents in ion separations by ion exchange resins has been summarized in detail in the earlier edition of this book [26] and need not to be repeated here, no new insight having been obtained since then.

8.4 APPLICATIONS TO CHEMICAL REACTION RATES

The rates of chemical reactions, and in some cases the compositions of their products, depend on the solvation of the reactants and of the activated complex among other factors. This applies to both inorganic and organic reactions, whether in a pure solvent or a solvent mixture. The "solvent effects in organic chemistry" have been discussed in detail in the recent edition of the book with this title by Reichardt and Welton [63]. Earlier reviews concerning organic reactions are those of Parker [64], Buncel and Wilson [65], Abraham [66], Reichardt [67], and Buncel et al. [68] among others, and some aspects are discussed by Marcus [26]. Regarding inorganic reactions, there are the reviews by Blandamer and Burgess [69] and by Wherland [70].

The effects of solvents on reaction rates have been studied most extensively on unimolecular solvolysis reactions (S_N1) and on bimolecular nucleophilic substitution reactions (S_N2). Absolute rate theory specifies that the activated complex in the transition state is at equilibrium with the reactants and is formed on provision of the activation (Gibbs) energy, $\Delta G^{\neq}$. In other words, the energy barrier that the reaction must pass to proceed has to be overcome. The specific rate constant is given by:

$$k/\left(\text{mol}\cdot\text{dm}^{-3}\right)^{1-n} = K_0 T \exp\left(\frac{-\Delta G^{\neq}}{RT}\right) \qquad (8.20)$$

where n is the molecularity of the reaction (its kinetic order) and $K_0 = k_B/h = 2.291\times10^{10}\ \text{K}^{-1}\cdot\text{s}^{-1}$ is a universal constant. The enthalpy and entropy of activation, yielding $\Delta G^{\neq} = \Delta H^{\neq} - T\Delta S^{\neq}$, may be cooperative or antagonistic, having the same sign, and eventually compensating, or the one or the other may dominate. The differences in the Gibbs energy of solvation of the reactants and of the activated complex on going from solvent S1 to Solvent S2 then determine the difference in the overall Gibbs energy of activation:

$$\Delta\Delta G^{\neq} = \Delta_{tr}G^{\neq}(\text{activated complex, S1} \rightarrow \text{S2}) - \Delta_{tr}G^{\circ}(\text{reactants, S1} \rightarrow \text{S2}) \quad (8.21)$$

where the $\Delta_{tr}G^{\neq}$ in this expression are the Gibbs energy of transfer of the specified moieties. For the reactants, be they ions or neutral entities, their (standard) molar Gibbs energies of transfer $\Delta_{tr}G^{\circ}(\text{reactants, S1}\rightarrow\text{S2})$ can be obtained from their molar Gibbs energies of solvation in solvents S1 and S2, but for the (transiently present) activated complex, this route is not possible. The value of $\Delta_{tr}G^{\neq}$ (activated complex, S1→S2) is then obtained from the kinetic data, Equation 8.20, yielding $\Delta\Delta G^{\neq} = \ln(k_{\text{in S2}}/k_{\text{in S1}})$ at a given temperature T, and is then rationalized in terms of the expected structure of the activated complex and its presumed abilities to be solvated by the two solvents.

Empirical linear solvation energy relationships have, however, been widely employed, relating reaction rates to one or more properties of the solvents. In common reactions involving ions, namely bimolecular nucleophilic substitution reactions (S_N2), it was shown that hydrogen bond–donating protic solvents increase the activation Gibbs energy and reduce the rate by increased solvation of the incoming anion, making it less available for the reaction as pointed out by Abraham [66]. The reactivity order in protic solvents is therefore $I^- > Br^- > Cl^- > F^-$, but in aprotic dipolar solvents, the order is reversed $Cl^- > Br^- > I^-$, because in such solvents the anions are poorly solvated. Relative rates (to that in methanol) of the reaction

$$CH_3I + Cl^- \rightarrow CH_3Cl + I^- \tag{8.22}$$

$$p- FC_6H_4NO_2 + N_3^- \rightarrow p-N_3C_6H_4NO_2 + F^- \tag{8.23}$$

in various solvents are shown in Table 8.6 adapted from Alexander et al. [71]. The values of $\log(k_{in\,S}/k_{in\,MeOH})$ roughly increase as the polarity of the solvents, measured by their E_T^N values (Table 3.9), diminishes.

Inorganic reactions exhibit their own trends, examples being outer-sphere electron transfer reactions of transition metal complexes reported by Wherland [70]. In this case the rate depends on the difference $n^{-2} - \varepsilon^{-1}$, because the high frequency solvent response, n^2, where n is the refractive index, represents the rapid response of the solvent to changes in the electric field produced by the electron transfer, whereas the permittivity responds much more slowly. Results for the electron exchange constants for ferrocene(0)/(I) studied by McManis et al. [72] and chromium bisdiphenyl(0)/(I) studied by Li and Brubaker [73] are shown in Table 8.7. The rate constants follow approximately the reverse order of the $n^{-2} - \varepsilon^{-1}$ function of the solvents.

First-order solvent exchange rates at hydrated metal cations have already been described in Section 4.5.2 and listed in Table 5.4. Some similar data exist also for nonaqueous solvents: methanol, ethanol, DMF, DMSO, and MeCN for di- and trivalent metal cations. The rate constants at 25°C and, where available, also the enthalpies and entropies of activation have been reported in the book by Burgess [74]. The rate

TABLE 8.6   Relative Rates $\log(k_{in\,S}/k_{in\,MeOH})$ of the Nucleophilic Reactions (8.22) and (8.23) at 25°C [71] Compared with the Normalized Polarity index E_T^N

Solvent	Reaction (8.22)	Reaction (8.23)	E_T^N
Methanol	0	0	0.762
Water	0.05		1.000
Formamide	1.2	0.8	0.775
N-Methylformamide	1.7	1.1	0.722
Tetramethylenesulfone		4.5	0.410
Nitromethane	4.2	3.5	0.481
Acetonitrile	4.6	3.9	0.460
Dimethylsulfoxide		3.9	0.444
N,N-Dimethylformamide	5.9	4.5	0.386
Acetone	6.2	4.9	0.355
N,N-Dimethylacetamide	6.4	5.0	0.377
N-Methylpyrrolidin-2-one	6.9	5.3	0.355
Hexamethyl phosphoric triamide		7.3	0.315

TABLE 8.7 Rate Constants, $10^{-7}k_{exch}/M^{-1}s^{-1}$, for the Electron Exchange Reactions of the Redox Couples $Fe(Cp)_2^{+/0}$ [72] and $Cr(Ph-Ph)_2^{+/0}$ [73] in Various Solvents and their Values of the Function $n_D^{-2}-\varepsilon^{-1}$

Solvent	$n_D^{-2}-\varepsilon^{-1}$	$Fe(Cp)_2^{+/0}$	$Cr(Ph-Ph)_2^{+/0}$
9 Benzene:1 MeOH (V/V)	~0.29		0.61
Nitrobenzene	0.388	3.0	
Benzonitrile	0.390	2.7	0.38
Dimethylsulfoxide	0.437	0.95	0.23
N,N-Dimethylformamide	0.462		0.21
Propylene carbonate	0.481	1.2	0.15
Acetone	0.496	0.8	
Nitromethane	0.498	1.2	
1 Benzene:4 MeOH (V/V)	0.501		0.16
1 Benzene:7 MeOH (V/V)	0.515		0.12
Acetonitrile	0.528	0.9	
Methanol	0.538	1.8	

TABLE 8.8 Rate Constants for Solvent Exchange, $\log(k_{exch}/s^{-1})$ at 25°C, at Solvated di- and Trivalent Metal Cations, Water From Table 5.4, Nonaqueous Solvents[a]

Cation	Water	MeOH	EtOH	DMF	DMSO	MeCN
Mg^{2+}	5.2	3.7	6.4	8.1 [75]		
Mn^{2+}	7.4	5.6				7.1
Fe^{2+}	6.5	4.7		6.2 [76]	6.0 [76]	5.7
Co^{2+}	6.4	3.6 to 4.3		5.4 to 5.6	5.2	5.5
Ni^{2+}	4.5	2.3 to 3.0	4.0	3.6 to 3.9	3.5 to 4.0	3.5
Cu^{2+}	9.3	~8		9.0 [77]		7.2 [78]
Al^{3+}	−0.8	3.6	3.5 [79]	−0.8	1.2	
Ti^{3+}	5.0	5.1				
V^{3+}	3.2	3.1				
Cr^{3+}	−6.3			−7.3	−7.5	
Fe^{3+}	4.3	3.4 to 3.7	4.3	1.5	1.7	<1.6
Ga^{3+}	3.3	4.0		0.2		

[a]From Ref. 74.

constants, $\log(k_{exch}/s^{-1})$, are shown in Table 8.8 and compared there with those for water exchange. In most cases, no information of whether a dissociative (positive volumes of activation, $V^{\neq} = (\partial \ln k_{exch}/\partial P)_T > 0$) or an associative ($V^{\neq} < 0$) mechanism take place. An exception is the case of Cr(III) in DMF and DMSO, where the latter, $V^{\neq} < 0$, was established [74].

The use of phase transfer catalysts (PTCs) is another area of reaction rates in which ion solvation plays a role. This term (PTC) was coined by Starks and Liotta [80] regarding biphasic reactions, in which a hydrophilic anion Y^- present in the aqueous phase is brought into an immiscible organic phase, where resides the substrate RX with which it is to react, by means of a suitable catalytic reagent, usually a quaternary ammonium salt, Q^+X^-. The general formulation of this process is:

$$Q^+X^-(\text{org}) + Y^-(\text{aq}) \rightleftarrows Q^+Y^-(\text{org}) + X^-(\text{aq}) \tag{8.24}$$

$$Q^+Y^-(\text{org}) + RX(\text{org}) \rightarrow RY(\text{org}) + Q^+X^-(\text{org}) \tag{8.25}$$

The homogeneous reaction (8.25) in the organic phase can proceed rapidly, whereas in the absence of the PTC, the heterogeneous reaction between the two immiscible phases will hardly proceed. The catalytic quaternary salt Q^+X^- should have altogether at least 24 carbon atoms in its alkyl chains that are of similar lengths in order to reside predominantly in the organic phase. Otherwise, with fewer carbon atoms or chains of grossly different lengths, the catalyst resides mainly in the aqueous phase or at the interface and partitions into the organic phase. The equilibrium (8.24) depends on the solvation of the ions X^- and Y^-, mainly their relative hydration, because they would not be appreciably solvated in water-immiscible organic solvents. The equilibrium quotients for the exchange reaction (8.24) where $X^- = Br^-$ for two quaternary cations—Dc_3MeN^+ in toluene employed by Starks and Liotta [80] and Ph_4As^+ in chloroform used by Bock and Jainz [81] are shown in Table 8.9. The quaternary ammonium salt carries with it

TABLE 8.9 Equilibrium Quotients, $\log K_{\text{exch}}$, for Exchange Reaction (8.24) for PTCs: $Dc_3MeN^+Br^-$ in Toluene [80] and $Ph_4As^+Br^-$ in Chloroform [81] with Aqueous Anions X^-

Anion X^-	$Dc_3MeN^+Br^-$	$Ph_4As^+Br^-$
OH^-	−3.22	
F^-	−2.92	<−2.68
Cl^-	−1.21	−1.40
Br^-	(0.00)	(0.00)
I^-	2.48	>1.78
CN^-	−1.22	
SCN^-		1.56[a]
NO_3^-		0.62
NO_2^-		−1.40
ClO_3^-		>1.48
BrO_3^-	−1.89	
IO_3^-		−3.10
ClO_4^-	1.48	>1.60
IO_4^-		−2.40
MnO_4^-	0.78	>1.78
HCO_2^-	−2.30[b]	
$CH_3CO_2^-$	−2.15	
$CH_3SO_3^-$	−1.82	
HCO_3^-	−2.52[b]	
HSO_4^-	−1.89[b]	
CO_3^{2-}	−3.52[b]	
SO_4^{2-}	−3.10[b]	

[a]The PTC is N-hexadecylpyridinium bromide.
[b]The PTC is $(C_{18}H_{37})_2Me_2N^+Br^-$.

some water into the organic phase: in the case of $Oc_4N^+X^-$ salts and toluene as the organic phase, the numbers of water molecules per salt formula are 1.5 for NO_3^-, 3.2 for Cl^-, 2.4 for Br^-, and 18 for SO_4^{2-} according to Heifets et al. [82].

The effects of various solvents in which HxBr is to react with solid KCl in order to convert it to HxCl with the PTC of Bu_4NBr at 70°C were studied by Danilova and Yufit [83]. Expressed as the rate constants $\log(k_r/s^{-1})$, they are shown in Table 8.10, reaction A. Also shown in Table 8.10 are the rate constants $\log(k_r/s^{-1})$ for fewer, but all water-immiscible solvents in biphasic reactions for the following: reaction B—oxidation of 5-vinylnorbornene with a combination of aqueous Na_2WO_4, H_3PO_4, and H_2O_2 with $Bu_4N^+HSO_4^-$ as the PTC at 30°C studied by Wang et al. [84]; reaction C—phenolysis of 1-bromoethylbenzene by aqueous KOH with $Bu_4N^+Br^-$ as the PTC at 50°C studied by Wu et al. [85]; reaction D—conversion of BuBr with aqueous Na_2S and the same PTC at 30°C studied by Wang et al. [86]; and reaction E—oxidation of 3,5-di-*t*-butyl-catechol by $KMnO_4$ catalyzed by dicyclohexyl-8-crown-6 (Section 8.3.1) at 25°C studied by Nakamura et al. [87]. Some authors attempted to relate the rates to the

TABLE 8.10   Rate Constants, $\log(k_r/s^{-1})$, for PTC Reactions A, B, C, D, and E Described in the Text in Solvents Arranged According to their Relative Pemittivities, with their Anion Solvating Abilities, the Acceptor Number *AN*, Also Shown

Solvent	ε	*AN*	A	B	C	D	E
n-Hexane	1.88	0.0			−2.84	1.81	4.17
Cyclohexane	2.02	1.6			−2.55	1.86	4.05
CCl_4	2.24	8.6	4.64				
Benzene	2.27	8.2			−2.84		
Toluene	2.38	6.8	4.85	0.85	−2.48	2.32	
Diethyl ether	4.20	3.9			−2.74		
Anisole	4.33		4.79				4.04
Chloroform	4.90	23.1		1.62		2.01	3.50
Chlorobenzene	5.62	11.9		0.63	−2.44		
Ethyl acetate	6.02	9.3	4.82				
THF	7.58	8.0	5.05				
Dichloromethane	8.93	20.4		1.64	−2.30		
1,2-Dichloroethane	10.36	16.7		1.62		1.96	3.40
Pyridine	12.91	14.2	6.05				
Pentan-2-one	15.38		5.14				
Acetophenone	17.39		5.27				
Ethanol	24.55	37.1	4.18				
Benzonitrile	25.20	15.5	5.06				
HMPT	29.30	9.8	5.97				
NMPyr	32.20	13.3	5.92				
Methanol	32.70	41.5	4.47				
Nitrobenzene	34.82	14.8	4.96				
Nitromethane	35.87	20.5	4.82				
Acetonitrile	35.94	18.9	4.94				
DMF	36.71	16.0	5.69				

solvent properties, such as the relative permittivities, but no clear trends are apparent for reaction A, the rate increases with ε for reaction C, but it decreases for reaction E. Neither can a clear trend be seen regarding the anion solvation abilities of the solvents, measured by their acceptor numbers AN (Table 3.9 and the compilation by Marcus [88]), that would be more appropriate for the kind of reactions dealt with here.

8.5 SOLVATED IONS IN BIOPHYSICAL CHEMISTRY

Ions play important roles in biophysical phenomena and their solvation, in particular in aqueous media but to some extent also in other media, is decisive in determining their behavior in biological systems. Three descriptors of the sequences of ions (separate sequences for cations and anions) pertaining to certain biophysical phenomena are commonly invoked: the *Hofmeister series*, the *kosmotropic* and *chaotropic* properties, and the *lyotropic numbers*, but these concepts are not interchangeable. However, some misuse of these descriptors pertaining to the specific effects of ions in solution is prevalent in biophysical publications, since for some biophysical phenomena alternative explanations, in terms of dehydration energies or direct interactions, should be better. Still, these properties do have their place in certain biophysical implications of aqueous ions. The Hofmeister series is a valuable phenomenological description when the aqueous ions are in the vicinity of a surface and at >0.1 molar concentrations rather than more dilute solutions. The notions of kosmotropic and chaotropic properties of ions describe their water structure ordering and destruction and are effective in dilute homogeneous solutions of the ions. Lyotropic numbers N_{lyo}, advocated by Voet and coworkers [89, 90] and reported in Table 8.11, were assigned to ions according to their effects on colloidal systems. The field of *specific ion effects* was recently reviewed by Kunz [91, 92].

TABLE 8.11 Lyotropic Numbers, N_{lyo}, Applicable to Colloidal Systems [89, 90]

Cation	N_{lyo}	Anion	N_{lyo}
Li$^+$	105.2	OH$^-$	5.5
Na$^+$	100.0	F$^-$	4.8
K$^+$	75.0	Cl$^-$	10.0
Rb$^+$	69.5	Br$^-$	11.3
Cs$^+$	60.0	I$^-$	12.5
Ca^{2+}	10.0	SCN$^-$	13.25
Sr^{2+}	9.0	NO$_2^-$	10.1
Ba^{2+}	7.5	NO$_3^-$	11.6
		ClO$_3^-$	10.65
		BrO$_3^-$	9.55
		IO$_3^-$	6.25
		ClO$_4^-$	11.8
		SO$_4^{2-}$	2.0
		H$_2$PO$_4^-$	8.2
		PO$_4^{3-}$	3.2

8.5.1   The Hofmeister Series

This concept was established by Hofmeister [93], pertaining to precipitation of egg albumin by a series of aqueous sodium salts with various anions, of which increasing concentrations were required with diminishing effectiveness along the series: $SO_4^{2-} \sim HPO_4^{2-} > CH_3CO_2^- > HCO_3^- > CrO_4^{2-} > Cl^- > NO_3^- > ClO_3^-$. Almost a century later, Collins and Washabaugh [94] listed nearly 1000 references involving Hofmeister series of anions and also of cations and many more have been added since then. Water is the key solvent for the phenomena treated under such headings and the specific ion effects are mediated by the solvent water. However, no universal underlying principle has so far emerged for the Hofmeister series of ions.

Studies of specific ion effects on such systems to which the Hofmeister series pertains generally involve salts at concentrations $\geq 0.1 M$. At lower salt concentrations, the electrostatic effects of the ions are of a general nature and no specific ion effects are to be expected. The binding of the water in the hydration shells of the ions at high salt concentrations makes less water available for hydrating other solutes, and this is, again, a general phenomenon that is outside the concept of specific ion effects. It is mainly the intermediate range of concentrations where the specific ion effects dealt with here are manifested.

The pH of the aqueous solutions in which biomacromolecules are dealt with has a strong effect on the biological function, such as enzyme activity. Such systems are, therefore, generally studied in buffer solutions in order to maintain a given pH value. Addition of neutral salts at molar concentrations sometimes used to make up the buffer tends to change the pH of buffer solutions according to Bauduin et al. [95, 96], causing effects beyond the specific ion effects on the biological activity. Such effects should be taken into account when the specific ion effects on biosystems and other colloidal systems are discussed.

8.5.1.1   The Anion Hofmeister Series   A comprehensive anion Hofmeister series is obtained by convolution of the various series proposed by several authors, Collins and Washabaugh [94], Miti et al. [97], Cacace et al. [98], Pinna et al. [99], and Zhang and Cremer [100], resulting in:

$$PO_4^{3-} \sim CO_3^{2-} > SO_4^{2-} > S_2O_3^{2-} > H_2PO_4^- > OH^- > F^- > HCO_2^- > CH_3CO_2^- >$$
$$Cl^- > Br^- > NO_3^- > I^- > ClO_4^- > SCN^-$$

$$(8.26)$$

The chloride anion has little effect on the relevant phenomena and is used as a reference point for the numerical effects of the anions (the value for Cl^- being near zero) that ranges from positive to negative or vice versa. The head of the series is the most effective anion, of which the lowest concentration of its salts with a given cation is needed for precipitation of a protein, henceforth called "head anion." The least effective one for precipitating proteins (weakly salting them out or even salting them in) at the end of the series is called hereafter the "tail anion." These epithets, "head anions" and "tail anions," are used here in order to avoid calling them kosmotropic and chaotropic and to distinguish the specific ion effects in the presence of a surface at appreciable concentrations from those in its absence and in dilute solutions.

Inversions of the positions of some anions in the series are noted when several related phenomena are used to establish it as reported by Cacace et al. and by Pinna et al. [98, 99]. For instance, F⁻ was placed at the head of the series according to Hochachka and Somero [101] and to Hall and Drake [102], although other authors relegated it to nearer its middle. A related series, whose lyotropic numbers are given in Table 8.11, also has several reversals with respect to the Hofmeister series as pointed out by Schott [103] and by Lo Nostro et al. [104].

A very broad range of phenomena is described by means of the Hofmeister anion series, ranging from the original one of protein precipitation to salting-out of other solutes, enzyme activity, macromolecular conformational transitions, and critical micelle concentrations. A "head anion," having a high propensity to precipitate a protein has also a small propensity to denaturalize it (unfold its tertiary and quaternary structures). On the contrary, a "tail anion" has a high ability to denature the protein, although it keeps it in solution as shown by Cacace et al. [98].

Anion properties that could be relevant to their ordering in the Hofmeister series, arbitrarily arranged according to their molar refractivities, R_{DI}, are shown in Table 8.12. The reader confronted with an anion Hofmeister series for some biophysical phenomenon may attempt to correlate it with the listed anion properties.

8.5.1.2   The Cation Hofmeister Series

The nature of the cations of the salts used for the denaturation of proteins also plays a role, subordinate to the dominating role of the anions, establishing for a given anion the series according to Cacace et al. [98]:

$$(CH_3)_4 N^+ > (CH_3)_2 NH_2^+ > NH_4^+ > K^+ \sim Na^+ > Cs^+ > Li^+ > Mg^{2+} > Ca^{2+} > Ba^{2+}$$
(8.27)

TABLE 8.12   Anion Parameters that Might be Relevant to their ordering in the Hofmeister Series

Anion[a]	R_{DI} (cm³ mol⁻¹)	r_I (Nm)	$-\Delta_{hyd}H_I$ (kJ mol⁻¹)	N_{lyol}	$B_{\eta I}$ (dm³ mol⁻¹)	$d\Delta\sigma/dc_I$ (mN m⁻¹ mol⁻¹ dm³)
F⁻	2.21	0.133	510	4.8	0.127	1.10
OH⁻	4.65	0.133	520	5.8	0.12	1.35
Cl⁻	8.63	0.181	367	10.0	−0.005	1.20
HCO₂⁻	9.43	0.204	432		0.052	0.15
NO₃⁻	10.43	0.200	312	11.6	−0.045	0.45
CO₃²⁻	11.45	0.178	1397		0.278	0.95
Br⁻	12.24	0.196	336	11.3	−0.033	0.95
ClO₄⁻	12.77	0.240	246	11.8	−0.060	−0.50
SO₄²⁻	13.79	0.230	1035	2.0	0.206	1.15
CH₃CO₂⁻	13.87	0.232	425		0.236	0.05
H₂PO₄⁻	14.6	0.200	522	8.2	0.34	1.25
PO₄³⁻	15.1	0.238	2879	3.2	0.59	2.00
SCN⁻	17	0.213	311	13.25	−0.032	0.20
I⁻	18.95	0.220	291	12.5	0.007	0.35
S₂O₃²⁻	23.2	0.250				1.45

[a] The anions are arranged arbitrarily according to their molar refractions, R_{DI}.

Sodium ions, near the midpoint of the series, have a minimal effect as have chloride ions in the anion series. The order for the monatomic alkali metal cations does not agree with their sizes, charge densities, or their lyotropic series (Table 8.11). However, cases where the lyotropic series is followed are also known. The rate of the penetration of the alkali metal cations through leaf cuticles decreases in the order $Cs^+ \geq Rb^+ > K^+ > Na^+ > Li^+$ according to McFarlaneand Berry [105] and the critical micelle concentration (*cmc*) of sodium dodecylsulphate increases in the reverse order according to Maiti et al. [97]. Variants of the cation Hofmeister series (8.27) have been proposed, for example, by Arakawa and Timasheff [106], where guanidinium, $C(NH_2)_3^+$, is added at the tail of the series:

$$
\begin{aligned}
&\left(CH_3\right)_4 N^+ > \left(CH_3\right)_2 NH_2^+ > K^+ \sim Na^+ > Cs^+ > Li^+ > NH_4^+ > \\
&Mg^{2+} > Ca^{2+} \sim Ba^{2+} > Mn^{2+} > Ni^{2+} > C\left(NH_2\right)_3^+
\end{aligned}
\tag{8.28}
$$

But many reversals in the order have been noted over the years by Fischer and Moore [107], Richter-Quitner [108], Carpenter and Lovelace [109], and by von Hippel and Wong [110].

Oligopeptides can be considered as protein analogs and the peptide groups are *salted-in* in increasing order along the series, $\left(CH_3\right)_4 N^+ < K^+ \sim Cs^+ \sim Li^+ < Na^+ < Ca^{2+}$, attributed to direct ion-peptide association by Nandi and Rolbinson [111]. On the other hand, ethyl esters of *N*-acetylamino acids are *salted-out* in the order $Cs^+ < K^+ \sim Na^+ \sim Li^+ < Ca^{2+}$, as expected from the charge density and the electrostriction. The observed salting-out of proteins is therefore attributed to the salting-out of the hydrophobic side groups. The cation order of stability of the enzyme halophilic malate dehydrogenase is reversed when the cations are examined at low (≤ 1 M) or at high concentrations according to Ebel et al. [112], due to the compensation between the general electrostatic interactions prevailing at low salt concentrations and direct association of the cations with the abundant carboxylate groups at high ones. Some series concerning enzyme activities do conform to the Hofmeister series (5.27) but others show reversals as discussed by Zhao [113]. The preferred interactions of weakly hydrated cations with weakly hydrated negatively charged side groups of the biomolecules and of strongly hydrated cations with strongly hydrated negatively charged groups, such as carboxylate and phosphate ones, should be taken into account according to Cacace et al. [98], Collins et al. [114], and Hess et al. [115].

Cation properties that could be relevant to their ordering in the Hofmeister series, arbitrarily arranged according to their surface charge densities, $z_I/4\pi r_I^2$, are shown in Table 8.13. The reader confronted with a cation Hofmeister series for some biophysical phenomenon may attempt to correlate it with the listed cation properties.

8.5.1.3 Interpretation of the Hofmeister Series   Many different explanations of the Hofmeister series of ions have been proposed over the years. Specific ion interactions with specific sites of the biomolecules must be taken into account and a subtle balance of several competing evenly matched interactions, such as differences in hydration strength, dispersion forces, polarization, ion size effects, and the impact on interfacial water structure, is involved according to Koelsch et al. [116] and Tobias and Hemminger [117].

TABLE 8.13 Cation Parameters that Might be Relevant to their ordering in the Hofmeister Series

Cation[a]	r_i (Nm)	$-\Delta_{hyd}H_i$ (kJ mol^{-1})	R_{Di} (cm^3 mol^{-1})	N_{lyol}	$B_{\eta l}$ (dm^3 mol^{-1})	$d\Delta\sigma/dc_1$ (mN m^{-1} mol^{-1} dm^3)
(CH$_3$)$_4$N$^+$	0.280	218	22.9		0.123	−0.40
(H$_2$N)$_3$C$^+$	0.20	602	11.21		0.058	−0.26
Cs$^+$	0.170	283	6.89	60.0	−0.047	0.50
NH$_4^+$	0.148	329	4.7		−0.008	0.40
K$^+$	0.138	334	2.71	75.0	−0.009	0.80
Na$^+$	0.102	416	0.65	100.0	0.085	0.90
Ba^{2+}	0.136	1332	5.17	7.5	0.229	0.50
Ca^{2+}	0.100	1602	1.59	10.0	0.298	1.50
Li$^+$	0.069	531	0.08	105.2	0.146	0.65
Mn^{2+}	0.083	1874	2.2		0.39	0.75
Ni^{2+}	0.072	2119	1.6		0.375	1.10
Mg^{2+}	0.072	1949	−0.7		0.385	1.65

[a]The cations are arranged arbitrarily according to their surface charges, $z_I/4\pi r_I^2$.

Collins and Washabaugh [94] stressed the analogy of the anion Hofmeister series with the surface tension and surface potential at the water/air surface, when the anions are ordered according to the sizes of $d\sigma/dc_1$ and of $\Delta\Delta\chi$ (Section 2.3.2.6). The authors postulated the same structure near the surface of a biomolecule as at the water/air surface, "head anions" interact with the layer near solutes more strongly than bulk water does, appropriating some of the water to themselves. For "tail anions" the binding of water is loose, so they do not dehydrate the solute. The underlying concept of Cacace et al. [98] is that biopolymer interactions are water-mediated and the anions affect both the surface of the biopolymer and the structure of the water. The water/protein interfacial area A_{WP} is diminished when a protein folds, aggregates, or adsorbs, but when a protein is denatured A_{WP} increases. The native conformation of a protein, assumed more compact than any non-native one, is stabilized by diminishing the A_{WP}, and the larger the work that needs to be done to exclude an anion from an interface, the greater the tendency to minimize that interface. On the other hand, the more strongly the anions are hydrated, the less readily they dehydrate when binding to positive sites of the protein. Baldwin [118] proposed that "head anions" affect the nonpolar (hydrophobic) groups of proteins, the more the larger these groups are, whereas "tail anions" interact with the peptide group according to Nandi and Rolbinson [111], losing their hydration water more readily than the former anions.

Collins [119] revised his previous view concerning analogy with surface tension, and stressed the surface charge density of the ions, $q = z_1 e/\pi r_1^2$, as determining the strength of their hydration relative to water–water interactions. According to their structural entropy (Section 5.1.1.7), monovalent anions having $r_1 < 0.178$ nm (near that of Cl$^-$) are more strongly hydrated, those with $r_1 < 0.178$ nm are more weakly hydrated. Direct interactions of "head anions" with "solute head cations" and of "tail anions" with "solute tail cations" were considered to be the primary mechanism for the specific ion effects. The view promoted by Collins and coworkers over the years [94, 114, 119, 120], that the cation hydration strength measured by the structural entropy compared with the water–water interaction strength is the main driving force for the cation effects, based solely on salts of the alkali metal cations, cannot be valid in view of the many order reversals in the cation Hofmeister series that have been noted.

Boström et al. [121–123] emphasized the importance of dispersion forces regarding the specific ion effects on both protein-related and non-protein-related Hofmeister effects. Silica membranes show Hofmeister series effects in pH measurements with a glass electrode in moderately concentrated salt solutions, 0.8 M. The order of decreasing pH is $NaCl > NaBr > NaNO_3 \sim NaClO_4$ ascribed to competition between the polarizability of the anions and electrostatic interactions at the surface of the glass electrode.

Artificial colloids permit a wider variation of surface polarity and charge than do proteins. The distinction between surface polarity (hydrophilic/hydrophobic balance) and net surface charge is of crucial importance, since it is related to reversals of the Hofmeister series according to Schwierz et al. [124]. Negatively charged colloids that are rather hydrophobic show the direct Hofmeister series but this is reversed for positively charged hydrophobic colloids. The opposite trends occur for very polar colloids as pointed out by Peula-Garcia et al. [125]. Leontidis et al. [126] measured the surface pressure at monolayers of dipalmitoyl phosphatidylcholine in the presence of sodium salts of a variety of anions. The ion penetration model presumes that Na^+ is excluded from the monolayer and that "tail anions" partition between it and the bulk solution. A model that emphasizes the dispersion forces is also able to fit the surface pressures well. A model that allows complexation of Na^+ with three lipid molecules is able to fit the results for univalent "head anions." The surface pressure is linear with $[(r_i + r_w)^2 - 0.029/(r_i + r_w)]$, the factor in square brackets being the difference between the cavity formation energy, proportional to the surface area of the hydrated ion, and an electrostatic energy, proportional to the reciprocal of the size of the hydrated ion and depends also on the permittivity difference between bulk water and the lipid layer [126].

The general conclusion that may be drawn from all these studies and many more not dealt with here is that not only the properties of the ions, listed in Tables 8.12 and 8.13, are relevant to the position of an ion in the Hofmeister series, but also the surface of the specific biomolecule or other colloid with which the ion interacts is relevant to a large extent. The hydrophilic–hydrophobic balance of this surface, the sign of the charge on it, if any, and the existence of strongly ion-binding groups at this surface, related to direct interaction of the ion with sites there, are as important as are general electrostatic and ion–water interactions that are independent of the surface of any given substrate. Although water-mediation does probably play a role in positioning ions in the series, whether direct or reverse, direct interactions with the surface of the solute may dominate. A multidimensional correlation of the properties of the substrate studies with several ion parameters appears to be needed for a given colloidal or biomolecular surface, but when other surfaces are involved, the same ion parameters may be operative but with changed weights.

8.5.2 Water Structure Effects of Ions

In the present context, the terms "water-structure-breaking" and "water-structure-making" (Section 5.1) are respectively synonymous with "chaotropic," coined by Hamaguchi and Geiduschek [127] as applied to anions, and "kosmotropic" popularized together with "chaotropic" in the biophysical literature by Collins and Washabaugh [94]. Such properties are manifested at infinite dilution and pertain to ions surrounded only by water being remote from interfaces. The effects on the structure of water persist also in homogeneous dilute solutions at finite concentrations containing other solutes than the ions, but when interionic forces become dominant or when the ions are in a non-homogeneous region in the solution, other considerations than water structure play a role.

The anions are ordered in dilute solutions from chaotropic to kosmotropic (see the viscosity B-coefficients in Table 8.12):

$$I^- \sim ClO_4^- > SCN^- \sim Br^- > NO_3^- \sim Cl^- > HCO_2^- \sim S_2O_3^{2-} > SO_4^{2-} > F^- \sim H_2PO_4^- >$$
$$OH^- \sim CH_3CO_2^- > CO_3^{2-} > PO_4^{3-}$$

$$(8.29)$$

Borderline anions, neither pronounced chaotropic nor kosmotropic, are F^- and $H_2PO_4^-$ and a reversed order according to some criteria may apply to anions preceded by approximately. The corresponding series for the cations, but reversed (from kosmotropic to chaotropic), is:

$$Mg^{2+} > Ca^{2+} > Li^+ > Na^+ > NH_4^+ > \left(CH_3\right)_4 N^+ > K^+ \sim Cs^+ \qquad (8.30)$$

A borderline position, being neither pronounced kosmotropic nor chaotropic is occupied by Na^+. The ions included in these two series are the same that are listed in the Hofmeister series in Section 8.5.1, but not necessarily in exactly the same order.

Some biophysical specific ion effects have been ascribed to the ions being classified as chaotropic or kosmotropic. However, few biophysical phenomena take place in homogeneous dilute aqueous solutions, since biomolecules tend to be large and colloidal, forming micro-heterogeneous domains when dispersed in water.

Koga and coworkers [128–130] suggested that the moderately hydrophobic solute 1-propanol in aqueous solutions, having a comparable ratio of hydrophobic and hydrophilic moieties to that of some soluble proteins, is sufficiently biomimetic to serve as a biophysical probe. Fairly dilute aqueous sodium salt solutions, $x_E \leq 0.02$, dilute also in 1-propanol were used to obtain thermodynamic excess functions and their concentration dependences, so that the premise for characterizing the anions as kosmotropic or chaotropic is fulfilled. In the three relevant studies, the orders were $CH_3CO_2^- > Cl^- \sim SO_4^{2-} > ClO_4^- \sim SCN^-$ [128] $Cl^- > Br^- > I^-$ [129], and $SO_4^{2-} > F^- > Cl^- > I^- \sim ClO_4^-$ [130]. These partial series correspond with Equation 8.29 with some significant reversals. Although the epithets kosmotropic and chaotropic have been applied by the authors to the ions, the molecular interactions involved may have little to do with the structure of the water and the effects of the ions on it. The effects of the ions also involve the number of water molecules immobilized in their hydration shells, their fitting into the hydrogen bonded network of the water, and their retarding its fluctuations (Section 5.1), and any direct interactions of them with the hydrophilic and hydrophobic parts of the 1-propanol may also be involved. All the five salts studied by Miki et al. [130] salt-out the 1-propanol, but the abilities of salts to salt-out or salt-in hydrophobic solutes are only remotely related to the water-structure effects of their constituent ions. For instance, the salting-out of hydrophobic gases or benzene in Table 7.6 shows that Li^+ ions do not conform to the sequence of Equation 8.30. Furthermore, salting-in is often due to direct interactions between the ions and the solute, which are not necessarily water-mediated. Indeed, Smith [131] studied the salting behavior of He, Ne, Ar, and methane in aqueous NaCl, $(NH_4)_2SO_4$, $CaCl_2$, $NH_4CH_3CO_2$, $(CH_3)_4NCl$, and $C(NH_2)_3Cl$ and concluded that the degree of preferential binding of the ions to the hydrophobic solutes was the dominant factor in the salting behavior.

The effect of various sodium salts on the lower consolute temperature (LCST) of aqueous dipropylene glycol monopropyl ether was studied by Bauduin et al. [132]. The LCST is $t_{LCST} = 14°C$ in the absence of salts and at a mole fraction of 0.11 of the ether in water. Salts change t_{LCST} linearly with their concentration, and the slopes (the units of which are K·(mmol salt/mole solvent mixture)$^{-1}$ and which have an uncertainty of ≤ 0.2 units) range from -14.4 for the salting-out Na_3PO_4 to $+3.2$ for the salting-in NaSCN. The sequence of the algebraic values of the slopes is: $Na_3PO_4 < Na_2SO_4 \sim Na_2CO_3 < NaOH \sim NaCH_3CO_2 < NaCl < NaBr < NaI < NaClO_4 < NaSCN$, similar to Equation 8.29 but deviates from it in several cases. Just below the LCST the solution is homogeneous, and the electrolytes change the chemical potential of the components toward phase separation, but there is no surface effect here and the words Hofmeister series should not have been mentioned in the title of the paper.

Thomas and Elcock [133] applied molecular dynamics simulations with approximately 500 water molecules to pure water and to 1 M solutions of alkali metal, alkaline earth, and tetraalkylammonium halides. The hydrogen bonding fraction between neighboring water molecules was defined as $\theta_{HB} = N_{HB}/N_{neighbors}$, using geometrical criteria for the formation of a hydrogen bond, finding for pure water $\theta_{HB} = 0.67$. There was a clear linear correlation between the Setchenow constants k_E for methane and neopentane and θ_{HB}, that ranged from salting-out, $\theta_{HB} < 0.67$, for the fluorides and salts of divalent cations through salts of the alkali metals and ammonium, water, and reaching the salting-in tetramethyl- and tetraethylammonium bromides, having $\theta_{HB} > 0.67$. Still, the correlation of k_E with θ_{HB} does not prove a causal relationship, as the authors themselves stressed.

Zangi [134] used molecular dynamics simulations with 1030 water molecules and 60 ions (cations and anions of equal charge) to investigate whether salting-out and salting-in ions can be classified as chaotropic or kosmotropic. The variable studied was the surface charge density q, ranging from 0.5 to 1.4 elementary charges e per $\pi\sigma^2$ surface area ($1\,e\,nm^{-2} = 16\,C\,m^{-2}$), with $\sigma = 0.50$ nm being the fixed Lennard–Jones diameter of the ions. For the association between two hydrophobic plates in the electrolyte solutions relative to that in neat water, the boundary between salting-in and salting-out ions was at $q = 0.71\,e\,nm^{-2}$, nearly coinciding with $q = 0.68e\,nm^{-2}$, the boundary between ions that enhance the viscosity of dilute aqueous solutions and those that reduce it. However, it was concluded that the hydrophobic association depends on the direct interaction of low-q ions (poorly hydrated ones) with the hydrophobic entities rather than on the water structural effects of the ions. A final conclusion from this study was that for small hydrophobic solutes, the water structure effects of the ions may be dominant, but for large (and polar) hydrophobic solutes, the direct interactions dominate.

8.5.3 Some Aspects of Protein Hydration

The hydration of the ionized side chains of proteins as well as that of the peptide moieties of the protein backbone and hydrophobic hydration of the side chains at the surface of proteins play important roles in the biophysics of such species. Native proteins in aqueous solutions are in their folded form and consist of a mainly hydrophobic core and a mainly hydrophilic periphery, only the latter being exposed to the solvent that borders it. Saito et al. [135] studied five representative globular proteins: bovine pancreatic trypsin inhibitor, bovine pancreatic ribonuclease, hen egg-white lysozyme, bovine milk β-lactoglobulin A, and bovine pancreatic α-chymotrypsinogen A. The relevant quantity was the solvent accessible surface area (*SASA*), obtained by rolling a spherical probe having the diameter of a water molecule, 0.28 nm, on the surface of the protein as stipulated by Sanner et al. [136]. The fraction of hydrophilic *SASA* ranged from 69 to 83%, of which a portion was ionic, comprising from 27 to 50% of the *SASA*, pertaining to the carboxylate and ammonium groups of the amino acid side chains, ionized at appropriate pH values.

Gerstein and Chothia [137] studied 22 proteins, the average number of surface atoms of per molecule of which was 420, and found them to be 47% hydrated, that is, in contact with water molecules. The peripheral hydrating water molecules had a molecular volume of $0.0245\,nm^3$ compared with the molecular volume of bulk water, $V^*/N_A = 0.0300\,nm$ at room temperature, that is, they were compressed by 22% with respect to it.

Svergun et al. [138] studied three proteins: lysozyme, *Escherichia coli* thioredoxin reductase, and *E. coli* ribonucleotide reductase protein R1, in aqueous solution, using x-ray and neutron scattering. The density of the water of the first hydration shell of these proteins differed from that of bulk water, the average relative densities were 1.08 ± 0.02, 1.16 ± 0.05, and 1.12 ± 0.06. These experimental values are smaller than those calculated from the packing, 1.22 according to Gerstein and Chothia [137], but still appreciable.

Menzel and Smith [139] computed by molecular dynamics simulation the small angle scattering profiles of the proteins in the Svergun et al. [138] study, concluding that the variation in this density is determined by both topological and electrostatic properties of the protein surface. The surface roughness with respect to a perfectly flat and smooth surface ranged from −0.6 nm for depressions (concave regions, grooves) to +0.4 nm for ridges (convex regions). In the 0.3 nm thick hydration shell, the density of the water increased by up to $7\pm2\%$ in the depressions but by only $2\pm2\%$ at the ridges. In spots where the electric field

perpendicular to the surface corresponded to a surface charge density σ of 0.05 to
$3.0\,e\,\text{nm}^{-2}$ ($1\,e\,\text{nm}^{-2} = 16\text{C m}^{-2}$), the water density was enhanced by up to $10 \pm 3\%$.
The topological effect was explained by a water molecule in a depression being
less exposed to randomizing, disorienting effects of neighboring water molecules,
and being in closer contact with ionic charges at the protein surface, than water
molecules at ridges.

Danielewicz-Ferchmin et al. [140] concluded that the compression of the water
near the protein surface, subscript σ, was due to the electrostriction of the water pro-
duced by the electric fields generated by the ions, $E_\sigma = \sigma/\varepsilon_0\varepsilon_\sigma$, being of the order of
GV·m^{-1}. The relative permittivity of water is drastically reduced by such high electric
fields: from $\varepsilon_w^* = 82$ (at 20°C) at zero charge density to $\varepsilon_\sigma = 63$ at 0.2 C m^{-2}, to
$\varepsilon_\sigma = 11$ at 0.3 C m^{-2}, and to $\varepsilon_\sigma = 4$ at 0.4 C m^{-2}. The values of σ required to produce
the experimental compression measured according to Svergun et al. [138] was
obtained by methods established for aqueous solutions of small ions by Danielewicz-
Ferchmin et al. [141]. There was no appreciable density enhancement up to surface
charge densities of 0.24 C m^{-2}, but beyond this value the surface densities were to a
good approximation given by:

$$\sigma/\text{C·m}^{-2} = -3.82 + 9.15\left(\frac{\rho_\sigma}{\rho_w}\right) - 6.83\left(\frac{\rho_\sigma}{\rho_w}\right)^2 + 1.74\left(\frac{\rho_\sigma}{\rho_w}\right)^3 \qquad (8.31)$$

where (ρ_σ/ρ_w) is the relative density of the hydration water. This expression was
valid up to 0.41C m^{-2}, where the compression yielded volumes that went down to
the van der Waals volume of the water molecules. The charge densities at the surfaces of the three proteins studied by Svergun et al. [138] were 0.294 ± 0.010,
0.326 ± 0.018, and $0.309 \pm 0.025\,\text{C m}^{-2}$, respectively, if the total compression of
the water was due to electrostriction, being within the range estimated by Menzel
and Smith [139].

The temperature and pressure dependences of the native-to-denatured transition
of proteins were related to local values of the charge density at the surface of the protein according to Danielewicz-Ferchmin et al. [142]. The water in the hydration shell
is at equilibrium with bulk water, and hence the orientation work caused by the field
of the surface charges of the molecular dipoles in the two environments is the same.
The charge density σ and the electrostrictive pressure yielded via equations-of-state
to temperature–pressure crossover points for the native-to-denatured transition
dependences for six proteins: ribonuclease, human interferon γ, chymotrypsinogen,
lysozyme, staphylococcal nuclease, and ribonuclease A, occurring in the range
$0.289 < \sigma/\text{C m}^{-2} < 0.306$ for these six proteins. The energy of dipole orientation is
commensurate with the energy for breaking a hydrogen bond between water molecules at these values of σ.

The rotational retardation factor of hydration shell water came from terahertz
spectroscopy applied by Ebbinghaus et al. [143], yielding hydrogen bond lifetimes,
up to 2.4 ps at 0.2 nm distance from the protein surface, that were larger than those in
bulk water, 1.6 pm, and were still detected up to a distance of 0.6 nm. The where-
abouts of sites on the surfaces of the proteins lysozyme and staphylococcal nuclease,

where the water binding is particularly strong, were sought by Priya et al. [144]. The criterion for strong binding was:

$$\eta = 3.4 + \ln\left[\frac{\tau_1}{(\tau_1 + \tau_0)}\right] > 2 \tag{8.32}$$

where τ_1 is the average time a site is occupied by a water molecule and τ_0 the average time it is vacant. Of the two proteins studied, the former has 150 sites of durable occupancy and the other 242 such sites, 15 and 10% of which are with kinetically bound water $(\eta > 2.7)$. Such sites appear to be the depressions in the rough topology of the protein surface, as suggested by Menzel and Smith [139] but could instead be at ionized amino acid side chains as proposed by Danielewicz-Ferchmin et al. [142].

It is still not clear how relatively important are the charge density and the surface roughness for affecting the water of hydration of protein surfaces. They contribute to the compression of the water, the orientation of the molecules, and the dynamics of their exchange and rotation, but it is still not known at what specific sites of the protein surface do they take place.

REFERENCES

[1] K. Xu, C. A. Angell, J. Electrochem. Soc. **145**, L70 (1998).

[2] X.-G. Sun, C. A. Angell, Electrochem. Commun. **7**, 261 (2005).

[3] Y. Abu-Lebdeh, I. Davidson, J. Electrochem. Soc. **156**, A60 (2009).

[4] J.-Y. Luo, W.-C. Cui, P. He, Y.-Y. Xia, Nat. Chem. **2**, 760 (2010).

[5] X.-P. Gao, H.-X. Yang, Energy Environ. Sci. **3**, 174 (2010).

[6] F. Beguin, V. Presser, A. Balducci, E. Frackowiak, Adv. Mater. **26**, 2219 (2014).

[7] E. Perricome, M. Chamas, L. Cointeaux, J.-C. Lepretre, P. Judeinstein, P. Azais, F. Beguin, F. Alloin, Electrichim. Acta **93**, 1 (2013).

[8] R. Vali, L. Laheäär, A. Jänes, E. Lust, Electrichim. Acta **121**, 294 (2014).

[9] A. K. Covington, R. G. Bates, R. A. Durst, Pure Appl. Chem. **57**, 531 (1985).

[10] H. B. Kristensen, A. Salomon, G. Kokholm, Anal. Chem. **63**, 885A (1991).

[11] R. P. Buck, S. Rondinini, A. K. Covington, F. G. K. Baucke, C. M. A. Brett, M. F. Camões, M. J. T. Milton, T. Mussini, R. Naumann, K. W. Pratt, P. Spitzer, G. S. Wilson, Pure Appl. Chem., **74**, 2169 (2002).

[12] R. G. Bates, Pure Appl. Chem. **54**, 229 (1982).

[13] Y. Marcus, Pure Appl. Chem. **61**, 1134 (1989).

[14] K. H. Khoo, R. W. Ramette, C. H. Culberson, R. G. Bates, Anal. Chem. **49**, 29 (1977).

[15] T. Mussini, P. Longhi, I. Marcolungo, P. R. Mussini, S. Rondinini, Fresenius J. Anal. Chem. **339**, 608 (1991).

[16] M. Rosès, J. Chromatogr. A **1037**, 283 (2004).

[17] C. Kalidas, G. Hefter, Y. Marcus, Chem. Rev. **100**, 819 (2000).

[18] Y. Marcus, Chem. Rev., **107**, 3880 (2007).

[19] B. G. Cox, A. J. Parker, W. E. Waghorne, J. Am. Chem. Soc. **95**, 1010 (1973).

[20] D. A. McInnes, The Principles of Electrochemistry, Dover, New York, 2nd ed. (1961).

[21] H. Strehlow, H. Wendt, Z. Phys. Chem. (NF) **30**, 141 (1961).

[22] G. Gritzner, Inorg. Chim. Acta **24**, 5 (1973).

[23] G. Gritzner, J. Mol. Liq. **156**, 103 (2010).

[24] H. D. Inerowicz, W. Li, I. Persson, J. Chem. Soc. Faraday Trans. **90**, 2223 (1994).

[25] B. G. Cox, A. J. Parker, W. Earle Waghorne, J. Phys. Chem. **78**, 1731 (1974).

[26] Y. Marcus, Ion Solvation, Wiley, Chichester, 1985.

[27] D. A. Owensby, A. L. Parker, J. W. Diggle, J. Am. Chem. Soc. **96**, 2682 (1974).

[28] B. G. Cox, G. R. Hedwig, A. J. Parker, D. W. Watts, Aust. J. Chem. **27**, 477 (1974).

[29] M. Johnsson, I. Persson, Inorg. Chim. Acta **127**, 15 (1987).

[30] Y. Marcus, J. Solution Chem. **13**, 599 (1984).

[31] R. Modin, G. Schill, Talanta **22**, 1017 (1975).

[32] D. S. Flett, J. Organomet. Chem. **690**, 2426 (2005).

[33] B. R. Reddy, D. N. S. V. Rao, P. Radhika, Hydrometallurgy **77**, 253 (2005).

[34] O. J. Solis-Marcial, G. T. Lapidus, Hydrometallurgy **131–132**, 120 (2013).

[35] G. Hefter, Y. Marcus, W. E. Waghorne Chem. Rev. **102**, 2773 (2002).

[36] A. J. Parker, D. M. Muir, Hydrometallurgy **6**, 239 (1981).

[37] A. J. Parker, Pure Appl. Chem. **53**, 1437 (1981).

[38] A. J. Parker, B. W. Clare, R. P. Smith, Hydrometallurgy **4**, 233 (1979).

[39] H. J. Gores, J. Barthel, Naturwissenschaften **70**, 495 (1983).

[40] D. Ghosh, S. Dasgupta, Metall. Mater. Trans. **39B**, 35 (2008).

[41] M. Yoshio, N. Ishibasi, J. Appl. Electrochem. **3**, 321 (1973).

[42] G. Eisenman, S. M. Ciani, G. Szabo, Fed. Proc. **27**, 1289 (1968).

[43] Y. Takeda, Bull. Chem. Soc. Jpn. **53**, 2393 (1980).

[44] M. Tanigawa, S. Nishiyama, S. Tsuruya, M. Masai, Chem. Eng. J. **39**, 157 (1988).

[45] Y. Marcus, L. E. Asher, J. Phys. Chem. **82**, 1246 (1978).

[46] Y. Marcus, L. E. Asher, J. Hormdaly, E. Pross, Hydrometallurgy **7**, 27 (1981).

[47] Y. Marcus, E. Pross, J. Hormadaly, J. Phys. Chem. **84**, 2708 (1980).

[48] M. Kyrš, J. Radioanal. Nucl. Chem. Lett. **187**, 185 (1994).

[49] J. Rais, B. Gruener, Ion Exch. Solv. Extr. **17**, 243 (2004).

[50] J. Rais, T. Okada, J. Alexova, J. Phys. Chem. B **110**, 8432 (2006).

[51] J. Hormodaly, Y. Marcus, J. Phys. Chem. **83**, 2843 (1979).

[52] Y. Marcus, J. Inorg. Nucl. Chem. **37**, 493 (1975).

[53] A. Jouyban, W. E. Acree, Jr., J. Pharm. Pharmaceut. Sci. **9**, 262 (2006).

[54] A. Jouyban-Gharamaleki, Chem. Pharm. Bull. **46**, 1058 (1998).

[55] Y. Marcus, J. Mol. Liq. **140**, 61 (2008).

[56] S. H. Yalkowski, T. Rosennab, in S. H. Yalkowski, ed., Solubilization of Drugs by Cosolvents, Dekker, New York, 91 (1981).

[57] E. Ruckenstein, I. L. Shulgin, Int. J. Pharm. **258**, 193 (2003); **260**, 283 (2003).

[58] G. M. Wilson, J. Am. Chem. Soc. **86**, 127 (1964).

[59] Y. Marcus, Chem. Rev. **109**, 1346 (2009).

[60] D. R. Delgado, A. R. Holguin, O. A. Almanza, F. Martinez, Y. Marcus, Fluid Phase Equil. **305**, 88 (2011).

[61] M. A. Ruidiaz, D. R. Delgado, F. Martinez, Y. Marcus, Fluid Phase Equil. **299**, 259 (2010).

[62] A. R. Holguin, D. R. Delgado, F. Martinez, Y. Marcus, J. Solution Chem. **40**, 1987 (2011).

[63] C. Reichardt, T. Welton, Solvents and Solvent Effects in Organic Chemistry, Wiley-VCH, Weinheim, 4th ed. (2011).

[64] A. J. Parker, Chem. Rev. **60**, 1 (1969).

[65] E. Buncel, H. Wilson, Accounts Chem. Res. **12**, 42 (1979).

[66] M. H. Abraham, Pure Appl. Chem. **57**, 1055 (1985).

[67] C. Reichardt, Chem. Rev. **94**, 2319 (1994).

[68] E. Buncel, R. Stairs, H. Wilson, The Role of Solvents in Chemical Reactions, Oxford University Press, Oxford (2003).

[69] M. J. Blandamer, J. Burgess, Pure Appl. Chem. **51**, 2087 1979.

[70] S. Wherland, Coord. Chem. Rev. **123**, 169 (1993).

[71] R. Alexander, E. F. C. Ko, A. J. Parker, T. J. Broxton, J. Am. Chem. Soc. **90**, 5049 (1968).

[72] G. E. McManis, R. M. Nielson, A. Gochev, M. J. Weaver, J. Am. Chem. Soc. **111**, 5533 (1989).

[73] T. T.-T. Li, C. H. Brubaker, Jr., J. Organomet. Chem., **216**, 223 (1981).

[74] J. Burgess, Metal Ions in Solution, Ellis Horwood, Chichester, (1978) Ch. 11.

[75] D. L. Pisanello, S. F. Lincoln, Aust. J. Chem. **32**, 715 (1979).

[76] S. Funahashi, R. B. Jordan, Inorg. Chem. **16**, 1301 (1977).

[77] D. H. Powell, P. Furrer, P.-A. Pittet, A. E. Merbach, J. Phys. Chem. **99**, 16622 (1995).

[78] R. J. West, S. F. Lincoln, J. Chem. Soc. Dalton Trans. 281 (1974).

[79] D. Richardson, T. D. Alger, J. Phys. Chem. **79**, 1733 (1975).

[80] C. M. Starks, C. Liotta, Phase Transfer Catalysis, Academic Press, New York (1978).

[81] R. Bock, J. Jainz, Z. Anal. Chem. **198**, 316 (1983).

[82] V. L. Heifets, N. A. Yakovleva, B. Ya. Krasil'shchik, Zh. Prikl. Khim. **46**, 549 (1973).

[83] O. I. Danilova, S. S. Yufit, Mendeleev Commun. (1993) 165.

[84] M.-L. Wang, T.-H. Huang, W.-T. Wu, Ind. Eng. Chem. Res. **41**, 518 (2002).

[85] H.-S. Wu, M.-C. Lu, J. Mol. Catal. A Chem. **104**, 139 (1995).

[86] M.-L. Wang, Y.-H. Tseng, J. Mol. Catal. A Chem. **203**, 79 (2003).

[87] K. Nakamura, S. Nishiyama, S. Tsuruya, M. Masai, J. Mol. Catal. **93**, 195 (1994).

[88] Y. Marcus, The Properties of Solvents, Wiley, Chichester (1998).

[89] B. H. Büchner, A. Voet, E. M. Bruins, Proc. Kong. Neder. Akad. Veten. **35**, 563 (1932).

[90] A. Voet, Chem. Rev. **20**, 169 (1937); Z. Koll **78**, 201 (1937).

[91] W. Kunz, Curr. Opinion Coll. Interf. Sci. **15**, 34 (2010).

[92] W. Kunz, Specific Ion Effects, World Scientific Publishing, Singapore (2010).

[93] F. Hofmeister, Arch. Exp. Pathol. Pharmakol. **24**, 247 (1888).

[94] K. D. Collins, M. W. Washabaugh, Quart. Rev. Biophys. **18**, 323 (1985).

[95] P. Bauduin, A. Renoncourt, D. Touraud, W. Kunz, B. W. Ninham Curr, Opinion Coll. Interf. Sci. **9**, 43 (2004).

[96] P. Bauduin, F. Nohmie, D. Touraud, R. Neuder, W. Kunz, B. W. Ninham, J. Mol. Liquids **123**, 14 (2006).

[97] K. Maiti, D. Mitra, S. Guha, A. P. Moulik. J. Mol. Liq. **146**, 44 (2009).

[98] M. G. Cacace, E. M. Landau, J. J. Ramsden, Quart. Rev. Biophys. **30**, 241 (1997).

[99] M. C. Pinna, A. Salis, M. Monduzzi, B. W. Ninham, J. Phys. Chem. B **109**, 5406 (2005).

[100] Y. Zhang, P. S. Cremer, Curr. Opinion Chem. Biol. **10**, 658 (2006).

[101] P. W. Hochachka, G. N. Somero, Biochemical Adaptation, Princeton University Press, Princeton (1984).

[102] D. L. Hall, P. L. Drake, J. Biol. Chem. **270**, 22697 (1995).

[103] H. Schott, Coll. Surf. **11**, 51 (1984).

[104] P. Lo Nostro, L. Fratoni, B. W. Ninham, P. Baglioni, Biomacromolecules **3**, 1217 (2002).

[105] J. C. McFarlane, W. L. Berry, Plant Phys. **53**, 723 (1974).

[106] T. Arakawa, S. N. Timasheff, Biochemistry **23**, 5912 (1984).

[107] M. H. Fischer, G. Moore, Am. J. Physiol. **20**, 330 (1907).

[108] M. Richter-Quitner, Biochem. Z. **121**, 273 (1921).

[109] D. C. Carpenter, F. E. Lovelace, J. Am. Chem. Soc. **57**, 2337 (1935).

[110] P. H. von Hippel, K-Y Wong, Science **145**, 577 (1964).

[111] P. K. Nandi, D. R. Rolbinson, J. Am. Chem. Soc. **94**, 1299, 1308 (1972).

[112] C. Ebel, P. Faou, B. Kernel, G. Zaccai, Biochemistry **38**, 9039 (1999).

[113] H. J. Zhao, Mol. Catal. B Enzym. **37**, 16 (2005).

[114] K. D. Collins, G. W. Neilson, J. E. Enderby, Biophys. Chem. **128**, 95 (2007).

[115] B. Hess, N. F. A. Van der Vegt, Proc. Natl. Acad. Sci. U.S.A **106**, 13296 (2009).

[116] P. Koelsch, P. Viswanath, H. Motschmann, V. L. Shapovalov, G, Brezesinski, H. Möhwald, D. Horinek, R. R, Netz, K. Giewekemeyer, T. Salditt, H. Schollmeyer, R. von Klitzing, J. Daillant, P. Guenoun, Coll. Surf. Sci. A Physicochem. Eng. Aspects **303**, 110 (2007).

[117] D. J. Tobias, J. C. Hemminger, J. Sci. **319**, 1197 (2008).

[118] R. L. Baldwin, Biophys. J. **71**, 2056 (1996).

[119] K. D. Collins, Biophys. J. **72**, 65 (1997).

[120] K. D. Collins, Proc. Natl. Acad. Scic. U.S.A **92**, 5553 (1995).

[121] M. Boström, D. R. M. Williams, B. W. Ninham, Biophys. J. **85**, 686 (2003).

[122] M. Boström, V. S. J. Craig, R. Albion, D. R. M. Williams, B. W. Ninham, J. Phys. Chem. B **107**, 2875 (2003).

[123] M. Boström, V. Deniz, B. W. Ninham, J. Phys. Chem. B **110**, 9645 (2006).

[124] N. Schwierz, D. Horineck, R. R. Netz, Langmuir **26**, 7370 (2010).

[125] J. M. Peula-Garcia, J. L. Ortega-Vinuesa, D. Bastos-Gonzalez, J. Phys. Chem. C **114**, 11133 (2010).

[126] E. Leontidis, A. Aroti, L. Belloni, J. Phys. Chem. B **113**, 1447 (2009); E. Leontidis, A. Aroti, J. Phys. Chem. B **113**, 1460 (2009).

[127] K. Hamaguchi, E. P. Geiduschek, J. Am. Chem. Soc. **84**, 1329 (1962).

[128] Y. Koga. P. Westh, J. V. Davies, K. Miki, K. Nishikawa, H. Katayanagi, J. Phys. Chem. A **108**, 8533 (2004).

[129] P. Westh, H. Kato, K. Nishikawa,Y. Koga, J. Phys. Chem. A **110**, 2072 (2006).

[130] K. Miki, P. Westh, Y. Koga, J. Phys. Chem. B **112**, 4680 (2008).

[131] P. E. Smith, J. Phys. Chem. B **103**, 525 (1999).

[132] P. Bauduin, L. Wattebled, D. Touraud, W. Kunz, Z. Phys. Chem. **218**, 631 (2004).

[133] A. S. Thomas, A. H. Elcock, J. Am. Chem. Soc. **129**, 14887 (2007).

[134] R. Zangi, J. Phys. Chem. B **114**, 643 (2010).

[135] H. Saito, N. Matubayashi, K. Nishikawa, H. Nagao, Chem. Phys. Lett. **497**, 218 (2010).

[136] M. F. Sanner, A. J. Olson, J.-C. Spehner, Biopolymer **38**, 305 (1996).

[137] M. Gersten, C. Chothia, Proc. Natl. Acad. Sci. U.S.A. **93**, 10167 (1996).

[138] D. I. Svergun, S. Richard, M. H. J. Koch, Z. Sayers, S. Kuprin, G. Zaccai, Proc. Natl. Acad. Sci. U.S.A. **95**, 2267 (1998).

[139] F. Merzel, J. C. Smith, Proc. Natl. Acad. Sci. U.S.A. **99**, 5378 (2002).

[140] I. Danielewicz-Ferchmin, E. Banachowicz, A. R. Ferchmin, Biophys. Chem. **106**, 147 (2003).

[141] I. Danielewicz-Ferchmin, A. R. Ferchmin, A. Szlaferek, Chem. Phys. Lett. **288**, 197 (1998).

[142] I. Danielewicz-Ferchmin, E. Banachowicz, A. R. Ferchmin, ChemPhysChem **7**, 2126 (2006).

[143] S. Ebbinghaus, S. J. Kim, M. Heyden, X. Yu, U. Heugen, M. Gruebele, D. M. Leitner, M. Hevenith, Proc. Natl. Acad. Sci. U.S.A. **104**, 20749 (2007).

[144] M. H. Priya, J. K. Shah, D. Asthagiri, M. E. Paulaitis, Biophys. J. **95**, 2219 (2008).

AUTHOR INDEX

SUBJECT INDEX

Ions in Solution and Their Solvation, First Edition. Yizhak Marcus.
© 2015 John Wiley & Sons, Inc. Published 2015 by John Wiley & Sons, Inc.

Printed and bound by CPI Group (UK) Ltd, Croydon, CR0 4YY

16/04/2025

14658344-0004